Optical Multidimensional Coherent Spectroscopy

Optical Multidimensional Coherent Spectroscopy

HEBIN LI

Florida International University

BACHANA LOMSADZE

Santa Clara University

GALAN MOODY

University of California Santa Barbara

CHRISTOPHER L. SMALLWOOD

San Jose State University

STEVEN T. CUNDIFF

University of Michigan

OXFORD

UNIVERSITY PRESS

OXFORD
UNIVERSITY PRESS

Great Clarendon Street, Oxford, OX2 6DP,
United Kingdom

Oxford University Press is a department of the University of Oxford.
It furthers the University's objective of excellence in research, scholarship,
and education by publishing worldwide. Oxford is a registered trade mark of
Oxford University Press in the UK and in certain other countries

Published in the United States of America by Oxford University Press
198 Madison Avenue, New York, NY 10016, United States of America

British Library Cataloguing in Publication Data
Data available

Library of Congress Control Number: 2023930434

ISBN 978-0-19-284386-9

DOI: 10.1093/oso/9780192843869.001.0001

Printed and bound by
CPI Group (UK) Ltd, Croydon, CR0 4YY

We dedicate this book to our families who had to put up with our absence while we wrote it.

Preface

Ultrashort optical pulses, with duration ranging from a few picoseconds down to a few femtoseconds, have been used to study dynamics in matter since pulsed lasers were first developed in the 1960s. Indeed, this research area has been one of the drivers for improvements in the field ultrafast optics, such as reductions in pulse duration. The field of using ultrashort light pulses to probe dynamical processes in matter is generally known as "ultrafast spectroscopy."

Beginning in the late 1990s, the field of ultrafast spectroscopy underwent a revolution due to the introduction of multidimensional coherent spectroscopy based on concepts originally developed in nuclear magnetic resonance spectroscopy. Given their power, multidimensional coherent methods are becoming the dominant ultrafast spectroscopic techniques.

This book presents optical multidimensional coherent spectroscopy methods and their application to systems and materials that fall primarily within the field of physics. The systems include atomic vapors and solids—particularly semiconductors and semiconductor nanostructures. Multidimensional coherent spectroscopy in the infrared and visible spectral regions has been more extensively used to study molecules. As the application of multidimensional coherent spectroscopy to molecular systems has been covered by other books, we have chosen to not repeat that coverage here. Rather, we seek to broaden the coverage by addressing applications that are largely not covered elsewhere.

We begin by providing an introduction of multidimensional coherent spectroscopy for researchers in all fields, whether or not they have a background in ultrafast spectroscopy, or even in optical spectroscopy more generally. We then focus on the use of the technique to probe systems that are primarily of interest in the fields of physics and materials science. Our goal is to illustrate the information that multidimensional coherent spectroscopy can provide and its advantages over other methods. To do so, we focus on several exemplary materials, but also aim to illustrate the technique's broader applicability.

Acknowledgements

We acknowledge all of those individuals who have worked or collaborated with us on developing and using multidimensional coherent spectroscopy including Diogo Almeida, Travis Autry, Wan Ki Bae, Manfred Bayer, Camelia Borca, Alan Bristow, Xingcan Dai, Matthew Day, Denis Karaiskaj, Irina Kuznetsova, Xiaoqin (Elaine) Li, Albert Liu, Torsten Meier, Richard Mirin, Eric Martin, Shaul Mukamel, Gaël Nardin, Lazaro Padilha, Marten Richter, Mark Siemens, Kevin Silverman, Rohan Singh, Bo Sun, Takeshi Suzuki, Peter Thomas, Andreas Wieck and Tianhao Zhang. STC would like to acknowledge the sustained funding for research on this topic from the DOE Atomic, Molecular and Optical Science program and NSF funding through the JILA Physics Frontier Center.

Table of symbols

Symbol	Description	Section
A	cross-sectional interaction area	3.1
a_0	Bohr radius	5.3
a_B	exciton Bohr radius	9.1
$\hat{B}/\hat{B}^\dagger$	exciton annihilation/creation operator	5.3
C	probability amplitude	1.4
C_i	i^{th} London dispersion coefficient	5.3
c	speed of light in vacuum	1
D	spatial diffusion coefficient	1.3
$\mathscr{D}_\alpha(x, y, z)$	double-sided Feynman diagram equation	3.1
$E(t)$	total electric field in the time domain	1.2
$\hat{E}(t)$	complex-valued, slowly varying component of $E(t)$	1.2
$\tilde{E}(t)$	phasor description of E	2.3.3
E_0	real-valued electric field amplitude	1.2
E_B	biexciton energy	9.1
E_e	single-particle electron energy	9.1
E_h	single-particle hole energy	9.1
E_n	energy difference between two quantum states	1
E_{signal}	radiated electric field from polarization	2.3.3
E_X	exciton energy	9.1
$\mathcal{F}\{...\}$	Fourier transform operation	1.2
$\mathcal{E}(\omega)$	total electric field in the frequency domain	1.2
$\mathcal{E}_0$	electric field pulse area	1.2
f_0	comb offset frequency	6.1
f_{CE}	carrier-envelope offset frequency	4.4
f_n	frequency of nth comb tooth	6.1
f_{rep}	laser repetition rate	4.4
$G(\omega)$	Fourier transform of $g(t)$ into the frequency domain	3.2
$g(t)$	Gaussian distribution function in the time domain	3.10
H	Hamiltonian operator of a system	1.4
H_0	unperturbed Hamiltonian operator of a system	1.4
h	Planck's constant, $h = 6.62607015 \times 10^{-34}$ J·s	1
$\hbar$	reduced Planck constant, $\hbar = h/2\pi$	1.4
I	optical intensity	1.2
I_{SI}	spectral interferogram intensity	4.3
$\mathcal{I}$	electric field power spectrum in the frequency domain	1.2
J_{mn}	one-exciton coupling strength between excitons m and n	5.3
K_{mn}	two-exciton coupling parameter	5.3

$\Delta\phi_{ce}$	carrier-envelope phase-slip	6.1
ΔX	exciton binding energy	7.1
$\delta(t)$	Dirac delta function	2.3.1
δt	temporal intensity pulse duration	1.2
δt_e	temporal intensity e^{-1} half width	1.2
δt_{e2}	temporal intensity e^{-2} half width	1.2
δt_{FWHM}	temporal intensity full-width at half maximum	1.2
δU	spectral intensity bandwidth in energy	1.2
$\delta\alpha_0$	change in absorption coefficient	1.3
$\delta\kappa$	spectral intensity bandwidth in wavenumbers	1.2
$\delta\lambda$	spectral intensity bandwidth in wavelength	1.2
$\delta\nu$	spectral intensity bandwidth in frequency	1.2
$\delta\omega$	spectral intensity bandwidth in angular frequency	1.2
$\delta\omega_e$	spectral intensity e^{-1} half width	1.2
$\delta\omega_{e2}$	spectral intensity e^{-2} half width	1.2
$\delta\omega_{\mathrm{FWHM}}$	spectral intensity full-width at half maximum	1.2
ϵ_0	vacuum permittivity	3.1
Γ	relaxation operator	1.4
Γ_{10}	excited state population decay rate	3.1
Γ_{gr}	grating relaxation rate	1.3
Γ_{pop}	population relaxation rate	1.3
γ	dephasing rate / homogeneous linewidth	1.3
γ_{10}	excited/ground state decoherence rate	3.1
γ_{ij}	imaginary part of resonance frequency between states i and j	2.3.1
γ^{ph}	pure dephasing rate	1.4
γ_t	single quantum dephasing rate	6.4
γ_τ	double quantum dephasing rate	6.4
$\Theta(x)$	Heaviside step function	1.3
θ	angle between beams 1 and 2	1.3
ϑ	Bloch sphere polar angle	1.5
Λ	spatial grating period	1.3
λ	wavelength of light in vacuum	1
μ	electric dipole moment	1.4
ν	frequency of light	1
ν^*	difference frequency between signal and CW reference laser	4.4
π	mathematical constant $\pi \approx 3.1415$	1.1
ρ	density matrix operator	1.4
σ	N-particle spectral distribution width	3.2
τ	delay between first and second pulses	1.3
τ_{ex}	two-level transition excited-state lifetime	1.3
χ	electric susceptibility	1.1

ϕ	optical phase	1.2
$\phi_{S,R}$	phase of signal/reference electric field	4.3
φ	Bloch sphere azimuthal angle	1.5
$|\psi\rangle$	quantum mechanical wave function	1.4
Ω_{ij}	complex resonance frequency between states i and j	2.3.1
Ω_r	Rabi frequency	1.4
ω	optical angular frequency	1.2
ω_0	resonant angular frequency of an optical transition	3.2.1
ω_c	carrier angular frequency of a laser	1.2
ω_{ij}	real part of resonance frequency between states i and j	2.3.1
ω_t	frequency conjugate to emission time, t	3.1
ω_τ	frequency conjugate to coherence time, τ	3.1

Table of acronyms

Acronym	Description
2DCS	two-dimensional coherent spectroscopy
2DES	two-dimensional electronic spectroscopy
AOM	acousto-optic modulator
BS	beamsplitter
°C	degrees Celsius
c.c.	complex conjugate
CAD	computer-aided design
CB	conduction Band
CCD	charge-coupled device
CP	compensation plate
CQD	colloidal quantum dot
CVD	chemical vapor deposition
CW	continuous wave
DBR	distributed Bragg reflector
DCM	dichroic mirror
DCS	dual-comb spectroscopy
DDS	direct-digital synthesizer
DFT	discrete Fourier transform
DQW	double quantum well
DSP	digital signal processor
EID	excitation induced dephasing
EIS	excitation induced shift
ESE	excited-state emission
FFT	fast Fourier transform
FPGA	field programmable gate array
FTIR	Fourier-transform infrared
FWHM	full-width at half-maximum
GNE	gold nanoelectrode
GSB	ground-state bleaching
HeNe	helium-neon
hh	heavy hole
HWP	half-wave plate
IFQD	interface fluctuation quantum dots
IR	infrared
lh	light hole
LO	local oscillator
LP	lower polariton

MAPI	Methylammonium Lead Iodide
MBE	molecular beam epitaxy
MDCS	multi-dimensional coherent spectroscopy
MONSTR	multidimensional optical nonlinear spectrometer
MQC	multi-quantum coherence
NA	numerical aperture
NMR	nuclear magnetic resonance
NW	narrow well
OD	optical density
PBS	polarizing beamsplitter
PL	photoluminescence
PLE	photoluminescence excitation
PZT	piezoelectric transducer
QD	quantum dot
QW	quantum well
RBM	radial breathing mode
RF	radio frequency
SAQD	self-assembled quantum dot
SERS	surface-enhanced Raman spectroscopy
SNR	signal-to-noise ratio
SR-TFWM	spectrally resolved transient four-wave mixing
STM	scanning tunneling microscope
SWNT	single-walled carbon nanotube
TBP	time-bandwidth product
TC	lock-in amplifier time constant
TCS	tri-comb spectroscopy
TFWM	transient four-wave mixing
TI-TFWM	time-integrated transient four-wave mixing
TMD	Transition Metal Dichalcogenide
Tr	trace of a linear operator
TR-TFWM	time-resolved transient four-wave mixing
UP	upper polariton
VB	valence band
WW	wide well
ZPL	zero phonon line

Contents

1
Basics of ultrafast spectroscopy

Starting from Issac Newton's use of a prism to observe the spectrum of sunlight, optical spectroscopists have been striving to further our understanding of matter by studying how it absorbs and emits light. Spectroscopic techniques remained fundamentally unchanged for centuries compared to Newton's method. Light was dispersed by a prism or diffraction grating and the intensity was measured as a function of angle, which can be mapped into wavelength. Changes in the spectral intensity can be related to either inherent properties of the source or the light passing through a medium that absorbs at specific wavelengths. The absorption or emission wavelength can be converted to a frequency through $c = \lambda \nu$ where c is the speed of light, λ is the wavelength of the light and ν is its frequency. Using the quantum mechanical relation between energy, E_n, and frequency, $E_n = h\nu$, where h is Planck's constant, the frequency can be understood as the energy difference between two quantum states such as electronic or vibrational states.

Before the invention of the laser, spectroscopic measurements were all performed in the linear regime where the material properties are independent of the intensity of the light. In this regime the electric field of the light is weak compared to the internal fields of the atom or molecule. A laser can produce light that is no longer weak compared to the internal fields of an atom or molecule, thus the invention of the laser brought the field of optical spectroscopy into a new era of nonlinear spectroscopy. In the nonlinear regime, the material properties are no longer independent of the light intensity, signals scale with a higher power of the laser intensity, and two laser beams can interact in a sample. For instance, a strong pump beam can saturate an absorption resonance and thus increase the transmission of a weaker probe beam. Also, wave-mixing of multiple light fields in the sample can lead to a signal beam with an entirely new direction and frequency.

Optical spectroscopic measurements can also be made not as a function of wavelength but rather of time. Time-domain spectroscopy is analogous to the idea of a stroboscope, in that short flashes of light can capture stop-action images of ultrafast dynamics such as a chemical reaction or charge carriers relaxing in a solid. Measurements made in the time domain using laser pulses can be converted into frequency-domain spectra using a Fourier transform. In general, the time-domain spectra can be functions of multiple time delays, so the resulting frequency-domain spectra are functions of multiple frequencies and thus are multidimensional. The concept of multidimensional Fourier transform spectroscopy was developed in nuclear magnetic resonance (NMR) [108] and is now transforming the field of ultrafast laser spectroscopy.

Multidimensional coherent spectroscopy has a number of advantages over other types of spectroscopy, including one-dimensional methods and multidimensional methods that are not coherent. At the same time, multidimensional coherent spectroscopy is challenging to experimentally implement in the optical portion of the electromagnetic spectrum due to the need to use phase-related light pulses to excite the sample and to measure the phase of the emitted light signals.

To set the stage before discussing multidimensional coherent methods, in this chapter we will review several ultrafast spectroscopic methods, both because they serve as the foundation from which multidimensional coherent methods were developed, and they provide context for describing the advantages of multidimensional coherent methods. In Chapter 2 we will introduce the basic concepts involved in multidimensional coherent spectroscopy, followed by an in-depth discussion of how to interpret multidimensional coherent spectra in Chapter 3. Chapter 4 will review several experimental implementations, describing how each overcomes the aforementioned challenges.

1.1 Basics of spectroscopy: linear versus nonlinear

The field of spectroscopy involves measuring a spectrum that displays the frequency (spectral) dependence of the interaction between matter and electromagnetic radiation. The electromagnetic radiation may be incident on the matter from an external source, or it may be emitted by the matter. In this book, we will discuss the former case.

When an electromagnetic field is incident on matter, it displaces the electrons or ions in the matter from their equilibrium positions, producing a polarization in the matter that in turn radiates a new electromagnetic field. Treating the polarization as a driving term in Maxwell's equations and taking the far-field limit gives the result that the phase of the reradiated field lags the phase of the polarization by 90° (a factor of i in complex phasor notation). The interference of this reradiated field with the incident field results in modification of the field due to propagation through the matter, which is usually attributed to material properties such as an index of refraction or absorption.

In linear spectroscopy, the incident electromagnetic field is weak and the induced polarization is linearly proportional to the incident field. For a continuous wave (CW) incident field, the polarization will have the same frequency and wavevector as the incident field. However, the phase of the polarization with respect to the incident field depends on frequency if resonances are present in the material. First, we consider only a single resonance as the simplest example.

If the frequency of the incident field is significantly below the resonant frequency, then the induced polarization will be in-phase with the incident field and the reradiated field will be 90° out of phase in the far field due to aforementioned far-field phase lag. Since the polarization has the same wavevector as the incident field, the reradiated field will propagate in the same direction, so a detector placed after the sample will detect the sum of the incident field and the reradiated field. If the reradiated field is weak compared to the incident field, the dominant result will simply be a phase shift of the transmitted field compared to the incident field, as illustrated in Fig. 1.1(a) [note that $e^{ikx} + i\delta e^{ikx} \approx e^{i(kx+\delta)}$ for small values of δ, where $k = 2\pi n/\lambda$]. This phase shift is consistent with a transparent material with an index of refraction n.

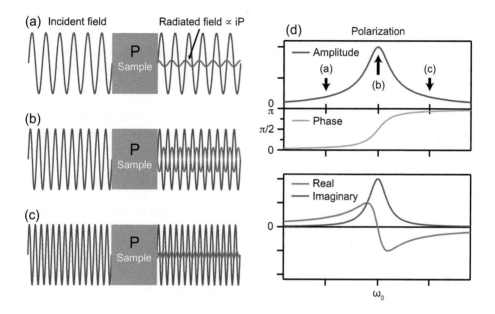

Fig. 1.1 Sketch of linear spectroscopy showing incident fields and reradiated fields in the vicinity of an optical resonance. **(a)** Below resonance. **(b)** On resonance. **(c)** Above resonance. **(d)** Frequency dependence of the complex polarization, in terms of amplitude and phase (top) and real and imaginary quadratures (bottom).

If the frequency of the incident field is tuned so that it is on the resonance in the sample, the phase of the polarization will lag that of the incident field by 90°. Together with the far-field phase shift, the net result will be that the reradiated field will be 180° out of phase with the incident field and they will experience destructive interference when measured together, as sketched in Fig. 1.1(b). This destructive interference means that a lower transmitted intensity is measured, as expected, because the incident field is now being absorbed by the sample since it is on resonance.

Tuning the frequency of the incident field to be above resonance results in a situation similar to below resonance, just a phase shift of the transmitted field, although the phase shift is in the opposite direction, as shown in Fig. 1.1(c).

Taken as a whole, this simplified picture of absorption spectroscopy is useful because it emphasizes the importance of the relative phase between the incident field and the induced polarization: it is all the difference between a material being transparent but causing a phase shift of the transmitted light and the material absorbing the light and resulting in less light being transmitted. The ability to measure the phase of the polarization with respect to the incident fields can be implemented in multidimensional coherent spectroscopy, and it yields important information about how the material is responding to the incident light.

The complete frequency dependence of the material polarization in the vicinity of the resonance is depicted in Fig. 1.1(d), divided into complex amplitude and phase components at top and into real and imaginary quadratures on bottom. The quadrature depiction, in particular, exhibits a number of important features that will serve as useful reference points for the nonlinear multidimensional spectra that will be discussed in much of the rest of this book. As mentioned above, when the polarization signal is both small and primarily real-valued, the dominant effect on the emitted signal is a change in phase. In fact, the real part of the polarization remains closely connected to phase shifts in the optical field regardless of what imaginary part if the overall reradiated signal amplitude remains small in comparison to the incident field. Because these phase shifts are different for light at different frequencies, the quadrature is very often referred to in spectroscopy literature as the polarization's "dispersive" component. Likewise, the polarization's imaginary component bears a strong resemblance to the material's absorption properties and as such is frequently termed the signal's "absorptive" component. In terms of the reradiated field—which is the most commonly examined quantity in MDCS—the dispersive component shifts into the signal's imaginary quadrature and the absorptive component shifts into the real quadrature.

This discussion has described a simple spectroscopy experiment, namely sweeping the frequency of an electromagnetic field incident on a sample and measuring the transmitted intensity. If a dip in the transmitted intensity is observed at a specific frequency it indicates the presence of resonance; the width of the resonance characterizes the damping of the resonance. However, there are important ambiguities in linear spectra, namely the inability to distinguish between homogeneous and inhomogeneous broadening and the inability to determine if two resonances are coupled or uncoupled.

Typically the sample is an ensemble of many systems, whether they are atoms undergoing electronic transitions, molecules undergoing vibrational transitions, or nuclei flipping their spins. If all the systems in the ensemble are identical, i.e., they have the same resonant frequency and same linewidth, the ensemble is designated as being "homogeneously broadened"; however, this may not necessarily be the case. In particular, there may be a distribution of resonant frequencies due to effects such as the Doppler shift in a vapor, random crystal fields in an ion-doped solid or structural disorder in a nanostructure. In this case, the linewidth of the measured resonance may have nothing to do with the linewidth of the individual members of the ensemble, but rather reflects the distribution of resonance frequencies. This case is known as "inhomogeneous broadening."

The linear spectrum of an inhomogeneously broadened ensemble will have a resonance feature (the absorption "line") that has a width that is characteristic of the inhomogeneous distribution, not the damping of the individual members of the ensemble. While both are useful to know, they provide quite different information. The width in the absence of the inhomogeneous broadening, often called the "homogeneous width" provides information about processes that interrupt the oscillations, for example collisions and radiative decay.

There is also an ambiguity in linear spectroscopy if two resonances are observed in a linear spectrum. There are two possible situations. One possibility is that the sample is heterogeneous, i.e., a mixture of two species with different resonance frequencies. The other possibility is that it is pure, i.e., a single substance, but that substance has two transitions. A good example of this latter case would be the D_1 and D_2 lines in the alkali metals, which correspond to the single outer electron making a transition to ground $S_{1/2}$ state the $P_{1/2}$ and $P_{3/2}$ states. Linear spectroscopy cannot distinguish between these two possibilities.

These ambiguities can be resolved by using some form of nonlinear spectroscopy. In nonlinear spectroscopy, as the intensity of the excitation field is increased, the dielectric polarization $P(\omega)$ of a material is no longer linearly proportional to the incident field, but rather higher-order terms must be considered. In the frequency domain, we can describe this in terms of a power series expansion of $P(\omega)$ as function of electric field strength $\mathcal{E}(\omega)$ as

$$P[\mathcal{E}(\omega)] = \epsilon_0 \left[\chi^{(1)}\mathcal{E} + \chi^{(2)}\mathcal{E}^2 + \chi^{(3)}\mathcal{E}^3 + \chi^{(4)}\mathcal{E}^4 + \dots \right], \qquad (1.1)$$

where the constant ϵ_0 is the vacuum permittivity and the coefficients $\chi^{(n)}(\omega)$ describe electric susceptibility parameters of the material at each of the different orders n. Linear spectroscopy corresponds to the situation where all the terms $\chi^{(n)}\mathcal{E}^n$ for $n > 1$ are small enough to be neglected, resulting in the relationship $P(\omega) = \epsilon_0 \chi^{(1)}(\omega)\mathcal{E}(\omega)$. Of course, this approximation depends on the strength of $\mathcal{E}$, because for large enough $\mathcal{E}$, the $\mathcal{E}^n$ factor can make $\chi^{(n)}\mathcal{E}^n > \chi^{(1)}\mathcal{E}$, no matter how small the ratio $\chi^{(n)}/\chi^{(1)}$. It is possible to show that for inversion symmetric systems, the second-order term in the expansion of Eq. (1.1), and indeed all of the even-valued higher-order terms, must be identically equal to zero.[1] Hence, in delving into the world of nonlinear spectroscopy, it is often the $\chi^{(3)}$ term that is actually the most important element governing nonlinear corrections to the polarization as a whole, and so it is upon this term that we will most heavily concentrate our attention in this book.

To understand how nonlinear spectroscopy can resolve the ambiguities in a linear spectrum, it is easiest to consider a simple frequency-domain method known as "spectral hole burning." In spectral hole burning, a continuous wave (CW) "pump" laser excites the sample, saturating its absorption. A second laser is then scanned to measure the absorption of the sample. If the sample is homogeneously broadened, the absorption of the entire line simply decreases. However, if it is inhomogeneously broadened, then the sub-ensemble that is resonant with the pump laser is most strongly saturated. In this case, the measured absorption spectrum is unchanged, except in the spectral region close to the pump, where the absorption is decreased, known as "burning a hole." The width of the spectral hole is proportional to the homogeneous width. Thus the observation of spectral hole burning shows that the system is inhomogeneously broadened and the width gives the homogeneous width.

Similarly, if two resonances are present in the spectrum, tuning the pump laser to one resonance and probing the other can determine if they are coupled. If they are coupled, then this situation will result in a change in the absorption, whereas if they are uncoupled it will not. This example was based on using CW lasers. Although there are some implementations of optical multidimensional coherent spectroscopy based on this approach [55–57, 443], most are based on using mutually coherent pulses and scanning their delays.

A time-domain multidimensional coherent spectroscopy (MDCS) measurement is made by illuminating a sample with a series of light pulses and measuring a signal

---

[1]See, for example, *Nonlinear Optics*, by Robert W. Boyd [40].

from the sample as a function of the delays between the pulses. Typically the pulses have duration of a few picoseconds or less, which is considered the domain of "ultrafast optics," where traditional photodetectors are too slow to directly measure the pulse duration. Due to their short duration, such pulses intrinsically have broad spectral bandwidth, thus spectral features can be measured without tuning them spectrally, but rather by spectrally resolving them. While this can be done using traditional spectrometers, it can also be realized using Fourier transform methods. Some MDCS approaches only use Fourier transforms, whereas others use a combination of Fourier transforms and a spectrometer to spectrally resolve the signal.

Before we introduce MDCS, we need to briefly review the properties of ultrashort pulses and introduce less related forms of spectroscopy that are based on ultrashort optical pulses.

1.2   Ultrashort pulses

A short optical pulse passing through a fixed point in space can be described by its electric field in the time domain

$$E(t) = \left|\hat{E}(t)\right| \cos\left(-\omega_c t + \phi(t)\right)$$

$$= \frac{1}{2}\hat{E}(t)e^{-i\omega_c t} + \text{c.c.}, \tag{1.2}$$

where ω_c is the carrier frequency, and where $\phi(t)$ is a time-dependent phase. The second line of the equation is expressed in phasor notation, with the complex-valued amplitude $\hat{E}(t) = |\hat{E}(t)|e^{i\phi(t)}$, and with the abbreviation "c.c." standing for "complex conjugate." For the discussion throughout the first three chapters of this book, we ignore the the fact that light has a polarization, and thus treat the electric field as a scalar. Note that the choice of ω_c is in principle arbitrary; the same pulse could be described using a different ω_c by adjusting the time dependence of the amplitude coefficients to include linearly ramping phase factors. Typically, however, ω_c is chosen to eliminate a linear ramp in $\phi(t)$. Throughout this book we use a caret ($\hat{X}$) placed over the top of a function X to indicate a value assumed to vary "slowly" in time (i.e., to be nearly constant over several periods of the optical frequencies).

Although slow in comparison to the carrier oscillations, time dependence of the amplitude factor $\hat{E}(t)$ as introduced in Eq. (1.2) is still required to shape the light into a pulse. Writing out the amplitude and phase components of $\hat{E}(t)$ explicitly and then expressing them in terms of optical intensity $I(t) \equiv n\epsilon_0 c \langle E(t)^2 \rangle$ leads to

$$E(t) = \frac{1}{2}\sqrt{\frac{2I(t)}{n\epsilon_0 c}}e^{-i[\omega_c t - \phi(t)]} + \text{c.c.}, \tag{1.3}$$

where n is refractive index, ϵ_0 is the vacuum permittivity, c is the speed of light, and the angle brackets in the definition of intensity specify the time average over an optical period. As shown in Fig. 1.2, the pulse described using Eq. (1.3) can be visualized as an oscillating carrier, of frequency ω_c, under an envelope proportional to $\sqrt{I(t)}$. The constant in the phase factor $\phi(t)$ allows the possibility of a shift in

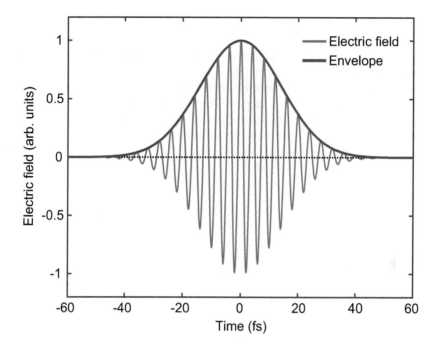

Fig. 1.2 The electric field (solid red line) and the envelope (blue line) of an ultrafast pulse.

alignment between the ripples of the carrier and the overall envelope position, known as the "carrier-envelope phase."

One reasonable mathematical description of an ultrafast pulse is Gaussian. The envelope of a Gaussian pulse is

$$\hat{E}(t) = E_0 e^{-\frac{t^2}{(\delta t_{e2})^2}} \tag{1.4}$$

where E_0 is the (real-valued) amplitude and the parameter δt_{e2} describes pulse duration. The pulse intensity associated with Eq. (1.4) is

$$I(t) \propto |\hat{E}(t)|^2 = E_0^2 e^{-\frac{2t^2}{(\delta t_{e2})^2}} = E_0^2 e^{-\frac{t^2}{(\delta t_{e2}/\sqrt{2})^2}}, \tag{1.5}$$

which demonstrates that for a Gaussian pulse, the duration parameter δt_{e2} corresponds to the half width at which the pulse intensity drops to $1/e^2$ (13.5%) of its peak. Beyond this, the equation shows that the pulse duration of the intensity function is a factor of $\sqrt{2}$ shorter than the pulse duration of the amplitude function. Although the $1/e^2$ half-width definition of pulse duration is common, the duration of a pulse can be defined according to any number of different conventions, including the intensity half width $1/e$ value (δt_e), the intensity full width at half maximum (FWHM) value (δt_{FWHM}), and various definitions associated with the intensity autocorrelation function. Thus, it is important to define the pulse duration carefully when introducing it as a parameter in order to avoid ambiguities. For Gaussian pulses, the duration parameters δt_e, δt_{e2}, and δt_{FWHM} are related to each other by

$$(\delta t_{\mathrm{FWHM}}) = \sqrt{2 \ln 2}\,(\delta t_{e2}) = 2\sqrt{\ln 2}\,(\delta t_e). \tag{1.6}$$

However, the conversion factors relating the various width definitions are different for different pulse shapes.

An ultrashort pulse can also be described in the frequency domain where the electric field is written as a function of the angular frequency or frequency. The conversion between the time and frequency domains is performed by the Fourier Transform of the electric field (not the intensity), where the forward Fourier Transform

$$\mathcal{E}(\omega) = \int_{-\infty}^{\infty} E(t) e^{i\omega t}\,dt \tag{1.7}$$

converts from a time-domain signal, $E(t)$, to its frequency-domain representation, $\mathcal{E}(\omega)$, and the inverse Fourier transform

$$E(t) = \frac{1}{2\pi} \int_{-\infty}^{\infty} \mathcal{E}(\omega) e^{-i\omega t}\,d\omega. \tag{1.8}$$

converts the other direction, giving the time-domain waveform from the frequency-domain representation. Note that there differing conventions for the definition of the Fourier Transform with regards to the sign of the exponent in the kernel ($e^{i\omega t}$ versus $e^{-i\omega t}$) and normalization ($\frac{1}{2\pi}$ in front of the inverse Fourier transform versus $\sqrt{\frac{1}{2\pi}}$ in front of both the forward and inverse transforms). Throughout this book we will use the conventions given in Eqs. (1.7) and (1.8). The Fourier transform of Eq. (1.3) gives

$$\mathcal{E}(\omega) = \frac{1}{2}\sqrt{\frac{2\mathcal{I}(\omega - \omega_c)}{n\epsilon_0 c}}\,e^{i\phi(\omega-\omega_c)} + \frac{1}{2}\sqrt{\frac{2\mathcal{I}(-\omega - \omega_c)}{n\epsilon_0 c}}\,e^{-i\phi(-\omega-\omega_c)}. \tag{1.9}$$

The field in the frequency domain has both positive and negative frequencies. The two frequency components are actually equivalent, but required to make the signal real. We usually take only the positive frequency and center the pulse at zero frequency (i.e., expressed in the rotating frame) in complex phasor notation as

$$\mathcal{E}(\omega) = \frac{1}{2}\sqrt{\frac{2\mathcal{I}(\omega)}{n\epsilon_0 c}}\,e^{i\phi(\omega)}. \tag{1.10}$$

Here $\mathcal{I}(\omega)$ is the spectral intensity function and the frequency-domain phase is

$$\phi(\omega) = \arctan\left\{ \frac{\mathrm{Im}[\mathcal{E}(\omega)]}{\mathrm{Re}[\mathcal{E}(\omega)]} \right\}. \tag{1.11}$$

The phase calculated from Eq. (1.11) can only vary from $-\pi$ to π so there could be 2π phase jumps. We can unwrap the phase by adding or subtracting 2π to avoid phase jumps and have a continuous phase. The phase defined by Eq. (1.11) has no meaning when the intensity is zero. The phase is noisy when the intensity is small at the wings of spectrum, in which case the phase is usually not plotted. According to the Fourier shift theorem, a delay in the time domain is a linear phase ramp in the frequency domain, i.e., $\mathcal{F}\{E(t-t_0)\} = \mathcal{E}(\omega)e^{i\omega t_0}$.

Applying the Fourier transform to the time-domain Gaussian pulse described by Eq. (1.4) gives the frequency-domain pulse in a frame rotating at the carrier frequency as

$$\mathcal{E}(\omega) = \sqrt{\pi}(\delta t_{e2})E_0 e^{-\frac{(\delta t_{e2})^2\omega^2}{4}}. \tag{1.12}$$

The intensity spectrum of the pulse is

$$\mathcal{I}(\omega) = \frac{n\epsilon_0 c}{2}\pi(\delta t_{e2})^2 E_0^2 e^{-\frac{(\delta t_{e2})^2\omega^2}{2}} \propto E_0^2 e^{-\frac{2\omega^2}{(2/\delta t_{e2})^2}}, \tag{1.13}$$

which demonstrates that the spectral profile of a time-domain Gaussian pulse is also a Gaussian. The $1/e^2$ half width intensity spectral bandwidth of this Gaussian is $(\delta\omega_{e2}) = 2/(\delta t_{e2})$.

As in the time-domain representation of the electric field, there are different kinds of conventions for describing spectral bandwidth, including the spectral intensity $1/e^2$ half width (δ_{e2}), the spectral intensity $1/e$ half width ($\delta\omega_e$), and the spectral intensity FWHM value ($\delta\omega_{\text{FWHM}}$). In similar fashion to the ways that time-domain pulse durations are related to each other, for Gaussian pulses these frequency-domain bandwidth descriptions are connected through the relationship

$$(\delta\omega_{\text{FWHM}}) = \sqrt{2\ln 2}\,(\delta\omega_{e2}) = 2\sqrt{\ln 2}\,(\delta\omega_e). \tag{1.14}$$

Different research fields also use different units in the frequency domain. The spectrum and bandwidth can be presented as a function of frequency (THz), angular frequency (rad/s), wavelength (nm), energy (eV), or wavenumber (cm^{-1}). The bandwidth and even the lineshape are different when different units are used. The bandwidth can be properly converted into different units. As an example, we consider a short pulse that has a spectral bandwidth in frequency as

$$\delta\nu = 5 \text{ THz}. \tag{1.15}$$

This bandwidth can be converted, for example, into units of angular frequency,

$$\delta\omega = 2\pi(\delta\nu) = 3.14 \times 10^{13} \text{ rad/s}, \tag{1.16}$$

wavenumber,

$$\delta\kappa = \frac{(\delta\nu)}{c} = 167 \text{ cm}^{-1}, \tag{1.17}$$

energy,

$$\delta U = h(\delta\nu) = 20.7 \text{ meV}, \tag{1.18}$$

or wavelength,

$$\delta\lambda = \frac{\lambda^2}{c}(\delta\nu), \tag{1.19}$$

where the last of these expressions is wavelength dependent. Assuming a wavelength of 800 nm, we have $\delta\lambda \approx 10$ nm.

Considering the intensity pulse duration $\delta t_{\text{FWHM}} = \sqrt{2\ln 2}\,(\delta t_{e2})$, the pulse duration and the spectral bandwidth are related as $\delta t_{\text{FWHM}} = 4\ln 2/(\delta\omega_{\text{FWHM}})$. Therefore,
a shorter pulse in the time domain requires a broader spectrum in the frequency do-
main and vice versa. The shortest possible pulse for a given bandwidth, known as the
Fourier transform-limited pulse, can be achieved when the spectral phase is constant
across the spectrum. The product of the bandwidth and the temporal duration of a
pulse is a dimensionless number known as the time-bandwidth product (TBP). The
TBP has a minimum value for a transform-limited pulse and the exact value depends
on the pulse shape. If consistent units are used for time and bandwidth, the TBP
is a unitless number. For example, a transform-limited Gaussian pulse has a TBP
of $(\delta t_{\text{FWHM}})(\delta\nu_{\text{FWHM}}) = 0.441$ while the TBP is $(\delta t_{\text{FWHM}})(\delta\nu_{\text{FWHM}}) = 0.315$ for
a transform-limited sech^2-shaped pulse. These relations imply that 100-fs Gaussian
pulses must have a bandwidth of 4.41 THz while 100-fs sech^2-shaped pulses need a
minimum bandwidth of 3.15 THz.

A pulse can have a carrier frequency that varies in time, in which case the pulse is
called a chirped pulse. A linearly chirped Gaussian pulse can be written as

$$E(t) = E_0 e^{-\frac{t^2}{(\delta t_{e2})^2}} e^{-i(\omega_c t + \beta t^2)}, \tag{1.20}$$

where ω_c is the carrier frequency and βt^2 is the chirp. The term βt^2 modifies the carrier
frequency and varies with time. It can be considered as a second-order phase. When
β is positive, the pulse increases its frequency linearly in time (from red to blue) and
is positively chirped. When β is negative, the pulse decreases its frequency linearly in
time (from blue to red) and is negatively chirped.

Fourier transforming Eq. (1.20) gives the frequency-domain expression of the chirped
pulse,

$$\mathcal{E}(\omega) = E_0 e^{-\frac{1/4}{(\delta\omega_{e2}/2)-i\beta}(\omega-\omega_c)^2} = E_0 e^{-\frac{(\delta\omega_{e2}/2)/4}{(\delta\omega_{e2}/2)^2+\beta^2}(\omega-\omega_c)^2} e^{-i\frac{\beta/4}{(\delta\omega_{e2}/2)^2+\beta^2}(\omega-\omega_c)^2}, \tag{1.21}$$

with $\delta w_{e2} = 2/(\delta t_{e2})$. Adding a chirp in the time domain changes the spectral width
but not the temporal width of the pulse, while adding a chirp in the frequency domain
changes the temporal width but not the spectral width of the pulse. In an experiment,
a chirp is usually created by propagating through a dispersive medium, which is to say
that the chirp is added in the frequency domain. As a result, in experiments, adding
chirp typically results in increasing the temporal width of the pulses.

1.3   Ultrafast nonlinear/coherent spectroscopy

Ultrafast pulses enable unique capabilities in spectroscopy. First, the time resolution
provided by ultrafast pulses can probe events that occur on fs to ps timescales. Second, the high instantaneous power and hence the strong electric field in ultrafast
pulses can lead to more efficient nonlinear effects for nonlinear spectroscopy. Finally,
a proper pulse sequence can be used to perform coherent spectroscopy. Incoherent
spectroscopy, such as time-resolved fluorescence/luminescence and transient absorption spectroscopy, is only sensitive to population relaxation and the results can be
interpreted by modeling with rate equations. In contrast, coherent spectroscopy, such

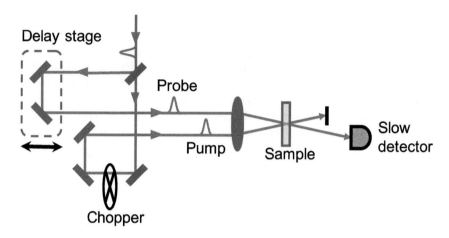

Fig. 1.3 Schematic of a typical pump-probe setup with two pulses and lock-in detection.

as transient four-wave mixing and multidimensional coherent spectroscopy, also probes phase relaxation and interpreting coherent spectra requires optical Bloch equations.

The simplest ultrafast technique is time-resolved fluorescence/luminescence spectroscopy. The technique uses only one pulse to excite the sample from the ground state to a high-lying excited state. The sample then relaxes to a lower excited state from which a fluorescence signal is spontaneously emitted. The fluorescence has a longer wavelength than the pump pulse so the signal can be distinguished from scattered pump photons. The signal can be recorded as a function of time by using time-resolved detection such as time-correlated photon counting or a streak camera, which has a typical time resolution of a few ps. The time resolution can be improved by cross correlating the fluorescence signal and a reference pulse through an upconversion nonlinear process, in which case the time resolution is only limited by the pulse duration. The measured fluorescence signal usually rises initially and then decays exponentially. The rise time of fluorescence is related to the relaxation time from the high-lying state to the lower excited state, while the decay dynamics are determined by the relaxation from the lower excited state to the ground state.

Pulse duration-limited time resolution can also be achieved by using a slow detector in the two-pulse transient absorption technique (also known as pump-probe). A typical pump-probe setup is depicted in Fig. 1.3. The sample is first excited by a pump pulse and the change in absorption due excitation created by the pump pulse is measured using a subsequent probe pulse. Typically the absorption decreases after the pump pulse due to bleaching of the transition by the pump-induced population of excited states and depletion of the ground state population. As the population relaxes from the excited state back into the ground state, the absorption will recover, which is monitored by changing the delay between pump and probe pulses. The time delay between the pump and probe pulses can be varied by translating a mirror or retroreflector using a delay stage, thereby changing the path length and hence time delay due to the changed time-of-flight. The average power of the transmitted probe pulse is measured by a slow detector. The change in the probe power due to the excitation by the pump is detected by using a lock-in amplifier as the pump beam is modulated by a optical chopper. The signal is recorded as a function of the time delay between the pump and probe pulses. The time resolution is determined by the minimum increment of the delay stage and the pulse duration. Assume that the unexcited sample has an absorption coefficient α_0 and the absorption coefficient decreases by $\Delta\alpha_0$ immediately after the excitation. For an excited state that decays exponentially with an excited-state lifetime of τ_{ex}, the change in the absorption coefficient at a delay time τ is given by

$$\Delta\alpha(\tau) = \Delta\alpha_0 e^{-\tau/\tau_{ex}}. \tag{1.22}$$

So the transmitted probe intensity and hence the average power depend on the delay τ and the lifetime τ_{ex} as

$$\begin{aligned} I_{tran}(\tau) &= I_{inc}e^{-[\alpha_0 - \Delta\alpha_0 e^{-\tau/\tau_{ex}}]L} \\ &\approx I_{tran}(0)(1 + \Delta\alpha_0 e^{-\tau/\tau_{ex}}L), \end{aligned} \tag{1.23}$$

where I_{inc} and I_{tran} are the incident intensity and the transmitted intensity, respectively, L is the sample length, and we assume $\Delta\alpha_0 L \ll 1$. The differential transmission, i.e., the relative change in transmitted intensity, or differential transmission, is

$$\begin{aligned} \frac{\Delta T(\tau)}{T_0} &= \frac{I_{tran}(\tau) - I_{tran}(0)}{I_{tran}(0)} \\ &\approx \Delta\alpha_0 e^{-\tau/\tau_{ex}}L, \end{aligned} \tag{1.24}$$

where T_0 is the transmission of unexcited sample and $\Delta T(\tau)$ is the change in the transmission at a delay τ. In the simplest case, the pump-probe signal features an exponential decay and measures the population decay dynamics of the pump-induced excitations. The pump-probe signal can be complicated in samples with more complex decay dynamics, for example, systems with multiple decay channels or intermediate states. The temporal behavior can deviate from a single exponential decay function and the differential transmission can even be negative. In the frequency-resolved version of pump-probe spectroscopy, known as spectrally resolved transient absorption, the transmitted probe is analyzed with a spectrometer to measure the wavelength dependence of the change in probe pulse. Spectral resolution can give more insight into the origin of the signal itself as well as helping to identify the reason for non-exponential dynamics.

A common coherent spectroscopic technique is transient four-wave mixing (TFWM), which can be performed with two or three pulses. The basic geometry for a two-pulse TFWM experiment is sketched in Fig. 1.4(a). Two pulses with wavevectors k_1 and k_2 are incident on the sample. The delay between the two pulses is τ and it is defined to be positive when pulse k_1 arrives first. The nonlinear interaction gives rise to a TFWM signal in the direction $k_s = 2k_2 - k_1$. A three-pulse TFWM experiment can be configured in different geometries. A planar geometry is shown in Fig. 1.4(b). The generated TFWM signal can be emitted in the direction $k_s = -k_1 + k_2 + k_3$. In the two-pulse and planar three-pulse geometries, it might be difficult to align the detector

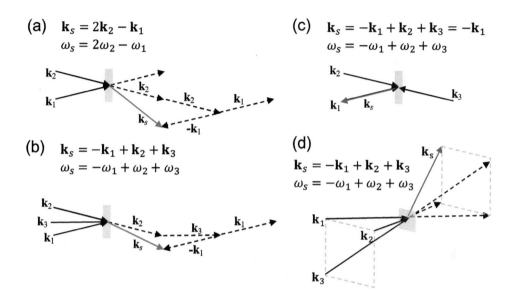

(a) $\mathbf{k}_s = 2\mathbf{k}_2 - \mathbf{k}_1$
$\omega_s = 2\omega_2 - \omega_1$

(c) $\mathbf{k}_s = -\mathbf{k}_1 + \mathbf{k}_2 + \mathbf{k}_3 = -\mathbf{k}_1$
$\omega_s = -\omega_1 + \omega_2 + \omega_3$

(b) $\mathbf{k}_s = -\mathbf{k}_1 + \mathbf{k}_2 + \mathbf{k}_3$
$\omega_s = -\omega_1 + \omega_2 + \omega_3$

(d) $\mathbf{k}_s = -\mathbf{k}_1 + \mathbf{k}_2 + \mathbf{k}_3$
$\omega_s = -\omega_1 + \omega_2 + \omega_3$

Fig. 1.4 Different geometries for TFWM experiments. (a) Two-pulse TFWM. (b) Three-pulse TFWM. (c) Three-pulse TFWM in the phase conjugate geometry. (d) Three-pulse TFWM in the box geometry.

to the signal direction when the TFWM signal is too weak to be visible. Neither of these geometries are fully phase-matched, i.e., the induced polarization has a different wavevector than the emitted signal. For thin samples the accumulated phase mismatch is small for small angles, and thus the lack of phase matching is not significant. However, for thick samples or large angles using a phase-matched geometry can be important.

A three-pulse TFWM experiment can be arranged in special geometries that are fully phase-matched and assist the alignment. One approach is the phase conjugate geometry, as shown in Fig. 1.4(c), where pulses $\mathbf{k}_2$ and $\mathbf{k}_3$ counter propagate, i.e., $\mathbf{k}_2 = -\mathbf{k}_3$, so the emission direction is $\mathbf{k}_s = -\mathbf{k}_1 + \mathbf{k}_2 + \mathbf{k}_3 = -\mathbf{k}_1$, i.e., the signal counter-propagates with respect to $\mathbf{k}_1$. Another approach is the box geometry, as shown in Fig. 1.4(d). In this geometry, three parallel beams with wavevectors $\mathbf{k}_1$, $\mathbf{k}_2$, and $\mathbf{k}_3$ are aligned to three corners of a square and focused by a lens to converge on the sample. The transmitted excitation beams point to three corresponding corners of the box on the other side of the sample. The TFWM signal direction $\mathbf{k}_s = -\mathbf{k}_1 + \mathbf{k}_2 + \mathbf{k}_3$ points to the fourth corner. For convenience, a fourth beam aligned to the fourth corner of the input box can be used as a reference beam to place the detector. The reference beam can be blocked during measurements. The box geometry is also commonly used in optical MDCS experiments. It is worth pointing out that the TFWM signal can emit in multiple phase-matched directions in addition to the direction $\mathbf{k}_s$ denoted in each geometry. However, the signals in other phase-matched directions contain the same information. We note that TFWM experiments can also be performed in the collinear geometry in which the excitation pulses co-propagate in one beam. In this case, different linear/nonlinear signals are not spatially separated and other methods such as frequency tagging are needed to isolate specific TFWM signals. The collinear geometry will be discussed in more detail in Chapter 4.

Based on how the signal is detected, TFWM experiment has three main variants: time-integrated (TI-TFWM), time-resolved (TR-TFWM), and spectrally resolved (SR-TFWM) TFWM. In TI-TFWM, a slow detector is used to integrate the signal over time. In TR-TFWM, the signal is measured with a sufficient time resolution to be resolved in time, using a reference pulse and up-conversion in a nonlinear crystal. The signal pulse is mapped out as a function of "real" time as the reference pulse is scanned across the signal in time. In SR-TFWM, a spectrometer is used to record the spectrum of the signal.

For a two-level system, the three-pulse TFWM experiment can be physically described as follows: The first pulse creates a coherence (also called a polarization) that oscillates at the absorption frequency. The second pulse converts the coherence to a population in either the excited or ground state, depending on the relative phase between the two pulses. For two crossing pulses, the relative phase changes with position to produce a population that is spatially modulated with a wavevector $\mathbf{k}_2 - \mathbf{k}_1$, forming a population grating, which in turn results in a grating in the sample absorption, analogous to the change in absorption for a pump-probe experiment. The third pulse then diffracts off the grating to generate the TFWM signal. Two-pulse TFWM is similar since the second pulse is considered as two pulses with a zero delay.

In two-pulse TFWM, the TFWM signal decays as a function of time delay τ between the pulses and the dynamics are determined by the decoherence rate in a sample of homogeneously broadened two-level systems [449]. In the approximation that the excitation pulses are infinitely short (i.e., a delta function in time) and within the Markovian approximation, the intensity of the TI-TFWM signal is a function of τ

$$I_s(\tau) \propto \Theta(\tau)e^{-2\gamma\tau}, \tag{1.25}$$

where γ is the dephasing rate (otherwise called the homogeneous linewidth) and $\Theta(\tau)$ is the Heaviside step function; however, in an inhomogeneously broadened system, the TFWM signal is a photon echo formed at a delay of τ after the second pulse. In the limit of strong inhomogeneous broadening, the TI-TFWM signal is

$$I_s(\tau) \propto \Theta(\tau)e^{-4\gamma\tau}. \tag{1.26}$$

Therefore, the TFWM technique can measure γ even in the presence of inhomogeneous broadening; however, to properly interpret the results, one must know whether or not the sample is inhomogeneously broadened. In some circumstances, comparing to the linear absorption linewidth can reveal whether Eq. (1.25) or (1.26) should be used, since the linear absorption spectrum is a convolution of the homogeneous and inhomogeneous lineshapes (a Voigt profile), which converges to the inhomogeneous lineshape, and hence a width determined by the inhomogeneous broadening, in the limit of strong inhomogeneous broadening. However, it can be ambiguous when the homogeneous and inhomogeneous linewidths are comparable. Additionally, the homogeneous linewidth can depend on the excitation intensity, making the comparison to an absorption linewidth more difficult. The TFWM signal can be temporally or spectrally resolved to help remove the ambiguity.

In three-pulse TFWM, the delay between the second and third pulse, T, can be varied. This gives the advantage to measure both the dephasing rate γ by scanning τ and the relaxation rate of the grating Γ_{gr} by scanning T [436]. For a homogeneous two-level system, the three-pulse TI-TFWM signal is

$$I_s(\tau, T) = \Theta(\tau)\Theta(T)e^{-2\gamma\tau}e^{-\Gamma_{gr}T}. \tag{1.27}$$

The factor of 2 multiplying γ becomes a 4 if the sample is homogeneously broadened. However, homogeneously and inhomogeneously broadened systems can be distinguished in three-pulse TFWM by measuring the signal around $\tau = 0$ for a fixed T where $T > 0$. The system is homogeneously broadened if the signal is symmetric about zero, whereas the system is inhomogeneously broadened if there is only signal for $\tau > 0$. In the presence of inhomogeneous broadening, three-pulse TFWM is also sensitive to spectral diffusion, a process by which an excitation initially at one energy can shift to another energy. Clear signatures of spectral diffusion can be observed by varying both T and τ. Three-pulse TFWM can also be used to study non-Markovian behaviors in a system where the Markovian approximation is not valid. Three pulse echo peak shift spectroscopy [85, 172, 436] can be used to extract the correlation function of the frequency fluctuations that give rise to dephasing.

The decay of the grating occurs at the rate Γ_{gr} is determined by the excited-state population decay rate Γ_{pop} and spatial diffusion, which washes out the population grating and leads to an exponential decay of the signal as a function of T. The grating relaxation rate is

$$\Gamma_{gr} = 2\Gamma_{pop} + 8\pi^2 D\Lambda^{-2}, \tag{1.28}$$

where D is the spatial diffusion coefficient and $\Lambda = n\lambda/2\sin\theta$ is the grating spacing for an angle θ between the beams $\mathbf{k}_1$ and $\mathbf{k}_2$. If there is no spatial diffusion or very slow diffusion compared to the population decay, γ gives a direct measurement of Γ_{pop}. In the presence of spatial diffusion, the effects of spatial diffusion and population decay can be separated by varying the angle between the first two pulses. Both Γ_{pop} and D can be determined by fitting the angle dependence of Γ_{gr} with Eq. (1.28).

1.4   The density matrix

Interpretation of laser spectroscopy relies on the theory of light-matter interaction. The semi-classical approach is usually used for laser spectroscopy in which matter is treated quantum mechanically while light is classical. Within this approximation, rudimentary scenarios can be treated using a wavefunction formalism in which the atoms are assumed to be coherent states at all times. Very quickly, however, it becomes helpful to expand this into a density matrix formalism capable of simulating incoherent states in addition to coherent ones. Here we illustrate the salient features of these two different formalisms in turn, using a two-level atom interacting with a monochromatic field as an illustrative model system.

We begin with the wavefunction formalism. As shown in Fig. 1.5, we consider a two-level atom with the ground state $|0\rangle$ and the excited state $|1\rangle$. An electromagnetic

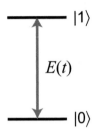

Fig. 1.5 Energy level diagram of a two-level atom interacting with a monochromatic field.

field with the angular frequency ω interacts with the atom. The electric field is $E(t) = |\hat{E}(t)| \cos(\omega_c t)$. The Hamiltonian of the system is $H = H_0 + V$, where H_0 is the unperturbed part,

$$H_0 = \hbar\omega_0 |0\rangle\langle 0| + \hbar\omega_1 |1\rangle\langle 1|, \tag{1.29}$$

here ω_0 and ω_1 are atomic frequencies such that $\hbar\omega_0$ and $\hbar\omega_1$ are the energies of states $|0\rangle$ and $|1\rangle$, respectively.

Under the dipole approximation, the atom is considered to be small compared to the wavelength of light, thus only its dipole moment needs to be considered. The interaction between the atom and the field is described by the interaction Hamiltonian

$$V = -\left(\mu_{10}|1\rangle\langle 0| + \mu_{01}|0\rangle\langle 1|\right)|\hat{E}(t)| \cos(\omega_c t)$$

$$= -\hbar\Omega_r \cos(\omega_c t)|1\rangle\langle 0| - \hbar\Omega_r \cos(\omega_c t)|0\rangle\langle 1|, \tag{1.30}$$

where $\mu_{10} = \mu_{01}^* = e\langle 1|x|0\rangle$ is the matrix element of the electric dipole moment, and $\Omega_r = \mu_{10}|\hat{E}(t)|/\hbar$ is the Rabi frequency.

The system dynamics are governed by the Schrödinger equation

$$i\hbar\frac{\partial}{\partial t}|\psi\rangle = H|\psi\rangle. \tag{1.31}$$

The state of the system can be expressed in the basis of the eigenstates of H_0 as

$$|\psi(t)\rangle = C_0(t)|0\rangle + C_1(t)|1\rangle. \tag{1.32}$$

Substituting Eqs. 1.29, 1.30, and 1.32 into Eq. 1.31, we obtain the equations of motion

$$\dot{C}_0 = -i\omega_0 C_0 + i\Omega_r \cos(\omega_c t)C_1, \tag{1.33}$$

$$\dot{C}_1 = -i\omega_1 C_1 + i\Omega_r \cos(\omega_c t)C_0. \tag{1.34}$$

Defining $C_0 = c_0 e^{-i\omega_0 t}$ and $C_1 = c_1 e^{-i\omega_1 t}$, the equations of motion can be rewritten

$$\dot{c}_0 = i\frac{\Omega_r}{2}\left(e^{i(\omega_c - \omega_{10})t} + e^{-i(\omega_c + \omega_{10})t}\right)c_1, \tag{1.35}$$

$$\dot{c}_1 = i\frac{\Omega_r}{2}\left(e^{-i(\omega_c - \omega_{10})t} + e^{i(\omega_c + \omega_{10})t}\right)c_0, \tag{1.36}$$

where $\omega_{10} = \omega_1 - \omega_0$. Applying the rotating-wave approximation in which the counter-rotating terms including $e^{\pm i(\omega_c + \omega_{10})t}$ are ignored yields

$$\dot{c}_0 = i\frac{\Omega_r}{2}e^{i(\omega_c - \omega_{10})t}c_1, \tag{1.37}$$

$$\dot{c}_1 = i\frac{\Omega_r}{2}e^{-i(\omega_c - \omega_{10})t}c_0. \tag{1.38}$$

The solutions to these equations describe the dynamics of the system. For simplicity, we consider the resonant case ($\omega_c = \omega_{10}$) and assume that the atom is initially in the ground state, i.e., $c_0(t = 0) = 1$ and $c_1(t = 0) = 0$. The solutions are

$$c_0(t) = -i\cos\left(\frac{\Omega_r}{2}t\right) \quad \text{and} \quad c_1(t) = \sin\left(\frac{\Omega_r}{2}t\right). \tag{1.39}$$

The population inversion between the excited and ground states is $W(t) = |c_1(t)|^2 - |c_0(t)|^2 = -\cos(\Omega_r t)$. The inversion oscillates between -1 and 1 at the Rabi frequency Ω_r, meaning that the population oscillates between the excited and ground states as a result of the field-atom interaction. Note that this solution has ignored any kind of dissipation or damping and thus assumes that the coherent superposition states induced in the atom do not experience any decoherence, either with respect to the driving field or within members of an ensemble. The inclusion of decoherence causes the amplitude of oscillations to decay with time after the field is turned on.

If the excitation field is a pulse instead of a CW field, the resulting inversion can be either 1 or -1 depending on the pulse area. The pulse area theorem states that the final state of the system does not depend on the shape of the pulse, but only its area. For pulses that are short compared to the system's relaxation time, the area of a pulse $\Omega_r(t)$ is

$$A = \int_{-\infty}^{+\infty} \Omega_r(t)dt. \tag{1.40}$$

A pulse with $A = \pi$, known as π-pulse, drives the system into the excited state and results in an inversion of 1. A 2π-pulse with $A = 2\pi$ drives the system into the excited state and then back to the ground state, leaving an inversion of -1.

We may ask ourselves at this point where the wavefunction formalism breaks down. The answer lies in the fact that state vectors such as $|\psi\rangle$ are ultimately incapable of capturing the physics of many practical situations involving dissipation or coupling to a thermal reservoir. Beyond this, we may want to consider an ensemble, and thus must take into account the possibility of finding different members of the ensemble existing within in a statistical mixture of different wavefunctions (this is a fundamentally different object from a quantum superposition), where the probability of being in state $|\psi\rangle$ is P_ψ. To capture these effects, it is necessary to expand our formalism into the language of density operators.

The density operator is defined as

$$\rho = \sum_\psi P_\psi |\psi\rangle\langle\psi|. \tag{1.41}$$

The state is called a pure state if the density operator can be written as $\rho = |\psi_0\rangle\langle\psi_0|$, that is the probability P_ψ is nonzero only for state $|\psi_0\rangle$. With the density operator, the expectation value of an operator O can be calculated as

$$\langle O \rangle = \sum_\psi P_\psi \langle\psi|O|\psi\rangle = \text{Tr}(O\rho). \tag{1.42}$$

The density operator can be written in a matrix form for a given basis $|n\rangle$, in which case the matrix elements are $\rho_{nm} = \langle n|\rho|m\rangle$ and the product $O\rho$ indicates matrix multiplication.

To write the Schrödinger equation, Eq. (1.31), in terms of the density operator, we take the derivative of Eq. 1.41

$$\dot{\rho} = \sum_\psi P_\psi(|\dot{\psi}\rangle\langle\psi| + |\psi\rangle\langle\dot{\psi}|). \tag{1.43}$$

Substituting Eq. (1.31) into Eq. (1.43) leads to the Liouville-von Neumann equation,

$$\dot{\rho} = -\frac{i}{\hbar}[H, \rho], \tag{1.44}$$

which determines the density matrix time evolution. The density matrix theoretical framework allows a greater breadth of phenomena to be captured than can be described using wave functions and the Schrödinger equation alone. For example, relaxation processes can be considered by including phenomenological decay terms in the equation of motion

$$\dot{\rho} = -\frac{i}{\hbar}[H, \rho] - \frac{1}{2}\{\Gamma, \rho\}, \tag{1.45}$$

where $\{\Gamma, \rho\} = \Gamma\rho + \rho\Gamma$ and Γ is the relaxation operator.

As an example of how the formalism associated with density matrices be concretely put into practice, we consider a two-level system in the state

$$|\psi\rangle = C_0|0\rangle + C_1|1\rangle. \tag{1.46}$$

The density operator can be written as

$$\rho = |\psi\rangle\langle\psi| = |C_0|^2|0\rangle\langle 0| + |C_1|^2|1\rangle\langle 1| + C_0 C_1^*|0\rangle\langle 1| + C_1 C_0^*|1\rangle\langle 0|. \tag{1.47}$$

The matrix elements are

$$\rho_{00} = |C_0|^2, \quad \rho_{11} = |C_1|^2, \quad \rho_{10} = C_1 C_0^*, \quad \rho_{01} = \rho_{10}^* = C_0 C_1^*. \tag{1.48}$$

In the matrix form, we can write the state vector as

$$|\psi\rangle = \begin{pmatrix} C_0 \\ C_1 \end{pmatrix}. \tag{1.49}$$

The density matrix is correspondingly

$$\rho = \begin{pmatrix} C_0 \\ C_1 \end{pmatrix} \begin{pmatrix} C_0^* & C_1^* \end{pmatrix} = \begin{pmatrix} |C_0|^2 & C_0 C_1^* \\ C_1 C_0^* & |C_1|^2 \end{pmatrix} = \begin{pmatrix} \rho_{00} & \rho_{01} \\ \rho_{10} & \rho_{11} \end{pmatrix}. \tag{1.50}$$

We can see that ρ_{00} and ρ_{11} are the probabilities of being in states $|0\rangle$ and $|1\rangle$, respectively, while ρ_{10} and ρ_{01} are related to the coherence or polarization. In general, in a density matrix the diagonal matrix elements ρ_{nn} represent the probabilities being in state $|n\rangle$ and the off-diagonal matrix elements ρ_{nm} $(n \neq m)$ represent a coherent superposition between $|n\rangle$ and $|m\rangle$.

In matrix form, the Hamiltonian given by Eqs. (1.29) and (1.30) for the interaction between light and a two-level system can be written

$$H = \underbrace{\begin{pmatrix} \hbar\omega_0 & 0 \\ 0 & \hbar\omega_1 \end{pmatrix}}_{H_0} + \underbrace{\begin{pmatrix} 0 & -\hbar\Omega_r^* \cos(\omega_c t) \\ -\hbar\Omega_r \cos(\omega_c t) & 0 \end{pmatrix}}_{V} \tag{1.51}$$

$$= \begin{pmatrix} 0 & -\hbar\Omega_r^* \cos(\omega_c t) \\ -\hbar\Omega_r \cos(\omega_c t) & \hbar\omega_1 \end{pmatrix}. \tag{1.52}$$

Substituting Eq. (1.52) into Eq. (1.45), we obtain equations of motion for the individual density matrix elements,

$$\dot{\rho}_{00} = -\gamma_{00}\rho_{00} - \frac{i}{\hbar}\mu_{01}E(t)(\rho_{01} - \rho_{10}), \tag{1.53}$$

$$\dot{\rho}_{11} = -\gamma_{11}\rho_{11} + \frac{i}{\hbar}\mu_{10}E(t)(\rho_{01} - \rho_{10}), \tag{1.54}$$

$$\dot{\rho}_{10} = -\gamma_{10}\rho_{10} - i\omega_{10}\rho_{10} - \frac{i}{\hbar}\mu_{10}E(t)(\rho_{11} - \rho_{00}), \tag{1.55}$$

where $\omega_{10} = \omega_1 - \omega_0$, and the relaxation matrix elements are $\gamma_{ij} = \frac{1}{2}(\Gamma_i + \Gamma_j) + \gamma_{ij}^{ph}$ with γ_{ij}^{ph} being the pure dephasing rate and $\gamma_{ij}^{ph} = 0$ for $i = j$.

There are several approaches to solving Eqs. (1.53)–(1.55). In cases where the electromagnetic field interactions are weak, the problem can be tackled using methods of perturbation theory. To illustrate, we show the first-order solution to the off-diagonal density matrix element $\rho_{10}(t)$ as described by Eq. (1.55). Introducing the parameter λ to keep track of bookkeeping (we can set $\lambda = 1$ when all is said and done to reproduce physical reality) and then replacing V by λV in the Hamiltonian matrix (1.51), we look to express $\rho_{ij}(t)$ in terms of a power series expansion in λ of the form

$$\rho_{ij} = \rho_{ij}^{(0)} + \lambda\rho_{ij}^{(1)} + \lambda^2\rho_{ij}^{(2)} + \ldots \tag{1.56}$$

Substituting Eq. (1.56) back into Eqs. (1.53)–(1.55) leads to a set of equations

$$\dot{\rho}_{ij}^{(0)} = -i\omega_{ij}\rho_{ij}^{(0)} - \gamma_{ij}(\rho_{ij}^{(0)} - \rho_{ij}^{(eq)}), \tag{1.57}$$

$$\dot{\rho}_{ij}^{(n)} = -i(\omega_{ij} + \gamma_{ij})\rho_{ij}^{(n)} - \frac{i}{\hbar}[V, \rho^{(n-1)}]_{ij} \quad \text{for} \quad n \in \mathbb{Z}^+, \tag{1.58}$$

that can be iteratively solved if the initially unperturbed states of the system are known. Assuming that the steady-state value of ρ is the same as its unperturbed value such that $\rho_{ij}^{(0)} = \rho_{ij}^{(eq)}$, and that the system relaxes entirely into the ground state under equilibrium conditions at time $t = \infty$, we have $\rho_{00}^{(0)} = 1$ and all other elements of $\rho_{ij}^{(0)} = 0$. This being the case, we can substitute the zeroth-order values of ρ into the first-order version of Eq. (1.58) and integrate. Writing $E(t)$ as two phasor components,

$$E(t) = \frac{1}{2}\left[\hat{E}(t)e^{-i\omega_c t} + \hat{E}^*(t)e^{+i\omega_c t}\right],\tag{1.59}$$

and defining

$$\rho_{10}(t) = \hat{\rho}_{10}(t)e^{-i\omega_{10}t},\tag{1.60}$$

we obtain

$$\hat{\rho}_{10}^{(1)}(t) = \frac{-iN\mu_{10}e^{-\gamma_{10}t}}{2\hbar}\int_{-\infty}^{t}\left[\hat{E}(t')e^{[\gamma_{10}-i(\omega_c-\omega_{10})]t'}\right.$$
$$\left. +\hat{E}^*(t')e^{[\gamma_{10}+i(\omega_c+\omega_{10})]t'}\right]dt'.\tag{1.61}$$

For a CW field turning on at time $t = 0$, we can set $\hat{E}(t) = \hat{E}^*(t) = E_0\Theta(t)$ and then analytically solve. The integral gives

$$\hat{\rho}_{10}^{(1)}(t) = \frac{-iN\mu_{10}E_0e^{-\gamma_{10}t}}{2\hbar}\left[\frac{e^{\gamma_{10}t'-i(\omega_c-\omega_{10})t'}}{\gamma_{10}-i(\omega_c-\omega_{10})} + \frac{e^{\gamma_{10}t'+i(\omega_c+\omega_{10})t'}}{\gamma_{10}+i(\omega_c+\omega_{10})}\right]_{t'=0}^{t}.\tag{1.62}$$

In the optical regime the resonance widths are generally narrow in comparison to the resonance frequencies, such that $\gamma_{10} \ll \omega_{10}$. At excitation frequencies near resonance then, the denominator of the second term is much larger than that of the first term, and the second term can be ignored. This is known as the rotating-wave approximation (RWA). In the case of a lower frequency regime such as infrared or microwave, the rotating-wave approximation begins to fail, and the second term causes a shift of the apparent resonance, known as Bloch–Siegert shift.

In the opposite limit of an infinitesimally short optical pulse arriving at time $t = 0$, we can set $\hat{E}(t) = \hat{E}^*(t) = \mathcal{E}_0\delta(t)$. The solution to Eq. (1.61) becomes

$$\hat{\rho}_{10}^{(1)}(t) = \frac{-iN\mu_{10}\mathcal{E}_0e^{-\gamma_{10}t}}{2\hbar}\left[\Theta(t) + \Theta(t)\right],\tag{1.63}$$

which decays exponentially with time after the initial perturbation at time $t = 0$. In the optical regime, the second of these two terms is typically suppressed down to unobservable values by the finite bandwidth of the laser pulses (an effect that delta-function pulse envelopes are unable to capture), and so the rotating-wave approximation is typically invoked by discarding this second term by hand.

1.5 Bloch sphere representation of quantum states

The quantum state of a two-level system and the state's evolution can be conveniently discussed by visualizing the state with a Bloch vector on the Bloch sphere. The Bloch sphere provides a geometrical visualization of any pure state in a two-level system. The north and south poles each correspond to one of the orthogonal basis states of a two-level system. For the two-level system shown in Fig. 1.5, the Bloch vector

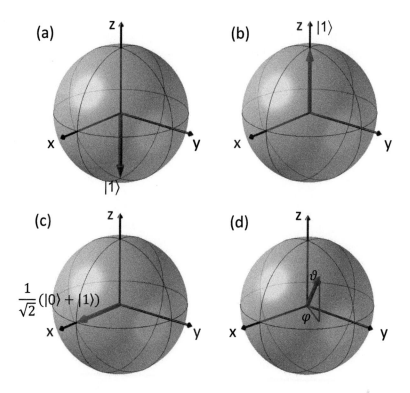

Fig. 1.6 Bloch spheres with (a) the Bloch vector representing state $|0\rangle$, (b) the Bloch vector representing state $|1\rangle$, (c) the Bloch vector representing state $\frac{1}{\sqrt{2}}(|0\rangle + |1\rangle)$, and (d) the Bloch vector representing an state $|\psi\rangle$.

pointing to the south pole corresponds to state $|1\rangle$, as shown in Fig. 1.6(a). The Bloch vector pointing to the north pole, as shown in Fig. 1.6(b), corresponds to state $|1\rangle$. A vector pointing to any other point on the surface of the sphere represents a coherent superposition of states $|0\rangle$ and $|1\rangle$. Particularly, a point on the equator corresponds to state $\frac{1}{\sqrt{2}}(|0\rangle + e^{i\varphi}|1\rangle)$, with φ being the azimuthal angle. In these states, the system has the maximum coherence or polarization. When $\varphi = 0$, as shown in Fig. 1.6(c), the state is $\frac{1}{\sqrt{2}}(|0\rangle + |1\rangle)$, representing an in-phase polarization. In general, a point defined by the polar angle ϑ and the the azimuthal angle φ, as shown in Fig. 1.6(d), corresponds to state

$$|\psi\rangle = \sin\frac{\vartheta}{2}|0\rangle + e^{i\varphi}\cos\frac{\vartheta}{2}|1\rangle. \tag{1.64}$$

As the quantum state evolves, the Bloch vector rotates and leaves a trajectory on the surface of the sphere. For example, if the system is initially in state $|0\rangle$ (south pole), a $\frac{\pi}{2}$-pulse drives the system from the south pole to the equator, i.e., from $|0\rangle$ to $\frac{1}{\sqrt{2}}(|0\rangle + |1\rangle)$, creating a coherence/polarization. A π-pulse can excite the system from the south pole to the north pole, i.e., from $|0\rangle$ to $|1\rangle$, or vice versa. A 2π-pulse can drive the system from the south (north) pole to the north (south) pole and then back to the south (north) pole. A Bloch vector can also precess about z-axis, corresponding to the phase evolution of a coherence/polarization.

2

Introduction to multidimensional coherent spectroscopy

Multidimensional coherent spectroscopy was first developed in nuclear magnetic resonance (NMR) spectroscopy [108], where it was known as "multidimensional Fourier transform spectroscopy." Key enabling technological advances were the development of digital computers with sufficient computing power and the development of the fast Fourier transform (FFT) algorithm [67]. This pair of advances meant that data could be acquired in the time domain and the spectrum calculated in a reasonable amount of time.

NMR relies on generating and measuring radio frequency (RF) waveforms with full phase information, which is routine. The wavelengths of the signals are typically on the order of a meter, thus the stability of the time delays does not need to be stringently controlled. In contrast, implementing similar methods at optical frequencies is challenging. In optics, detectors typically measure the intensity of the light, not its underlying electric field waveform. Thus, heterodyne/homodyne methods are needed to retrieve phase information. In addition, the orders-of-magnitude shorter wavelength means that fluctuations in path lengths must be tightly controlled to avoid corruption of the phase information.

The idea of implementing 2D coherent spectroscopy in the optical regime was first proposed by Tanimura and Mukamel in 1993 [389]. The technical challenges hindered experimental implementation initially and it took over six years to demonstrate the first realizations. Since then, many different approaches of optical 2D coherent spectroscopy have been developed. The technique is now well established in many ultrafast spectroscopic laboratories and even became commercially available in recent years.

The initial proposal by Tanimura and Mukamel [389] uses a Raman excitation scheme in which molecular vibrations are excited by a five-pulse sequence. The experiments were initially complicated by lower-order signals that emit in the same direction as the desired nonlinear signal, making it difficult to separate the two [32]. This difficult led to using IR to access molecular vibrations directly for 2D spectroscopy, which was demonstrated independently by the groups of Hochstrasser in 1999 [146] and Tokmakoff in 2001 [136]. In the IR regime, the requirement of the timing precision and stability can be satisfied with only passive stability due to the longer wavelength.

Implementing 2D coherent spectroscopy to access electronic transitions, often called "two-dimensional electronic spectroscopy (2DES)," requires visible or near-IR light and thus poses greater challenges with respect to path length stability due to the shorter wavelengths. The first demonstration of 2DCS in these wavelength regimes relied on measuring the delay for each time-step and resampling the data [161]. To eliminate the need for resampling, methods based on diffractive optics were demonstrated by Miller [68] and Fleming [45] in 2004. Subsequently, many implementations extended the technique into the near-IR and visible regime, realizing 2D spectroscopy for electronic transitions in atoms, molecules, or semiconductors. These experiments used more elaborate approaches to control the timing at shorter wavelengths. Subsequently approaches based on active stabilization of the path lengths were demonstrated [34, 461].

Multidimensional coherent spectra can be difficult to understand and interpret when they are first encountered. This chapter will provide an introduction into the basic concepts of multidimensional coherent spectroscopy and provide a primer for interpreting the spectra. A common tool for analyzing the spectra, double-sided Feynman diagrams, will also be introduced.

2.1 Concepts of multidimensional coherent spectroscopy

A multidimensional coherent spectrum is a map of coherent light-matter interactions plotted across a domain of two or more frequencies. Such a spectrum is obtained by illuminating a material with a series of electromagnetic pulses and analyzing the frequency-dependent way in which excitation of the sample by the initial or intermediate pulses affects the material's response to the subsequent pulses.

Perhaps the simplest way to understand the technique is as an extension of transient absorption spectroscopy, as illustrated by Fig. 2.1. In a transient absorption experiment [Fig. 2.1(a)], a sample is illuminated by two pulses. The first of these (the "pump") drives the sample into a nonequilibrium state, which is then measured by subtracting the sample's response to the second pulse (the "probe") in the pump's presence from the response to the probe in the pump's absence. The resulting differential signal can be measured as a spectrally and temporally integrated quantity using a photodetector or, by sending the transmitted probe into a spectrometer, as a spectrally resolved quantity plotted against the probe frequency [Fig. 2.1(b)].

In order for a signal to be observed, the optical response of the sample must be nonlinear, i.e., it depends on the intensity of light. If the response is not nonlinear, then the presence of the pump pulse will not affect the interaction of the sample with the probe pulse. The most straightforward of these is saturation, in which the pump pulse decreases the sample's net absorption of the probe pulse, thereby increasing the transmission of the probe and generating a positive signal in differential transmission. For interacting many-body systems, there are also other possibilities. For example, the pump can broaden the linewidth of the sample's response to the probe pulse (excitation-induced dephasing, or EID), which tends to produce positive central peaks with negative wings. It can also shift the center frequency of the resonance (excitation-induced shift, or EIS), creating an asymmetric pump-probe spectrum. The effects of both EIS and EID are visible in Fig. 2.1(b), which corresponds to an asymmetric $In_{0.05}Ga_{0.95}As$ double quantum well where many-body effects are known to dominate the optical response.

Although it improves upon the information available through linear absorption, transient absorption is still incomplete because spectral dependence of the sample's

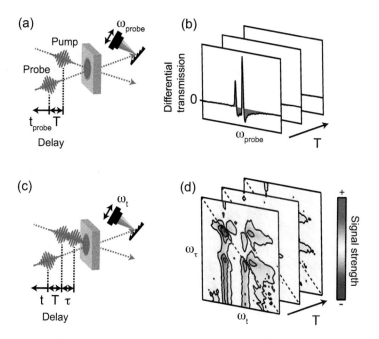

Fig. 2.1 Comparison between transient absorption spectroscopy and MDCS. (a) In transient absorption spectroscopy, the differential transmission of an optical pulse (the "probe") is measured in response to the application of an earlier pulse (the "pump"). The pulses are delayed with respect to each other by a time T. (b) For a given value of T, the response may be plotted as a spectrally resolved function of the probe frequency ω_{probe}. (c) In a simple implementation of MDCS, the pump pulse from (a) is divided into sub-pulses, defining an additional time delay τ. d) Data may be acquired as a function of varied τ and/or T, and Fourier transformed to produce spectra plotted against a multidimensional frequency domain. Adapted from Ref. 363.

response to the pump pulse remains unknown even as its differential probe-pulse response is spectrally resolved. In part because of this limitation, results are often obscured in the presence of inhomogeneity resulting, for example, from Doppler broadening in atomic vapors, sample imperfections, size variations in nanostructures mapping into variations in confinement potentials, or uneven strain. To remedy the situation, the pump may be divided into a pair of sub-pulses [Fig. 2.1(c)]. The inter-pulse delay between these, (τ), can then be systematically varied, and the data can be numerically Fourier transformed to obtain a spectrally resolved excitation axis. The delay between the second and third pulses (T) tracks the pump-probe delay. The third pulse continues to act as a probe, leading to a differential response that is either emitted from the sample over an elapsed time (t) as coherent four-wave mixing, or converted by a fourth pulse (not shown) after time t into a signal in photoluminescence, photocurrent, or photoemission. The resulting MDCS plot is a time-resolved signal that is also frequency-resolved as a function of both excitation (pump) and emission (probe) frequency [Fig. 2.1(d)]. As presented throughout this book, unfolding the spectral response across these multiple frequency dimensions facilitates a wealth of new capabilities, including the ability to disentangle microscopic dephasing in the midst of sample inhomogeneity, the ability to identify coherent coupling mechanisms between resonances, and the ability to more thoroughly constrain the mechanisms behind many-body interactions.

2.1.1   Coherent spectroscopy

One might wonder, based on the schematic depicted in Fig. 2.1(c), how coherent information related to the sample's response to the first pulse can possibly be extracted from the final measurement. After all, the sample's response to the probe pulse is ultimately the only signal measured. The issue is resolved by noting that if time delays between pulses are controlled with sub-wavelength precision, then coherent interactions from the earlier pulses will be written into the phase of measured signal.

A simple illustration of the effect is shown in Fig. 2.2, depicting a Bloch sphere illustration [8] of a two-pulse correlation measurement in which $\pi/2$ pulses are applied to a two-state system. As illustrated by Fig. 2.2(a), the system is initially in the ground

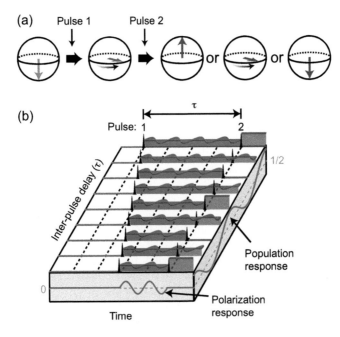

Fig. 2.2 Coherent spectroscopy of a two-level system in the $\pi/2$-pulse limit. **(a)** Bloch sphere illustration of the quantum state's evolution. The system is initialized in the ground state (left), driven into a superposition between ground and excited states by a $\pi/2$ pulse (center), and converted into a potential variety of population states and superposition states by a second $\pi/2$ pulse (right). **(b)** Varying the time delay between the two pulses maps the intermediate quantum state's coherent evolution (horizontal red oscillations, center) onto the final state's excited-state population (vertical blue projections and red oscillations, right). Adapted from Ref. 363.

state such that the Bloch vector points downward for times $t < 0$ [Fig. 2.2(a), left]. At $t = 0$, a $\pi/2$ pulse is applied, driving the system into a coherent superposition of the ground and excited states. The superposition evolves with time [Fig. 2.2(a), center], rotating around the Bloch sphere's equator until it dephases and relaxes back down into the ground state, or until (as illustrated) a second $\pi/2$ pulse is applied. Depending on the relative delay and phase difference between these two pulses, the second pulse can drive the system into a ground-state population, an excited-state population, or anywhere in between [Fig. 2.2(a), right].

When the delay and phase relationship between pulses 1 and 2 is controlled with sufficient precision, the model illustrates a striking feature of the response, as shown in Fig. 2.2(b). Even though the second pulse fundamentally alters the intermediate quantum state, and in cases converts the superposition between the ground and excited state into a population state without oscillatory motion, the information about the intermediate state still gets written onto the final state's dependence on inter-pulse delay. This dependence can be seen in the red trace on the right side of Fig. 2.2(b), which tracks oscillations in the population component of the final quantum state.

It is worth noting that the Bloch sphere example differs from a typical optical coherent spectroscopy experiment in one significant respect. Whereas the example from Fig. 2.2 employs $\pi/2$ pulses, the pulses much more commonly employed in MDCS are perturbative (much smaller pulse area than $\pi/2$). Nevertheless, the key aspects of the illustrated pathway remain a valid description of the measured portion of the perturbative response function. The connections are solidified in Section 2.3.1.

2.1.2   Multidimensional coherent spectroscopy

With two-pulse correlation spectroscopy established, the technique can be readily extended to encompass light-matter interactions involving three, four, and in some cases even more pulses, producing spectral information against multiple frequency dimensions. A characteristic example of an MDCS plot is depicted in Fig. 2.3, showing coupled exciton resonances in an asymmetric InGaAs double quantum well.

As shown by Fig. 2.3(a), the spectrum is generated by exciting the sample with three optical pulses to create a nonlinear polarization response (green oscillations). The response can either be measured as heterodyne-detected four-wave mixing, as was the case for the data in Fig. 2.3, or it can be converted by a readout pulse into a higher-order population response that can be detected as a photocurrent, photoluminescence, or photoemission signal. Regardless of the detection scheme, the time delay between pulses 1 and 2 is then varied to produce a two-dimensional plot of the response as a function of excitation time τ and emission time t as displayed in Fig. 2.3(b), plotted in a rotating frame. The frequency-domain data [Figs. 2.3(c)–2.3(e)] correspond to the Fourier transform of the time-domain data in Fig. 2.3(b). The excitation frequency axes have negative frequency units for reasons to be explained in Section 2.2.

Figures 2.3(c)–2.3(e) exhibit a number of features that illustrate the power of MDCS as an experimental technique. Prominent features are two peaks along the diagonal, where $|\omega_\tau| = |\omega_t|$. Such features are analogous to what might be observed using a simpler method such as one-dimensional absorption or photoluminescence spectroscopy, and they each originate from energetically degenerate pump-probe interactions

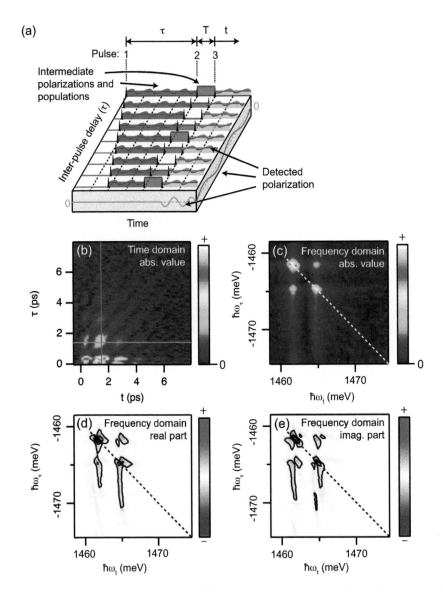

Fig. 2.3 Mechanistic illustration of MDCS. (a) Varying the time delay τ (between pulses 1 and 2) while keeping the delays T (between pulses 2 and 3) and t (following pulse 3) fixed writes coherent pump-pulse absorption characteristics (red oscillations, center left) onto the output signal (red oscillations, bottom right). Tracking t at fixed τ and T gives coherent emission characteristics of the output signal (green oscillations, bottom front). (b) The result may be plotted as a two-dimensional function of four-wave mixing vs. τ and t. (c–e) Performing the Fourier transform of (c) produces a coherent frequency-resolved two-dimensional spectrum. Adapted from Ref. 363.

where the pump-pulse excitations induce a change in the probe-pulse interaction at the same frequency. The frequencies of the diagonal peaks can be used to determine the quantity and spacing of a system's excited states. Ratios between peak heights can be used to determine relative optical absorption strength.

The diagonal peaks in an MDCS experiment provide more information than their one-dimensional counterparts in absorption, photoluminescence, and transient absorption. For example, the resonances in Fig. 2.3(c) are elongated along the diagonal, which is a consequence of the fact that sample inhomogeneities create variations in the local environment. In a one-dimensional measurement, these variations smear resonance characteristics out, making it hard to ascertain the microscopic dephasing times within the larger macroscopic ensemble. By contrast, an inhomogeneously broadened feature in an MDCS spectrum remains narrow along the perpendicular "cross-diagonal" direction almost independently of the system's inhomogeneity, making quantitative measurements of intrinsic dephasing rates possible [28, 354, 362].

A more striking contrast between MDCS and one-dimensional measurements is that degenerate interactions can be separated from nondegenerate ones. For example, the Fig. 2.3(c) spectrum exhibits two "cross peaks" at (1462,-1464.9) meV and (1464.9,-1462) meV, which can arise for a number of reasons. If the two resonances share a common ground state, then the pump-pulse interaction with one of the resonances can result in a ground-state bleaching, affecting the probe pulse's interaction at the frequency of the second resonance. The pump can also drive the system into a coherent superposition between the two excited states. In both cases, the detection of cross peaks constitutes a direct signature of coupling between the two resonances that would be more difficult to observe using simpler spectroscopic methods. The presence of cross peaks is particularly important in the spectrum of a sample that is a mixture of multiple species, as a cross peak indicates that the corresponding diagonal peaks are from the same species.

Finally, as shown by Figs. 2.3(d) and 2.3(e), the frequency spectrum is also phase-resolved. Analyses of the real and imaginary components of the signal and comparison to theoretical models can reveal important information about many-body effects and excitation-related interactions within the material sample. As discussed in Chapters 7–10, the study of many-body effects has been perhaps the single greatest contribution of MDCS to the field of semiconductor physics.

2.2 Spectrum classification

Depending on pulse ordering and signal detection filtering choices, different types of multidimensional spectra can be obtained from a material. Among the most common of these is the "rephasing" spectrum, which was depicted in Fig. 2.3 and which is illustrated for a simulated three-state "v" system in Fig. 2.4(a). Rephasing spectra are generated using an experimental configuration in which a material's interaction with the first pulse is conjugate to its interaction with the second and third pulses. This can be achieved by arranging the excitation pulses to have distinct momentum vectors (e.g., $\mathbf{k}_1$, $\mathbf{k}_2$, and $\mathbf{k}_3$) and isolating a four-wave mixing signal at $\mathbf{k}_s = \mathbf{k}_I \equiv -\mathbf{k}_1+\mathbf{k}_2+\mathbf{k}_3$ [162, 252], by tagging the pulse trains with different carrier frequencies (ω_1, ω_2, and ω_3) to obtain a four-wave mixing signal at $\omega_s = \omega_I \equiv -\omega_1 + \omega_2 + \omega_3$ [207, 257, 285, 391],

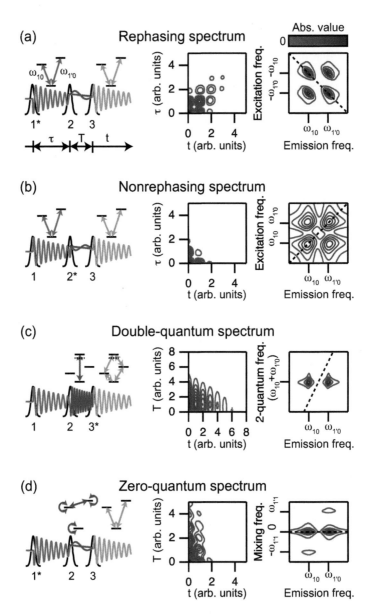

Fig. 2.4 MDCS classifications with simulated data. Spectra correspond to the signal's absolute value, with linearly spaced contours. 2D time plots are in the rotating frame. (a) Single-quantum rephasing spectrum for an inhomogeneously broadened three-state "v" system. (b) Single-quantum nonrephasing spectrum for an inhomogeneously broadened three-state "v" system. (c) Double-quantum spectrum for a four-state diamond system. (d) Zero-quantum spectrum for an inhomogeneously broadened three-state "v" system.

or by phase cycling [108, 394]. In such a spectrum, the coherence generated by the first pulse is opposite in frequency to that generated by the third pulse, which is why the excitation frequencies in Figs. 2.3(c)–2.3 (e) and in the right panel of Fig. 2.4(a) are negative. In turn, the pulse sequence tends to bring different resonances of an inhomogeneously broadened system into phase with each other during the emission process, resulting in the "photon echo" signature [1] in the time-domain spectrum of Fig. 2.4(a) (note the clustering of spectral intensity near emission times $t = \tau$), as well as the narrowed cross-diagonal lineshapes in the Fig. 2.4(a) frequency domain spectrum.

Spectra can also be collected from a conjugate second-pulse interaction relative to the interactions of the first and third pulses in a "nonrephasing" interaction [Fig. 2.4(b)], which can be achieved by collecting a four-wave mixing signal at $\mathbf{k}_{II} \equiv \mathbf{k}_1 - \mathbf{k}_2 + \mathbf{k}_3$ and $\omega_{II} \equiv \omega_1 - \omega_2 + \omega_3$. In this case, the phase difference between resonant interactions at different frequencies increases monotonically with time, resulting in a time-domain signal without a photon echo and in time-domain and frequency-domain signals that are typically weaker than their rephasing counterparts. Nevertheless, the nonrephasing pulse sequence is still preferred over the rephasing pulse sequence in certain applications because the resonant peaks of a nonrephasing spectrum arise from slightly different physical origins and interfere differently with each other in close proximity than do the peaks of a rephasing spectrum [59, 133, 190]. Beyond this, if the real parts of rephasing and nonrephasing spectra are added together, one obtains a purely "absorptive" spectrum, which is perhaps the closest physical analog to the signals generated in transient absorption spectroscopy [184].

Figure 2.4(c) shows the spectrum from a pulse sequence in which the third pulse is conjugate relative to nonconjugate first and second pulses, achieved by collecting a four-wave mixing signal at $\mathbf{k}_{III} \equiv \mathbf{k}_1 + \mathbf{k}_2 - \mathbf{k}_3$ and $\omega_{III} \equiv \omega_1 + \omega_2 - \omega_3$. The spectrum is commonly termed a double-quantum (or two-quantum) spectrum because it produces no signal except in the presence of a system with a doubly excited state, for which a direct optical transition is often dipole-forbidden, but nevertheless the action of two pulses can create a coherence between the ground state and the doubly excited state, known as a "double-quantum coherence." Correlating the evolution time of the double-quantum coherence residing within the time interval T and the single-quantum coherence generated during time t is often powerful because it gives a background-free view into many-body interactions [453].

In analogy to double-quantum spectroscopy, it is also possible to generate correlation plots probing the mixing interaction between nearly degenerate quantum states using rephasing or nonrephasing pulse-ordering sequences, which is often just as inaccessible to optical techniques as the dipole-forbidden double-quantum coherence. Figure 2.4(d) shows an example of such a "zero-quantum" spectrum, for the rephasing pulse sequence from Fig. 2.4(a). The language of "n-quantum spectra" originates from NMR [108] and refers to the number of quanta of the electromagnetic field that must be absorbed to make a transition between the corresponding levels.

Finally, one can even generate three-dimensional coherent spectra, in which all three inter-pulse delays are varied to probe for correlations across three independent frequency domains [74]. Such 3D spectra provide perhaps the clearest separation of quantum pathways possible in an optical spectroscopic measurement [214], and although less common than their 2D counterparts, have been acquired on atomic vapors and semiconductor quantum wells [83, 412].

2.3 Density matrix formalism and double-sided Feynman diagrams

Much of the preceding discussion becomes more concrete when framed in the context of semiclassical perturbation theory as applied to the Bloch model [40, 449]. Although the treatment of excitonic systems using a discrete level system such as this is an approximation to theoretical treatments based on first-principles calculations involving fermionic creation and annihilation operators; the two pictures can be reconciled as discussed in Ref. 418.

2.3.1 Interpreting MDCS in the perturbative limit

As mentioned briefly in the context of a two-level system back in Section 1.4, the basic mathematical object of importance within this framework is the material system's density matrix ρ, which evolves in time according to the Liouville-von Neumann equation

$$\dot{\rho}_{ij} = -\frac{i}{\hbar}[H, \rho]_{ij}. \tag{2.1}$$

In the Bloch model treatment, we tack on phenomenological relaxation terms to the right-hand side of the equation to accommodate randomized collision and relaxation events, yielding

$$\dot{\rho}_{ij} = -\frac{i}{\hbar}[H, \rho]_{ij} - \gamma_{ij}(\rho_{ij} - \rho_{ij}^{(eq)}) \tag{2.2}$$

where γ_{ij} are damping constants and there is no implied summation of repeated indices. Having established this master equation, we proceed to solve the system by methods of perturbation theory. We make an assumption that an interaction term V can be split off from the total Hamiltonian as $H = H_0 + V(t)$ leaving H_0 as a diagonal and time-independent Hamiltonian matrix consisting of elements $H_{0,ij} = E_i\delta_{ij}$ with energies E_i being associated with unperturbed eigenstates. Substituting this expression into Eq. (2.2) gives

$$\dot{\rho}_{ij} = -i\omega_{ij}\rho_{ij} - \frac{i}{\hbar}[V, \rho]_{ij} - \gamma_{ij}(\rho_{ij} - \rho_{ij}^{(eq)}), \tag{2.3}$$

where the frequency term $\omega_{ij} \equiv (E_i - E_j)/\hbar$. We can expand out the density matrix into a perturbative series $\rho = \rho^{(0)} + \rho^{(1)} + \rho^{(2)} + \rho^{(3)} + ...$ which, upon being substituted back into Eq. (2.3), transforms the equation into an infinite set of order-specific equations of the form

$$\dot{\rho}_{ij}^{(n)} = -(i\omega_{ij} + \gamma_{ij})\rho_{ij}^{(n)} - \frac{i}{\hbar}[V, \rho^{(n-1)}]_{ij}. \tag{2.4}$$

In slightly more compact form, these can be written

$$\dot{\rho}_{ij}^{(n)} = -i\Omega_{ij}\rho_{ij}^{(n)} - \frac{i}{\hbar}[V, \rho^{(n-1)}]_{ij} \tag{2.5}$$

with

$$\Omega_{ij} \equiv \omega_{ij} - i\gamma_{ij}, \quad \text{and} \quad \omega_{ij} \equiv \frac{E_i - E_j}{\hbar}. \tag{2.6}$$

In turn, the expressions can be integrated to obtain expressions such as

$$\rho_{ij}^{(n)}(t) = \int_{-\infty}^{t} -\frac{i}{\hbar} [V(t'), \rho^{(n-1)}]_{ij} e^{-i\Omega_{ij}(t-t')} dt' \tag{2.7}$$

from which each element of the series expansion of ρ can be iteratively determined once the lower-order elements are known. For optical perturbations, the interaction Hamiltonian

$$V(t) = -\frac{\mu}{2} \sum_m \left[\hat{E}_m(t) e^{i(\mathbf{k}_m \cdot \mathbf{r} - \omega_m t)} + \text{c.c.} \right] \tag{2.8}$$

is a sum of products between the material system's electric dipole moment operator μ and the relevant optical modes, indexed by the subscript m.

The nth-order perturbative regime satisfies the condition $\sum_{m=n+1}^{\infty} \rho^{(m)} \ll \rho^{(n)}$, which is a condition typically realizable at a desired order n by controlling the amplitudes $\hat{E}_m(t)$ of the interacting fields in Eq. (2.8). The task of modeling a multidimensional spectrum then amounts to characterizing ρ to the highest relevant perturbative order and discarding all orders beyond this.[1] Because MDCS requires a minimum of three distinct optical interactions, the task reduces to the calculation of a subset of contributions to $\rho^{(n \geq 3)}$ where the successive orders in perturbation theory leading up to the nth order correspond to exactly one interaction with each of the excitation pulses. These n^{th}-order density matrix corrections ultimately emerge as measurable quantities such as the sample's macroscopic polarization (for off-diagonal elements) or excited-state population (for diagonal elements).

In general, the tabulation of the constituent elements—termed Liouville space pathways—of an element of an individual perturbative order is quite large, and grows exponentially with interaction order. Fortunately, time ordering rules, the rotating-wave approximation, wave vector selection rules, and frequency selection rules substantially reduce the number of pathways that are actually relevant.

An illustrative example of these simplifications is the determination of the lower-left peak of the spectrum shown in Fig. 2.4(a), replicated in Fig. 2.5, corresponding to the interaction of a rephasing pulse sequence with a three-state "v" system. (We focus in on just one of the system's four prominent spectral peaks to begin with; a more complete analysis will be presented in Section 3.3.) The measured four-wave mixing signal has wave vector $\mathbf{k}_{sig} = -\mathbf{k}_1 + \mathbf{k}_2 + \mathbf{k}_3$ and carrier frequency $\omega_{sig} = -\omega_1 + \omega_2 + \omega_3$, which requires the signal to emerge from at least one interaction with each of the excitation pulses 1, 2, and 3. Because $\rho_{10}^{(3)}$ is a third-order correction, however, the interaction can also contain *no more* than one interaction each with each of these pulses. Time ordering dictates that the first-order interaction be the interaction with pulse 1, that the second-order interaction be the interaction with

---

[1]For collective systems the perturbative regime of the density matrix is more restrictive than the equivalent perturbative regime of the measured polarization. In many cases, however, the two approximations yield identical results, allowing experiments to be interpreted using only the density matrix ground state and first few excited states without difficulty [15].

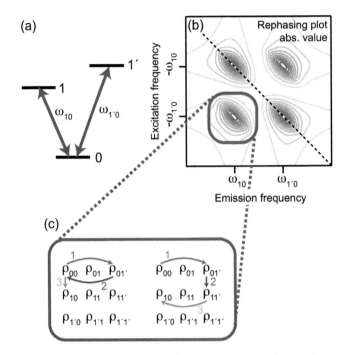

Fig. 2.5 Graphical representations of Liouville space pathways in density matrix perturbation theory. **(a)**–**(b)** Diagrams are derived from a rephasing interaction between light and a three-state "v" system as depicted in (a), and correspond to the lower left peak in (b). **(c)** Matrix representations of the relevant Liouville space pathways where the red, blue, and green arrows correspond to first-, second-, and third-order interactions with the excitation beams. Adapted from Ref. 363.

pulse 2, and that the third-order interaction be the interaction with pulse 3. Finally, the fact that the material's interaction with pulse 1 is conjugate to its interaction with pulses 2 and 3, in combination with the rotating-wave approximation, means that if the first-order interaction traverses a row of the density matrix, then the third-order interaction must traverse a column (and vice versa). Combining these restrictions with a selection of peaks centered on the coordinate $(\omega_t, \omega_\tau) = (\omega_{10}, -\omega_{1'0})$ leaves only two possible peak contributions, illustrated schematically in Fig. 2.5(c). The pathway on the left corresponds to a bleaching interaction in which the excitation pulse interaction at $\omega_{01'} = -\omega_{1'0}$ affects the detection pulse interaction at ω_{10}. The pathway on the right corresponds to a quantum beat interaction in which the second pulse drives the system's second-order correction to ρ into a coherent superposition of states $|1\rangle$ and $|1'\rangle$.

The restriction of quantum pathways to taxicab-geometry patterns in Fig. 2.5(c) (that is, patterns where n^{th}-order density matrix corrections are prohibited from incrementing diagonally upon corrections at order $n-1$) provides a convenient illustration of the fact that the structure of Eq. (2.7) requires the element $\rho_{ij}^{(n)}$ to share at least one of its indices with each of the elements of the lower-order density matrix correction

$\rho^{(n-1)}$ upon which $\rho_{ij}^{(n)}$ is generated. In turn, the illustration invites an intuitive understanding of MDCS in which the technique can be described as a process where the initial pulse creates a coherence in the sample, the second pulse converts the coherence into a population, and the third pulse converts the population back into a coherence.

2.3.2 Double-sided Feynman diagrams

To make the arguments laid out in the previous section more concrete, we can explicitly expand the commutator product depicted on the right-hand side of Eq. (2.7) as

$$\rho_{ij}^{(n)}(t) = \sum_{\alpha} \rho_{ij,\alpha}^{(n)}(t), \qquad (2.9)$$

where each of the different elements $\rho_{ij,\alpha}^{(n)}(t)$ corresponds to a different Liouville space interaction pathway.

We have seen that such pathways may be graphically depicted as in Fig. 2.5(c) [4, 21]. However, a far more common visualization is that of double-sided Feynman diagrams, where successive orders of density matrix elements are vertically stacked and adorned on the left and right with a collection of "vertices" involving intersecting arrows representing various optical interactions, as depicted in Fig. 2.6.

In the perturbative solution to the Bloch model, at each order of the interaction with the field, the state represented by density matrix elements evolves from one matrix element to an adjacent matrix element, as shown in Fig. 2.6(a). A double-sided Feynman diagram tracks the time evolution of the density matrix elements explicitly. As shown in Fig. 2.6(b), the diagrams use two vertical lines with two symbols in between to represent the "bra" and "ket" of the density matrix operator, where the left (right) line represents the "ket" ("bra"). The time increases upward. The interaction with a field is described by the vertex of an arrow with a vertical line. An arrow represents a field that can change one side of the density matrix element. There are four possible vertices, as shown in the box containing Eqs. (2.10)–(2.13), and these vertices mathematically connect back to the positive and negative terms of the commutator embedded within Eq. (2.7) as well as the complex conjugate terms of Eq. (2.8).

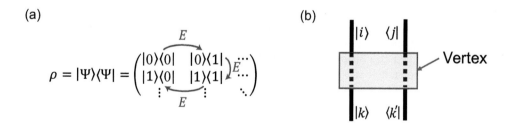

(a)

$$\rho = |\Psi\rangle\langle\Psi| = \begin{pmatrix} |0\rangle\langle 0| & |0\rangle\langle 1| \\ |1\rangle\langle 0| & |1\rangle\langle 1| \\ \vdots & \vdots \end{pmatrix} \overset{E}{\underset{E}{\cdots}}$$

(b)

Fig. 2.6 (a) Density matrix representation of tracking excitation pathways. Each excitation of the field changes a density matrix element to an adjacent matrix element. (b) Each interaction with the field can be represented by a vertex in double-sided Feynman diagrams.

Double-sided Feynman diagram vertices for density matrix perturbations

$$\rho_{ij,\alpha}^{(n)} = \frac{i\mu_{ik}}{2\hbar}e^{i\mathbf{k}_n\cdot\mathbf{r}}\int_{-\infty}^{t}\hat{E}_n(t')e^{-i\omega_n t'}e^{-i\Omega_{ij}(t-t')}\rho_{kj,\alpha}^{(n-1)}(t')dt' \quad (2.10)$$

$$\rho_{ij,\alpha}^{(n)} = \frac{i\mu_{ik}}{2\hbar}e^{-i\mathbf{k}_n\cdot\mathbf{r}}\int_{-\infty}^{t}\hat{E}_n^*(t')e^{i\omega_n t'}e^{-i\Omega_{ij}(t-t')}\rho_{kj,\alpha}^{(n-1)}(t')dt' \quad (2.11)$$

$$\rho_{ij,\alpha}^{(n)} = -\frac{i\mu_{kj}}{2\hbar}e^{-i\mathbf{k}_n\cdot\mathbf{r}}\int_{-\infty}^{t}\hat{E}_n^*(t')e^{i\omega_n t'}e^{-i\Omega_{ij}(t-t')}\rho_{ik,\alpha}^{(n-1)}(t')dt' \quad (2.12)$$

$$\rho_{ij,\alpha}^{(n)} = -\frac{i\mu_{kj}}{2\hbar}e^{i\mathbf{k}_n\cdot\mathbf{r}}\int_{-\infty}^{t}\hat{E}_n(t')e^{-i\omega_n t'}e^{-i\Omega_{ij}(t-t')}\rho_{ik,\alpha}^{(n-1)}(t')dt' \quad (2.13)$$

As can be seen in both the figures and equations, each vertex describes the time evolution from some component of the $(n-1)^{\text{th}}$-order density matrix element $\rho_{kj}^{(n-1)}$ or $\rho_{ik}^{(n-1)}$ to the nth order density matrix element $\rho_{ij}^{(n)}$. A right-pointing arrow corresponds to a "nonconjugate" interaction involving the first of the two terms listed within the integrand of Eq. (2.8) [i.e., a term containing $\hat{E}_m(t)e^{i(\mathbf{k}_m\cdot\mathbf{r}-i\omega_m t)}$]. A left-pointing arrow, by contrast, corresponds to a "conjugate" interaction involving the second of these two terms [i.e., a term containing $\hat{E}_m^*(t)e^{-i(\mathbf{k}_m\cdot\mathbf{r}-i\omega_m t)}$]. An arrow situated on the left side of a vertex indicates a "positive" vertex and corresponds to the first half of the commutator embedded within Eq. (2.7). An arrow situated on the right side of a vertex indicates a "negative" vertex and corresponds to the second half of this commutator product. In the rotating-wave approximation, these vertices carry an additional and physically intuitive association: For arrows pointing toward the vertical lines [Eqs. (2.10) and (2.12)], light is absorbed and drives the density matrix element from a lower energy state to a higher energy state. For arrows pointing away from the vertical lines [Eqs. (2.11) and (2.13)], light is emitted and the density matrix element goes from a higher energy state to a lower energy state.

Having defined the properties of double-sided Feynman diagram vertices in isolation, we can keep track of possible excitation pathways from 0th order to higher orders in perturbation calculations by using vertices and stacking them along the vertical lines. This process is demonstrated in Fig. 2.7 up to the 3rd order in the rotating-wave approximation for a two-level system with the ground state $|0\rangle$ and the excited state $|1\rangle$ ($E_1 > E_0$). At the 0th-order, the system is initially in the ground state, represented by density matrix element $\rho_{00}^{(0)}$. The 1st-order interaction has to absorb a photon and has two possibilities, raising the index on the right or the left from 0 to 1. The 1st-order interaction generates a linear polarization in the material by means of contributions

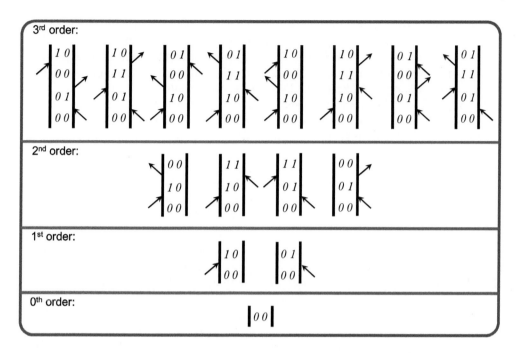

Fig. 2.7 The process for constructing all possible double-sided Feynman diagrams from the 0th order to the 3rd order for a two-level system interacting with three pulses in the rotating-wave approximation. At each order, the diagrams at the lower-order branch into more diagrams at the higher order. There are eight double-sided Feynman diagrams at the 3rd order when the pulses are not specified.

to the off-diagonal density matrix element corrections $\rho_{10}^{(1)}$ and $\rho_{01}^{(1)}$. The 2nd-order interaction can either absorb a photon or emit a photon, changing the corresponding index from 0 to 1 or from 1 to 0, respectively. The two 1st-order diagrams branch into four possible diagrams at the 2nd order, generating excited-state populations in the material by means of contributions to the diagonal density matrix element corrections $\rho_{00}^{(2)}$ and $\rho_{11}^{(2)}$. Continuing on to the 3rd-order interaction, each 2nd-order diagram gives rise to two possible 3rd-order diagrams, resulting in eight 3rd-order diagrams and generating 3rd-order polarizations in the material by means of contributions to the off-diagonal density matrix element corrections $\rho_{01}^{(3)}$ and $\rho_{10}^{(3)}$. We also need to specify the pulses. Considering a three-pulse experiment with three excitation pulses A, B, and C, the three pulses can be arranged in various different three-pulse sequences. If we do not consider third harmonic generation with one pulse acting three times or the cases where only two pulses act, the permutations of pulses A, B, and C give $3! = 6$ possible combinations of three excitation pules. For each pulse sequence, there are eight double-sided Feynman diagrams at the 3rd order as shown in Fig. 2.7. Therefore, there are a total of 48 diagrams in the rotating-wave approximation at the 3rd order for a two-level system interacting with three pulses each acting only once. This process can be extended to more complex energy-level structures and higher-order excitations to systematically construct all possible double-sided Feynman diagrams.

As mentioned previously in Section 2.3.1, in an experiment, the number of possible double-sided Feynman diagrams can be significantly reduced for a given pulse ordering and signal direction. In the example discussed above, if pulse A is applied first, pulse B second, pulse C third, and the signal is measured in the signal direction $-\mathbf{k}_A + \mathbf{k}_B + \mathbf{k}_C$ (i.e., pulse A is considered conjugate with a negative wave vector), then there are only four possible double-sided Feynman diagrams, corresponding to the left half of the diagram set shown at the top of Fig. 2.7. Moreover, the third and fourth of these diagrams are nothing more than the complex conjugates of the first and second diagrams, leaving only two physically distinct signal contributions. We shall return to a more quantitative discussion of this scenario shortly, in Section 2.3.5.

2.3.3 Measured observables

It remains to be explained how these different phase-matched elements connect to measured observables. Two different avenues are common: (1) detection of coherently emitted light, and (2) detection of intensity beatnotes or phase-cycled components of incoherent signals such as photoluminescence or photocurrent.

Detection of coherently emitted light is the more common form of extracting a signal between the two of these, and the signal emerges in this case by means of the fact that a material's dielectric polarization P and its off-diagonal density matrix elements ρ_{ij} are related by

$$P = (N/V)\operatorname{Tr}(\rho\mu). \tag{2.14}$$

The polarization, in turn, generates a coherently emitted beam of light, which is the ultimately measured quantity. The details of where exactly the emitted radiation goes are dependent on the geometrical arrangement of the emitters in the solid. Among the more commonly treated examples is the case of an infinite planar sheet of dipoles,[2] in which case polarization and emitted electric field amplitude are related to each other in the time domain according to

$$E(t)_{\text{signal}} = -\frac{(N/A)}{2c\epsilon_0}\frac{\partial}{\partial t}P(t)_{\text{signal}} \tag{2.15}$$

and are related to each other in the frequency domain according to

$$\mathcal{E}(\omega_t)_{\text{signal}} = \frac{(N/A)}{2c\epsilon_0}i\omega_t P(\omega_t)_{\text{signal}}. \tag{2.16}$$

Note that in the frequency domain, the radiated electric field $\mathcal{E}(\omega_t)$ has a factor of i difference from the polarization $P(\omega_t)$. In the narrow-band limit, this represents a simple $\pi/2$ phase difference between the radiated field and the polarization, as also

---

[2]Chapter 30-7 of *The Feynman Lectures on Physics*, vol. 1, provides an intuitive description of how Eqs. (2.15) and (2.16) come about [116]. See also Appendix A of *Laser Physics* by Sargent, Scully, and Lamb [328].

discussed in Section 1.1 and illustrated in Fig. 1.1. In experimentally reported plots of 2D spectra, the emitted electric field profile is treated much more frequently than the originating material polarization, and so the additional $\pi/2$ phase shift must be accordingly remembered.

As was the case in the perturbative expansion of the density matrix, the final coherent emission process can be usefully codified in terms of double-sided Feynman diagram vertices. Because the process is different from that involved in the density matrix expansion itself, however, a different kind of vertex notation is generally used, and the most common illustration of the use of a wavy line pointing out from the double-sided Feynman diagram as illustrated in the figure accompanying Eqs. (2.17) and (2.18), with the equation itself depicting the substitution of Eq. (2.14) into Eq. (2.15).

Double-sided Feynman diagram vertices for the coherent emission of light

$$\tilde{E}_{ji,\alpha}^{(n)}(t) = -\frac{N}{2Ac\epsilon_0}\frac{\partial}{\partial t}\mu_{ij}\,\rho_{ji,\alpha}^{(n)}(t) \tag{2.17}$$

$$\tilde{E}_{ij,\alpha}^{(n)}(t) = +\frac{N}{2Ac\epsilon_0}\frac{\partial}{\partial t}\mu_{ji}\,\rho_{ij,\alpha}^{(n)}(t) \tag{2.18}$$

An alternate detection scheme to the one in which coherently emitted light is measured is a scheme in which an incoherent fourth-order signal is measured, for example a fourth-order beatnote in photoluminescence or photocurrent intensity or a phase-cycled component of photoemitted electrons. In this case, double-sided Feynman diagrams can be built up entirely out of the vertices described by Eqs. (2.10)–(2.13) subject to a requirement that the fourth-order density matrix being examined be a population state ρ_{kk}. The measured signal can then be constructed out of a weighted Feynman diagram sum, for example, a summation of photoluminescence signal contributions

$$I_{kk,\alpha}^{(n)}(t) = k\,\rho_{kk,\alpha}^{(n)}(t) \tag{2.19}$$

in which the relative weights of different diagrams are specified according to whether the associated population states are singly-occupied, doubly-occupied, etc.

2.3.4 Putting it all together

Having laid out the equations for all of these different double-sided Feynman diagram vertices in isolation, we can now start to build up the equations corresponding to more complete double-sided Feynman diagrams, and we can ultimately then combine these diagrammatic equations into complete multidimensional spectroscopy simulations. It may help, before we start looking into specifics, to generalize the above formalism into slightly more compact form so that overall themes and commonalities can be accentuated. Beginning with the first-order interaction, we can write

$$\rho_{fg,\alpha}^{(1)}(t) = \frac{\pm i\mu_{1,\alpha}}{2\hbar} e^{i\eta_1 \mathbf{k}_1 \cdot \mathbf{r}} \int_{-\infty}^{t} dt' \, \hat{E}_{1,\eta_1}(t') \, e^{-i\eta_1\omega_1 t'} e^{-i\Omega_{fg}(t-t')}, \qquad (2.20)$$

where the symbol η_1 specifies whether an interaction is nonconjugate ($\eta_n = 1$) or conjugate ($\eta_n = -1$), where the plus or minus symbol in front of the interaction is respectively associated with whether the interaction pertains to the left or right side of the commutator product in Eq. (2.7), where $\Omega_{fg} = \Omega_{f0}$ or Ω_{0g} depending respectively on whether the interaction pertains to the left or right side of the commutator product in Eq. (2.7), where $\mu_{1,\alpha} = \mu_{f0}$ or μ_{0g} depending respectively on whether the interaction pertains to the left or right side of the commutator product, and where the electric field amplitudes $\hat{E}_{n,\eta_n}(t)$ are connected back to Eq. (2.8) and defined as

$$\hat{E}_{n,\eta_n}(t) \equiv \begin{cases} \hat{E}_n(t) & \text{for } \eta_n = 1 \\ \hat{E}_n^*(t) & \text{for } \eta_n = -1. \end{cases} \qquad (2.21)$$

In this way, all of the four possibilities in Eqs. (2.10)–(2.13) are encapsulated within a single equation.

Having thus worked this out, in accordance with the discussion of Eq. (2.7) [from which Eq. (2.20) is derived], all higher-order interactions can be obtained by repeatedly subsituting Eq. (2.20) back into itself. A common situation relevant to MDCS is the case of third-order density matrix correction terms of the form

$$\rho_{jk,\alpha}^{(3)}(t,T,\tau) =$$

$$\frac{\pm i\mu_{3,\alpha}}{2\hbar} e^{i\eta_3 \mathbf{k}_3 \cdot \mathbf{r}} \int_{-\infty}^{t} dt''' \, \hat{E}_{3,\eta_3}(t''') e^{-i\eta_2\omega_3(t''')} e^{-i\Omega_{jk}(t-t''')}$$

$$\left\{ \frac{\pm i\mu_{2,\alpha}}{2\hbar} e^{i\eta_2 \mathbf{k}_2 \cdot \mathbf{r}} \int_{-\infty}^{t'''} dt'' \, \hat{E}_{2,\eta_2}(t''+T) e^{-i\eta_2\omega_2(t''+T)} e^{-i\Omega_{hi}(t'''-t'')} \right.$$

$$\left. \left[\frac{\pm i\mu_{1,\alpha}}{2\hbar} e^{i\eta_1 \mathbf{k}_1 \cdot \mathbf{r}} \int_{-\infty}^{t''} dt' \, \hat{E}_{1,\eta_1}(t'+T+\tau) e^{-i\eta_1\omega_1(t'+T+\tau)} e^{-i\Omega_{fg}(t''-t')} \right] \right\}, \quad (2.22)$$

where the higher-order frequencies Ω_{hi} and Ω_{jk} and the higher-order dipole moments $\mu_{2,\alpha}$ and $\mu_{3,\alpha}$ obey the rules of Eqs. (2.10)–(2.13), and where the temporal arguments of the electric field functions $\hat{E}_{n,\eta_n}(t)e^{i\eta_n(\mathbf{k}_n \cdot \mathbf{r} - \omega_n t)}$ have been suggestively chosen to reflect a three-pulse interaction in which the pulses are time-delayed by values of τ and T, as schematically illustrated back in Figs. 2.1(c) and 2.3(a). (The origin is chosen such that time $t = 0$ coincides with the arrival of the third pulse.)

We can incorporate radiative emission processes in the cases where it is appropriate by substituting our obtained value of $\rho_{jk,\alpha}^{(3)}(t,T,\tau)$ back into Eqs. (2.17) or (2.18). Introducing the symbol $\mathscr{D}_\alpha$ for generalized double-sided Feynman diagram equations (and thereby allowing us to expand our formalism beyond density matrix perturbation theory alone and to incorporate radiative emission processes into the formalism), we have

$$\mathscr{D}_\alpha(t,T,\tau) = \mp \frac{N\mu_{4,\alpha}}{2Ac\epsilon_0} \frac{\partial}{\partial t} \Bigg\{ \tag{2.23}$$

$$\frac{\pm i\mu_{3,\alpha}}{2\hbar} e^{i\eta_3 \mathbf{k}_3 \cdot \mathbf{r}} \int_{-\infty}^{t} dt''' \hat{E}_{3,\eta_3}(t''') e^{-i\eta_2\omega_3(t''')} e^{-i\Omega_{jk}(t-t''')}$$

$$\Bigg[\frac{\pm i\mu_{2,\alpha}}{2\hbar} e^{i\eta_2 \mathbf{k}_2 \cdot \mathbf{r}} \int_{-\infty}^{t'''} dt'' \hat{E}_{2,\eta_2}(t''+T) e^{-i\eta_2\omega_2(t''+T)} e^{-i\Omega_{hi}(t'''-t'')}$$

$$\Bigg(\frac{\pm i\mu_{1,\alpha}}{2\hbar} e^{i\eta_1 \mathbf{k}_1 \cdot \mathbf{r}} \int_{-\infty}^{t''} dt' \hat{E}_{1,\eta_1}(t'+T+\tau) e^{-i\eta_1\omega_1(t'+T+\tau)} e^{-i\Omega_{fg}(t''-t')} \Bigg) \Bigg] \Bigg\}.$$

The overall signal is given by the sum of a subset of the possible double-sided Feynman diagrams that are experimentally filtered by signal direction or other means, resulting finally in

$$S(t,T,\tau) = \sum_{\alpha \text{ (selected)}} \mathscr{D}_\alpha(t,T,\tau). \tag{2.24}$$

Within the Bloch model framework, Eqs. (2.22) and (2.23) are almost entirely general. Note that not even do the time delays τ and T impose any limiting constraints, because the envelope functions can always be adjusted to negate their effect. Moreover, no restrictions have been imposed in these equations on whether or not the rotating-wave approximation is invoked (although we have generally been operating within the rotating-wave approximation from Section 2.3.2 onward). In the cases where the rotating-wave equation *is* invoked, there is an important mapping between the selected Liouville space pathways associated with different signal collection schemes and the vectors $\boldsymbol{\eta} \equiv [\eta_1, \eta_2, \eta_3]$ that can be constructed out of the different conjugate/non-conjugate symbols η_n. We have

$$S_I \text{ (rephasing pathways)} \iff \boldsymbol{\eta} = [-1,1,1] \text{ or } [1,-1,-1] \tag{2.25}$$
$$S_{II} \text{ (nonrephasing pathways)} \iff \boldsymbol{\eta} = [1,-1,1] \text{ or } [-1,1,-1] \tag{2.26}$$
$$S_{III} \text{ (double-quantum pathways)} \iff \boldsymbol{\eta} = [1,1,-1] \text{ or } [-1,-1,1]. \tag{2.27}$$

In the impulsive limit, this mapping must be explicitly applied before summing up diagrams, because non-rotating-wave terms [e.g., contributions to $S_I(t,T,\tau)$ from diagrams with $\boldsymbol{\eta} \neq [-1,1,1]$ or $[1,1,-1]$] have unphysically large amplitudes [280]. With finite pulses, by contrast, these terms are attenuated by pulse bandwidth, which helps justify the rotating-wave approximation in situations where it is valid, and illustrates the flaws of the approximation in situations where it is not. See Section 3.10 and also Ref. 362 for additional discussion of the ways in which finite-pulse effects can influence spectra.

2.3.5 Case study: Two-level system

Returning to the example of the two-level system introduced in Section 2.3.2 and discussed in the context of Fig. 2.7, we can now quantitatively analyze the specific case of a coherent-emission-detected rephasing MDCS experiment conducted on this

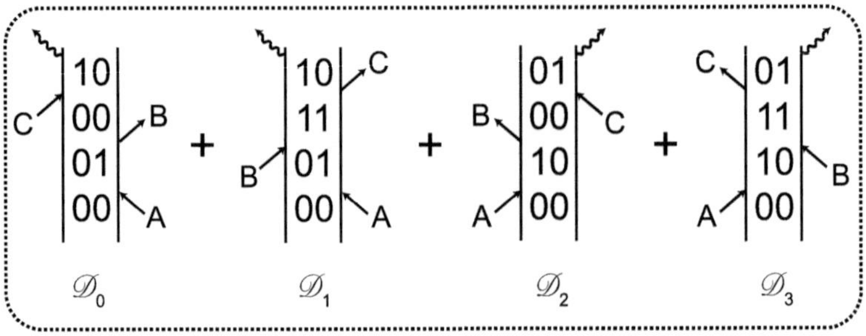

Fig. 2.8 Double-sided Feynman diagrams relevant to an MDCS measurement of a two-level system interacting with a specific pulse sequence A, B, and C and acquired in the signal direction $-\mathbf{k}_A + \mathbf{k}_B + \mathbf{k}_C$.

system in the limit where the excitation pulses are assumed to be of infinitesimally short duration and equal intensity. The relevant double-sided Feynman diagrams are as displayed in Fig. 2.8.

The laser pulse envelopes in this case can be mathematically modeled using the functions $\hat{E}_{n,\eta_n}(t) = \mathcal{E}_0 \delta(t)$, which correspond to Dirac delta functions of area $\mathcal{E}_0$. We assume as well that the system is initially in the ground state such that $\rho_{00}^{(0)} = 1$. The integrals in Eq. (2.23) can then be re-written as

$$
\mathcal{D}_\alpha(t,T,\tau) = -\frac{N\mu^* \mathcal{E}_0^3}{2Ac\epsilon_0} \frac{\partial}{\partial t} \Bigg\{
$$

$$
\frac{\pm i\mu}{2\hbar} e^{i\mathbf{k}_3 \cdot \mathbf{r}} \int_{-\infty}^{t} dt''' \delta(t''') e^{-i\omega_3(t''')} e^{-i\Omega_{10}(t-t''')}
$$

$$
\left[\frac{\mp i\mu}{2\hbar} e^{i\mathbf{k}_2 \cdot \mathbf{r}} \int_{-\infty}^{t'''} dt'' \delta(t''+T) e^{-i\omega_2(t''+T)} e^{-i\Omega_{\alpha\alpha}(t'''-t'')} \right.
$$

$$
\left. \left(\frac{-i\mu^*}{2\hbar} e^{-i\mathbf{k}_1 \cdot \mathbf{r}} \int_{-\infty}^{t''} dt' \delta(t'+T+\tau) e^{i\omega_1(t'+T+\tau)} e^{-i\Omega_{01}(t''-t')} \right) \right] \Bigg\} \quad (2.28)
$$

for $\alpha = 0$ and 1, and as $\mathcal{D}_\alpha(t,T,\tau) = \mathcal{D}_{\alpha-2}^*(t,T,\tau)$ for $\alpha = 2$ and 3. The symbol $\mu = \mu_{10} = -\langle 1|e\hat{r}|0\rangle$ is the electric dipole moment. All of these equations can be easily integrated to give equations of the form

$$
\mathcal{D}_\alpha(t,T,\tau) = \underbrace{\left(-\frac{\mu^*}{2Ac\epsilon_0} \frac{\partial}{\partial t} \right)}_{\text{Emission}} \underbrace{\left(\frac{\pm i\mu\mathcal{E}_0 e^{i\mathbf{k}_3 \cdot \mathbf{r}}}{2\hbar} \Theta(t) e^{-i\Omega_{10}t} \right)}_{3^{rd}\text{-order interaction}}
$$

$$
\times \underbrace{\left(\frac{\mp i\mu\mathcal{E}_0 e^{i\mathbf{k}_2 \cdot \mathbf{r}}}{2\hbar} \Theta(T) e^{-i\Omega_{\alpha\alpha}T} \right)}_{2^{nd}\text{-order interaction}} \underbrace{\left(\frac{-i\mu^*\mathcal{E}_0 e^{-i\mathbf{k}_1 \cdot \mathbf{r}}}{2\hbar} \Theta(\tau) e^{-i\Omega_{01}\tau} \right)}_{1^{st}\text{-order interaction}} \quad (2.29)
$$

for $\alpha = 0$ and 1, and (again) $\mathscr{D}_\alpha(t, T, \tau) = \mathscr{D}^*_{\alpha-2}(t, T, \tau)$ for $\alpha = 2$ and 3. The function $\Theta(x)$ is the Heaviside step function.

In this way we have constructed perhaps the simplest possible description of the expected results of an MDCS experiment, which is happily in this case an analytical function. Frequency-domain versions of the spectrum can be obtained by Fourier-transforming Eq. (2.29) with respect to the three time delays τ, T, and t. In general, the spectrum in the time domain can be constructed by stepping these different time delays and then Fourier transforming the result. Further aspects of the Bloch model treatment of MDCS experiments conducted on two-level systems, as well as the Bloch model treatment of MDCS experiments conducted on other archetypal systems, are discussed in detail in Chapter 3.

While double-sided Feynman diagrams are useful in determining what signals may exist, and their relative strengths, they only describe signals due to Pauli blocking, i.e., the fact that only one electron can occupy a given state. In many systems, particularly semiconductors, there are more complex interactions between excitations (electron-hole pairs or excitons) that are not captured by the optical Bloch equations or double sided Feynman diagrams. The optical Bloch equations can be phenomenologically modified to include some of these effects such as local fields [433], excitation-induced dephasing [158, 426, 427] or excitation-induced shift [343, 344]. However, there is no corresponding modification to the double-sided Feynman diagrams. To capture many-body effects diagrammatically, the traditional Feynman diagrams used to describe many-body interactions must be employed. The equivalent to a double-sided Feynman diagram is known as a "loop diagram." They have been used to describe multidimensional coherent spectra of semiconductors [419, 420] and molecules [100, 282].

2.4   Phase matching

The concept of "phase matching" is well defined in nonlinear optics, see for example the text by Shen [349]. Specifically it refers the situation where the phase evolution of the nonlinear polarization is identical to that of the emitted electromagnetic field. In this case the emission due the nonlinear polarization will be in phase with the field emitted at other locations along the propagation direction and the total emitted field will build up as square of the propagation distance. In the absence of phase matching, there will be a phase slip between the polarization and the field as they propagate and ultimately the newly emitted field can actually begin to cancel the existing field, resulting in a decrease in the total field strength.

Determining in general if a given geometry of excitation beams and signal beam is phase matched can be complicated if the beams, either the excitation beams or the signal beam, have different frequencies. In this case the dispersion of the index of refraction of the medium has to be taken into account. Often birefringence is exploited to realize full phase matching.

For MDCS, the situation is typically greatly simplified in that the excitation fields and signal field all have the same frequency, thus dispersion does not matter. In this case, the phase-matching condition can simply be expressed as $|\mathbf{k}_{sig}| = |\mathbf{k}_i|$ where $i = 1, 2, 3$, i.e., the magnitude of the signal wavevector is equal to the magnitude of the excitation wavevectors.

Note that simply summing up the wavevectors of the incident fields does not necessarily define a phase-matched direction, it only determines the direction of the wavevector for the nonlinear polarization. Whether or not that direction is phase matched depends on the specific arrangement as discussed in Section 1.3. For the example of a three-pulse experiment, the signal direction $\mathbf{k}_{sig} = -\mathbf{k}_1 + \mathbf{k}_2 + \mathbf{k}_3$ is phase matched if the incident beams are arranged in the box geometry, whereas it is not if the beams are arranged in a planar geometry. Unfortunately, the language in the MDCS community has become somewhat confusing, with the generic signal direction sometimes being referred to as the "phase-matched" direction although that is not the correct terminology from the perspective of general nonlinear optics.

2.5 Two-dimensional infrared (2D IR) spectroscopy

The experiments of 2D spectroscopy by using a five-pulse Raman excitation, as initially proposed by Tanimura and Mukamel [389], can be complicated by lower-order signals [32]. Alternatively, molecular vibrations can be directly excited by infrared (IR) fields and thus 2D IR spectroscopy. The groups of Hochstrasser [146] and Tokmakoff [136] independently demonstrated 2D IR spectroscopy to study molecular vibrations. In the IR regime, despite the challenges of IR sources and detectors, the requirement of the timing precision and stability for implementing 2D spectroscopy can be satisfied with only passive stability due to the longer wavelength.

Vibrational levels of molecules can be directly accessed by 2D IR spectroscopy. One of the first molecules that were studied by 2D IR spectroscopy was dicarbonyl-acetylacetonato rhodium [184, 185]. The spectrum shows two diagonal peaks at 2015 and 2084 cm^{-1}, corresponding to the symmetric and asymmetric stretching modes of the carbonyl (CO) groups. There are also two cross-diagonal peaks at the same frequencies due to the coupling of the two modes. More interestingly, the 2D IR spectrum also shows a set of four offset peaks due to anharmonicity of molecular vibrations. In contrast, these peaks would be too closely spaced to be distinguished in a 1D linear absorption spectrum thus the anharmonicity cannot be determined.

The time resolution of short pules in 2D IR spectroscopy allows the ability to probe ultrafast reaction dynamics. The kinetics of a chemical reaction can be probed by taking a series of 2D spectra at different waiting time T. For example, the formation and dissociation of solute-solvent complexes have been studied by using 2D IR spectroscopy [466]. In this reaction, phenol (the solute) forms hydrogen-bonded complexes with benzene (the solvent). At short T ($T < 10$ ps), the 2D spectrum does not have cross peaks since the the vibrational resonances of the complex and those of the free phenol are uncoupled. Cross peaks appear in 2D spectra at longer T ($T > 10$ ps). The creation and dissociation rates can be measured from the variation of cross-peak volume as the waiting time T varies. These complexes are now known to have lifetimes on the order of 10 ps. Such picosecond dynamics in chemical reactions can be probed with 2D IR spectroscopy but not with 2D NMR whose time resolution is limited to milliseconds.

3

Interpretation of multidimensional coherent spectra

Multidimensional spectra can be intimidating when they are first encountered. There are multiple peaks positioned on a plane (or higher-dimensional space), often with nontrivial shapes, which is not intuitive for spectroscopists more familiar with one-dimensional spectra or at most a series of one-dimensional spectra as a function of some other parameter. In Chapter 2, we gave an intuitive introduction to the different types of two-dimensional spectra. In this chapter, we will provide a more rigorous description of these various spectral forms as well as their analysis.

The analysis is largely based on the Bloch model of a multi-level atomic system, in which discretely separated electronic energy levels are connected through quantum mechanical dipole matrix elements and exhibit exponentially decaying coherence and population dynamics as is consistent with the Markov approximation [40, 449]. Such a model comprises an excellent approximation of the physics relevant to MDCS measurements of atomic gases and—to a certain extent—the physics of quantum dots and other forms of localized resonances in solid-state systems at low temperature.

3.1 Isolated two-level system

Perhaps the simplest possible MDCS example is the spectrum resulting from an isolated two-level system exhibiting an excited-state population lifetime T_1 and associated population decay rate $\Gamma_{10} = 1/T_1$, as well as an excited-state/ground-state coherence time T_2 corresponding to a decoherence rate $\gamma_{10} = 1/T_2$. In the coherent limit, these two timescales are related to each other according to $\gamma_{10} = \Gamma_{10}/2$. In many practical systems of interest, however, this equality is violated and more generally

$$\gamma_{10} = \frac{1}{2}\Gamma_{10} + \gamma_{10}^{ph} \qquad (3.1)$$

where (as in Section 1.4) the parameter γ_{10}^{ph} captures the effect of collisional or "pure" dephasing processes in which the coherence of the superposition of a two-level system's ground state and excited state is lost without any corresponding change in population. To preserve this generality, we shall allow γ_{10} and Γ_{10} to remain independent both here and below.

For the sake of concreteness, we consider the system's single-quantum rephasing spectrum derived from an experiment in which three laser pulses of wave vectors $\mathbf{k}_n$ and carrier frequencies ω_n (the subscript $n = 1, 2, 3$ is a pulse identification index) are arranged to produce a coherently emitted four-wave mixing signal in the direction $\mathbf{k}_I = -\mathbf{k}_1 + \mathbf{k}_2 + \mathbf{k}_3$ and carrier frequency $\omega_I = -\omega_1 + \omega_2 + \omega_3$. We consider an impulsive scenario in which all three excitation pulses are of identical shape and have envelopes $\hat{E}_n(t) = \mathcal{E}_0 \delta(t - t_n)$ corresponding to a collection of Dirac delta functions of area $\mathcal{E}_0$. To avoid the complexities of vector notation, we also restrict our analysis to the projection of electric field and dipole moments onto a particular direction in space, and we examine the output signal at position $\mathbf{r} = 0$ (so that phase factors of the form $e^{i\mathbf{k}\cdot\mathbf{r}} \to 1$). The coherently emitted signal in this case can be expressed in the time domain as $S_I(t, T, \tau) = E_I^{(3)}(t; T, \tau)$, or in the frequency domain as $S_I(\omega_t, T, \omega_\tau)$ by Fourier-transforming with respect to the first-order and third-order interaction variables τ and t.

In the lowest-order perturbative theoretical treatment of such an experiment, the signal arises from the sum of exactly two distinct double-sided Feynman diagrams

$$S_I = \mathscr{D}_0 + \mathscr{D}_1 \tag{3.2}$$

as illustrated in Fig. 3.1. (Complex conjugate diagrams are also generally relevant and required to produce a real-valued time-domain signal as discussed back in Section 2.3.5. However, they are not considered here or below because the information they provide is redundant.) The first of these diagrams can be physically interpreted as bleaching resulting from the third pulse's reduced probability of being able to interact with the remaining unperturbed population in the ground state after the arrival of the first two pulses. The second diagram can be interpreted as stimulated emission resulting from the third pulse's interaction with the population in the excited state created by the first two pulses.

In the Bloch model, both of these signal contributions are exactly the same, and can be written in the time domain (see Section 2.3.5) as

$$\mathscr{D}_\alpha(t, T, \tau) = \underbrace{\left(-\frac{\mu^*}{2Ac\epsilon_0}\frac{\partial}{\partial t}\right)}_{\text{Emission}} \underbrace{\left(\frac{\pm i\mu\mathcal{E}_0}{2\hbar}\Theta(t)e^{-i\Omega_{10}t}\right)}_{3^{rd}\text{-order}} \underbrace{\left(\frac{\mp i\mu\mathcal{E}_0}{2\hbar}\Theta(T)e^{-i\Omega_{\alpha\alpha}T}\right)}_{2^{nd}\text{-order}} \tag{3.3}$$

$$\times \underbrace{\left(\frac{-i\mu^*\mathcal{E}_0}{2\hbar}\Theta(\tau)e^{-i\Omega_{01}\tau}\right)}_{1^{st}\text{-order}}$$

$$= \frac{i|\mu|^4\mathcal{E}_0^3}{16Ac\epsilon_0\hbar^3}\frac{\partial}{\partial t}\Theta(t)e^{-i(\omega_{10}-i\gamma_{10})t}\Theta(T)e^{-\Gamma_{10}T}\Theta(\tau)e^{-i(-\omega_{10}-i\gamma_{10})\tau} \tag{3.4}$$

and in the frequency domain as

$$\mathscr{D}_\alpha(\omega_t, T, \omega_\tau) = \frac{\omega_t|\mu|^4\mathcal{E}_0^3}{16Ac\epsilon_0\hbar^3}\Theta(T)e^{-\Gamma_{10}T}$$

$$\times \left(\frac{i}{\omega_t - [\omega_{10} - i\gamma_{10}]}\right)\left(\frac{i}{\omega_\tau - [-\omega_{10} - i\gamma_{10}]}\right) \tag{3.5}$$

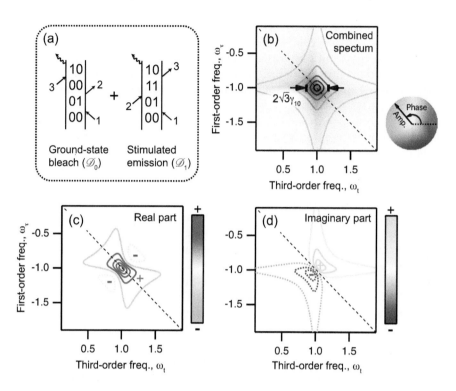

Fig. 3.1 MDCS rephasing plot corresponding to an isolated two-level system with exponential dephasing. (a) Relevant double-sided Feynman diagrams as described by Eqs. (3.4) and (3.5). (b)–(d) Resultant complex spectrum and projections into real and imaginary components.

where $\alpha = 0$ or 1, $\mu = \mu_{10} = -\langle 1|e\hat{r}|0\rangle$ is the electric dipole moment, A is the cross-sectional interaction area, c is the speed of light in vacuum, ϵ_0 is the vacuum permittivity, $\hbar$ is the reduced Planck's constant, $\Theta(x)$ is the Heaviside step function, $\gamma_{10} = \gamma_{01} = 1/T_2$, $\gamma_{11} = \gamma_{00} = \Gamma_{10} = 1/T_1$, and ω_t and ω_τ are the frequencies conjugate to the emission time t, and the delay between the first and second pulse τ, respectively. As discussed previously in Section 2.3.1, we define $\Omega_{ij} \equiv \omega_{ij} - i\gamma_{ij}$ with $\omega_{ij} \equiv (E_i - E_j)/\hbar$ [Eq. (2.6)], giving rise to the particular case at hand where ω_{10} corresponds to the two-level system's resonant frequency. As discussed in Section 2.3.3, values of $\mathscr{D}_\alpha(t, T, \tau)$ and $\mathscr{D}_\alpha(\omega_t, T, \omega_\tau)$ correspond to the field contribution of an electric dipole within an extended sheet of polarized material instead of the field emitted by a localized dipole in isolation, and so the system under consideration is perhaps most accurately imagined as a delocalized quantum state of planar geometry and not a localized point particle. Some mental acrobatics may be involved in visualizing this at the one-particle level, but the extended-sheet picture turns out to be simpler than the point-source picture from the point of view of phase-matching considerations, and it is also more straightforwardly generalized into the description of inhomogeneously broadened ensembles that will be discussed in Section 3.2.

The identical contributions of the two double-sided Feynman diagrams may strike the reader as curious given the diagrams' differing physical origins and the fact that one might naively expect the relevant ground state's lifetime to be of infinite duration while the excited state's lifetime is finite. This paradox is resolved by noting that the decay rate Γ_{00} corresponds *not* to the population dynamics of ρ_{00} dropping from its initial value down to zero, but rather to the dynamics associated with recovering back to equilibrium. For a two-level system, the timescale of this recovery must necessarily be the same as the dynamics of the excited-state population ρ_{11} by laws of probability conservation. In fact, under certain circumstances the decay rates Γ_{11} and Γ_{00} may not even correspond to overall population dynamics at all, but rather to other system parameters depending on the experimental configuration. In the case in which the MDCS signal is extracted by making use of spatially varied excitation-pulse characteristics, for example, the decay rate may be more indicative of diffusion processes that wash these spatial variation signatures out than it is of population dynamics specifically. These experimental considerations highlight the potential advantages of building MDCS systems capable of extracting MDCS signals via multiple avenues.

Among the spectrum's most prominent features is a characteristic star-shaped profile that breaks isotropic rotational symmetry and has an amplitude with a full-width at half-maximum (FWHM) linewidth in both the horizontal and vertical directions corresponding (in angular frequency units)[1] to $\Delta\omega_{\mathrm{FWHM}} = 2\sqrt{3}\,\gamma_{10}$. [The identity can be directly read off of Eq. (3.5) after computing its absolute value.] The broken symmetry can be understood as a consequence of the fact that the originating time-domain spectrum breaks rotational symmetry as well: in general, there is no reason why a lineshape other than highly symmetric profiles such as the Gaussian lineshape should maintain rotational symmetry when unfolded into higher-dimensional spaces. Beyond this, the spectrum exhibits an interesting characteristic "phase twist" evolving from the its lower-left to upper-right corners. This phase twist results in a pair of negative lobes in the real part of the spectrum that may confuse the uninitiated upon looking at multidimensional features of this sort for the first time, but which is nevertheless ubiquitous in multidimensional Fourier-transform spectroscopy, appearing at frequencies ranging from the ultraviolet regime all the way down to the radio waves of 2D-NMR. If desired, the twist can be unwound in certain cases by summing up the different rephasing and nonrephasing signal components, as described in more detail in Section 3.9.

Our aim here in discussing the isolated two-level system has been primarily pedagogical, but it might be noted before we move on to more complicated systems that— even in this extremely simple system—MDCS has the ability to elucidate properties that cannot be easily accessed by other means. A linear absorption measurement performed on a two-level system, for example, is sensitive to the dephasing time T_2 but

---

[1] We can translate Eq. (3.5) into ordinary frequency units by revising the identity $\gamma_{10} = 1/T_2$ into $\gamma_{10}^{(\mathrm{Hz})} = 1/(2\pi T_2)$, or we can translate into photon-energy units under the stipulation that $\gamma_{10}^{(\mathrm{eV})} = \hbar/T_2$.

not the population decay time T_1, whereas MDCS is sensitive to both of these quantities. T_2 can be measured (as explained above) through the Fig. 3.1 spectral linewidth. T_1 can be measured through the relaxation characteristics of the Fig. 3.1 peak height as a function of the waiting time T.

3.2 Inhomogeneously broadened ensemble of two-level systems

Having established the properties of an isolated two-level system, it is immediately possible to generalize to the more commonly encountered case of an ensemble of non-interacting two-level systems featuring a distribution of many *different* resonance frequencies that are simultaneously present, producing a spectral effect known as inhomogeneous broadening.

We can model the properties of such a system by generalizing the previous section's resonance frequency ω_{10} into a more complete set of frequencies $\{\omega_{10}\}$ with relative prominence given by the probability distribution function $Q(\omega_{10})$. The overall signal in this case will be given by a weighted sum of each of the individual frequency components, expressible in integral form as

$$S_{\text{Ensemble}} (\omega_t, \omega_\tau) \propto \int_{-\infty}^{\infty} Q(\omega'_{10}) \, S_{\text{1-atom}} (\omega_t, \omega_\tau, \omega'_{10}) \, d\omega'_{10}. \tag{3.6}$$

The most striking consequence of incorporating the distribution function $Q(\omega_{10})$ is that the width of spectral lineshapes is no longer merely governed by the intrinsic one-particle dephasing rate γ_{10}, but arises rather from a combination of this and the characteristic width of the inhomogeneous distribution function $Q(\omega_{10})$. In linear spectroscopy, this spectral inhomogeneity necessitates the introduction of a newly defined ensemble dephasing time T_2^* that can be orders of magnitude shorter than the single-particle dephasing time $T_2 = 1/\gamma_{10}$ discussed in the previous section, and which can hamper the extraction of parameters of fundamental interest from spectroscopic data.[2] Among the great advantages of MDCS is its ability to unambiguously disentangle these various parameters even in the presence of macroscopic ensembles and (perhaps especially) when T_2 and T_2^* are of comparable but not quite equal magnitude.

In principle, Eq. (3.6) can be directly applied to physical situations without need for any further formal development, and it can be applied in this form to zero-quantum, double-quantum, and nonrephasing single-quantum spectra just as well as it can be applied to the rephasing single-quantum spectrum that has been initially introduced; however, the integral equation can be cumbersome to treat from a computational and fitting perspective, and so it behooves the spectroscopist in many circumstances to rely on approximations. We discuss two of the more useful of these below.

3.2.1 Gaussian inhomogeneity, constant homogeneous linewidth

Among the most important special cases of Eq. (3.6) is the one in which the distribution function $Q(\omega_{10})$ is Gaussian and the homogeneous linewidth γ_{10} is assumed

---

[2]The parameter T_2^* has been defined somewhat inconsistently in spectroscopy literature. We adopt a definition matching that of Siegman [353] and Ernst, Bodenhausen, and Wokaun [108] but distinct from the definition introduced by Hamm and Zanni [147]. As such, we note that T_2^* is *not* the same thing here as the pure dephasing rate's reciprocal.

to be constant within the ensemble. This approximation is essentially exact for the case of a Doppler-broadened vapor, and the assumptions are also reasonable in many other situations, for example an ion-doped solid where the inhomogeneity comes from random crystal fields for which a normal (Gaussian) distribution is a good assumption.

Codifying the model in terms of an N-particle distribution function of width σ centered about an average resonance frequency ω_0, we have

$$Q(\omega_{10}) = G(\omega_{10} - \omega_0) \equiv \frac{N}{\sigma\sqrt{2\pi}} \exp\left[\frac{-(\omega_{10} - \omega_0)^2}{2\sigma^2}\right]. \tag{3.7}$$

Eq. (3.6) then takes on the form of a convolution integral between $G(\omega_0)$ and the one-atom two-dimensional Lorentzian function,

$$S_I(\omega_t, T, \omega_\tau) = \frac{\omega_t |\mu|^4 \mathcal{E}_0^3}{8Ac\epsilon_0\hbar^3} \Theta(T) e^{-\Gamma_{10}T}$$

$$\times \left(\frac{i}{\omega_t - \omega_0 + i\gamma_{10}}\right)\left(\frac{i}{\omega_\tau + \omega_0 + i\gamma_{10}}\right) * G(\omega_0). \tag{3.8}$$

Representative special cases of the Gaussian-distribution inhomogeneous ensemble are displayed in Fig. 3.2, and its one-dimensional equivalent is known as a Voigt profile.

Simplified descriptions of Eq. 3.8 exist. For optical-frequency resonances, for example, the peak center frequency, ω_0, is almost always much greater than both the homogeneous linewidth γ and the inhomogeneity parameter σ, and so the leading factor of ω_t can be replaced by ω_0. Fourier transforming into the time domain after making this substitution leads to the product expression

$$S_I(t, T, \tau) = \frac{N\omega_0 |\mu|^4 \mathcal{E}_0^3}{8Ac\epsilon_0\hbar^3} \Theta(T) e^{-\Gamma_{10}T}$$

$$\times \underbrace{\Theta(t)e^{-i(\omega_0 - i\gamma_{10})t}\Theta(\tau)e^{-i(-\omega_0 - i\gamma_{10})\tau}}_{\text{Isolated 2-level system response}} \underbrace{e^{\frac{-\sigma^2(t-\tau)^2}{2}}}_{\text{Inhomogeneity}}. \tag{3.9}$$

The projection-slice theorem of two-dimensional Fourier transforms permits an analytical frequency-domain formulation of Eq. 3.8 [28, 354], which is expressible in terms of exponentials and complementary error functions as

$$S_I(\omega_t, T, \omega_\tau) = \frac{N\omega_0 |\mu|^4 \mathcal{E}_0^3}{8Ac\epsilon_0\hbar^3} \Theta(T) e^{-i\Omega_{hi}T} \frac{\sqrt{2\pi}}{2\sigma(2\gamma_{10} - i(\omega_t + \omega_\tau))}$$

$$\times \left\{ \begin{array}{l} e^{\frac{(\gamma_{10} - i(\omega_t - \omega_0))^2}{2\sigma^2}} \operatorname{erfc}\left(\frac{\gamma_{10} - i(\omega_t - \omega_0)}{\sqrt{2}\sigma}\right) \\ + e^{\frac{(\gamma_{10} - i(\omega_\tau + \omega_0))^2}{2\sigma^2}} \operatorname{erfc}\left(\frac{\gamma_{10} - i(\omega_\tau + \omega_0)}{\sqrt{2}\sigma}\right) \end{array} \right\}. \tag{3.10}$$

Either equation 3.9 or 3.10 can be directly converted into fit functions used to extract γ_{10} and σ from actual data, and indeed two-dimensional functions of this nature are exceptional in terms of the number of constraints they provide. At times, however, such fits can be cumbersome to implement, and they can also fail in the presence

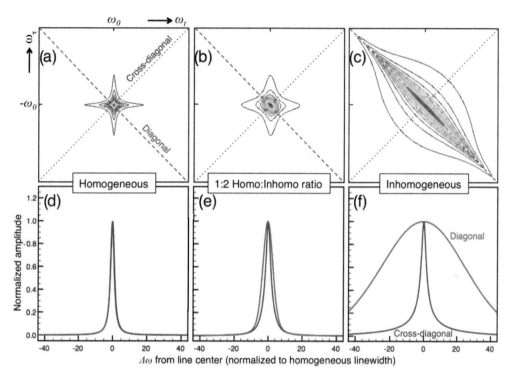

Fig. 3.2 MDCS rephasing amplitude lineshapes for an inhomogeneous ensemble of noninteracting two-level systems with a Gaussian distribution of spectral width σ and a frequency-independent single-particle dephasing rate γ_{10}. (a) $\sigma \ll \gamma_{10}$. (b) $\sigma = 2\gamma_{10}$. (c) $\sigma \gg \gamma_{10}$. (d)–(f) Diagonal lineouts (red traces) and cross-diagonal lineouts (blue traces) corresponding to the cases depicted in (a)–(c). Adapted from Ref. 354.

of congested spectral features or large amounts of background noise. It is therefore occasionally advantageous to examine selected slices of Eq. 3.10. A diagonal slice through the center of the resonance generated by varying ω_t and setting $\omega_\tau = -\omega_t$ yields

$$S_{\mathrm{I}\atop\mathrm{diag}}\left[\omega_t, \omega_\tau(\omega_t)\right] = \frac{N\omega_0|\mu|^4\mathcal{E}_0^3}{8Ac\epsilon_0\hbar^3}\Theta(T)e^{-i\Omega_{hi}T}$$

$$\times \left\{\frac{\pi}{\gamma_{10}}\mathrm{Voigt}(\omega_t - \omega_0; \sigma, \gamma)\right\}. \tag{3.11}$$

A cross-diagonal slice through the center as parameterized according to $\omega_\tau = \omega_t - 2\omega_0$ yields

$$S_{\underset{\text{cross-diag}}{\text{I}}} \left[\omega_t, \omega_\tau(\omega_t) \right] = \frac{N\omega_0 |\mu|^4 \mathcal{E}_0^3}{8Ac\epsilon_0 \hbar^3} \Theta(T) e^{-i\Omega_{hi}T}$$

$$\times \left\{ \frac{\sqrt{2\pi} e^{\frac{(\gamma_{10} - i(\omega_t - \omega_0))^2}{2\sigma^2}} \operatorname{erfc}\left(\frac{\gamma_{10} - i(\omega_t - \omega_0)}{\sqrt{2}\sigma} \right)}{2\sigma(\gamma_{10} - i(\omega_t - \omega_0))} \right\}. \tag{3.12}$$

Data sets are typically fit to Eqs. (3.11) and (3.12) in simultaneous fashion so that the values of γ_{10} and σ are shared between the two fits.

More generally within the Gaussian-inhomogeneity scenario, Eqs. (3.9)–(3.12) illustrate the important ability of MDCS to facilitate the independent extraction of the parameters γ_{10} and σ while imposing no requirements on their relative values.

This capability is lacking from other types of measurements. In a one-dimensional absorption measurement, for example, the lineshape would be expected to be a Voigt profile similar to Eq. (3.11), and attempting to extract the values of γ_{10} and σ from such a feature without the additional constraints supplied by Eq. (3.12) will typically lead to an unreliable estimate of the smaller of the two parameters. A one-dimensional photon-echo measurement, meanwhile, produces the integral of Eq. (3.9) with respect to t, resulting in a different set of limitations: in the homogeneous limit where $\sigma \ll \gamma_{10}$ the function $S_I(\tau)$ exponentially drops off with decay rate γ_{10}, and in the inhomogeneous limit where $\sigma \gg \gamma_{10}$ it exponentially drops off with decay rate $2\gamma_{10}$, but in the intermediate regime where $\sigma \approx \gamma_{10}$ it drops off with an unspecified decay rate somewhere in between these two bounds.

We have restricted our analysis here to rephasing spectra, but many of the equations can be easily expanded to incorporate nonrephasing pulse sequences as well. A more complete formal development of the model can be found in Refs. 28 and 354, to which the interested reader is referred. In the original treatment of Ref. 354, the expressions for the diagonal and cross-diagonal slices through an inhomogeneously broadened spectrum were derived in the context of a rotated, flipped, and squeezed coordinate system defined by variables $\omega_{t'} = (\omega_t + \omega_\tau)/2$ and $\omega_{\tau'} = (\omega_t - \omega_\tau)/2$. The dimensions and scaling of the homogeneous and inhomogeneous broadening parameters in the expressions in Ref. 354 and those of the parameterization described by Eqs. (3.11) and (3.12) are the same.

3.2.2   Large inhomogeneity

A second important analytically tractable scenario is the situation in which the probability distribution function $Q(\omega_{10})$ is much broader than the characteristic homogeneous linewidth $\gamma_{10}(\omega_{10})$, and in which the dependence of $\gamma_{10}(\omega_{10})$ on the parameter ω_{10} is small in comparison to its overall functional value. We shall refer to this scenario as the "large-inhomogeneity" limit, and it is a distinctively useful scenario because it is capable of capturing inhomogeneous dephasing situations in which the linewidth $\gamma_{10}(\omega_{10})$ is freed from the limitation of having to remain perfectly constant. Inhomogeneous dephasing does not occur in atomic vapors, but it can occur for electronic states in nanostructures, where the inhomogeneous distribution of resonance frequencies arise due to size fluctuations through the concomitant change in quantum confinement energy. The change in size can also change the coupling of the electronic states to the bath resulting in dephasing, as typically observed with a phonon bath.

If the signal is acquired at a sufficiently brief waiting time T to prevent frequency-dependent variations in Γ_{10} from having an observable impact, then for a given value of ω'_{10} we can expand the functions $\gamma_{10}(\omega'_{10})$ and $Q(\omega'_{10})$ about a selected third-order measurement frequency ω_t as

$$\gamma_{10}(\omega'_{10}) = \gamma_{10}(\omega_t) + \partial_{\omega'}\gamma_{10}|_{\omega_t} (\omega'_{10} - \omega_t) + \ldots \tag{3.13}$$

$$\text{and} \quad Q(\omega'_{10}) = Q(\omega_t) + \partial_{\omega'}Q|_{\omega_t} (\omega'_{10} - \omega_t) + \ldots \tag{3.14}$$

Substituting Eqs. (3.2) and (3.5) into Eq. (3.6), incorporating expansions (3.13) and (3.14) into the result, and then examining this new result under the stipulations of $\partial_{\omega'}\gamma_{10}|_{\omega_t} \ll 1$ and $\partial_{\omega'}Q|_{\omega_t} \ll Q(\omega_t)/\gamma_{10}(\omega_t)$ and in the vicinity of $|\omega_\tau + \omega_t| \sim \gamma_{10}$ (i.e., close to resonance) leads to a situation in which the terms in these expansions beyond zeroth-order can be ignored. The result is

$$S_I(\omega_t, T, \omega_\tau) \approx \frac{N\omega_0|\mu|^4\mathcal{E}_0^3}{8Ac\epsilon_0\hbar^3}Q(\omega_t)\Theta(T)e^{-\Gamma_{10}T} \tag{3.15}$$

$$\times \int_{-\infty}^{\infty} \left(\frac{i}{\omega_t - [\omega'_{10} - i\gamma_{10}(\omega_t)]}\right)\left(\frac{i}{\omega_\tau - [-\omega'_{10} - i\gamma_{10}(\omega_t)]}\right) d\omega'_{10}$$

$$= \frac{N\omega_0|\mu|^4\mathcal{E}_0^3}{8Ac\epsilon_0\hbar^3}Q(\omega_t)\Theta(T)e^{-\Gamma_{10}T}\left[\frac{i\pi}{(\omega_t + \omega_\tau)/2 + i\gamma_{10}(\omega_t)}\right]. \tag{3.16}$$

We see from this that a diagonal spectral slice along the lineout parameterized according to $\omega_\tau = -\omega_t$ traces out the lineshape

$$S_I \underset{\text{diag}}{[\omega_t, \omega_\tau(\omega_t)]} \propto \frac{Q(\omega_t)}{\gamma_{10}(\omega_t)}, \tag{3.17}$$

and that a cross-diagonal slice parameterized according to $\omega_\tau = \omega_t - 2\omega_{10}$ [where, in contrast to the case of Eq. (3.12), the diagonal-slice intersection point ω_{10} can be arbitrarily selected] traces out the lineshape

$$S_I \underset{\text{cross-diag}}{[\omega_t, \omega_\tau(\omega_t)]} \propto \left[\frac{i}{(\omega_t - \omega_{10}) + i\gamma_{10}(\omega_t)}\right] \tag{3.18}$$

$$\approx \left[\frac{i}{(\omega_t - \omega_{10}) + i\gamma_{10}(\omega_{10})}\right]. \tag{3.19}$$

Crucially, Eq. (3.19) takes on the same Lorentzian form as a horizontal lineout of the isolated two-level system, and as such exhibits an amplitude spectrum of linewidth $\Delta\omega_{\text{FWHM}} = 2\sqrt{3}\gamma_{10}(\omega_{10})$ that is independent of $Q(\omega_{10})$. An example simulated spectrum in this limit is depicted in Fig. 3.3.

Two other aspects of this scenario are worth noting. First, Eqs. (3.17) and (3.19) can be readily verified to be synonymous with Eqs. (3.11) and (3.12) in the $\sigma \rightarrow$

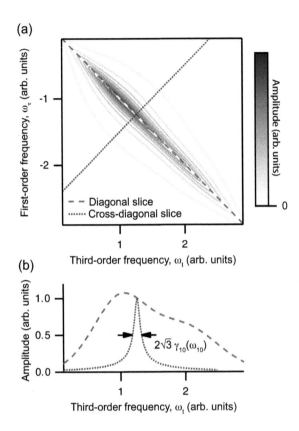

Fig. 3.3 Simulated example of an MDCS measurement of an inhomogeneously broadened ensemble of noninteracting two-level systems in the large-inhomogeneity limit. (a) Absolute value of the system's one-quantum rephasing spectrum. (b) Representative diagonal and cross-diagonal lineouts.

∞ limit, which establishes a common regime of applicability for these two different approximations and also highlights some of the ways in which the large-inhomogeneity approximation might be expected to fail. The appearance in a measured resonance of negative lobes flanking the central peak in the real part of the spectrum, for example, would be predicted by Eq. (3.12) but not by Eq. (3.19).

Second, and related to this, is the interesting fact that that phase twists disappear from the spectrum in the large-inhomogeneity limit. The effect can be understood by considering the phase-twisted shape of the isolated two-level resonance depicted in Fig. 3.1(c) and noting that the real part of the resonance has a butterfly-shaped profile exhibiting extended positive wings that are the same distance away from the diagonal line at $\omega_\tau = -\omega_t$ as are the negative lobes on the resonance's upper right and lower left sides. In the large-inhomogeneity limit, multiple butterfly resonances are superposed on top of one another at various different resonant frequencies while variations in peak height and homogeneous linewidth remain minimal. This being the case, the positive wings of a given resonance within the ensemble tend to cancel out the negative lobes of nearby resonances, thereby rendering the real part of the spectrum strictly positive while simultaneously narrowing its cross-diagonal linewidth. (The imaginary part of the spectrum is dispersive in nature and remains wide.)

An alternate way of understanding the same effect may be to note that the peaks of resonances at different frequencies tend to *constructively* interfere with each other in a rephasing spectrum [as can be seen by the fact that the isolated-system resonance depicted in Figs. 3.1(c) and 3.1(d) is strictly positive along the diagonal line at $\omega_\tau = -\omega_t$], while tending to *destructively* interfere with each other in a nonrephasing spectrum [the isolated-atom spectral peaks of a nonrephasing spectrum would be the same shape as the spectrum depicted in Figs. 3.1(c) and 3.1(d), but it would be superposed on top of one another along a line going from lower left to upper right instead of from upper left to lower right]. In the large-inhomogeneity limit, the overall magnitude of the rephasing spectrum becomes large while the magnitude of the nonrephasing spectrum goes to zero, and so the sum of the two spectra as rendered in a purely absorptive spectrum (see Section 3.9) becomes essentially identical to the rephasing spectrum collected in isolation. The effect justifies the fact that, in highly inhomogeneous systems, data corresponding to nonrephasing spectral plots are rarely collected.

3.3   Coherent coupling signatures

With a quantitative model for diagonal resonances established in the previous section, we proceed now to analyze coherent coupling. Among the simplest material systems in which this phenomenon is relevant is a three-state "v" system, as illustrated in Fig. 3.4 and previously introduced in Chapter 2.

In all, there are eight different double-sided Feynman diagrams corresponding to the formation of this spectrum, which can all be classified into different sorts of ground-state bleach and stimulated emission interactions as depicted in Fig. 3.4(c). Mathematically speaking, these diagrams can all be written in the form

$$\mathcal{D}_\alpha(\omega_t, T, \omega_\tau) = \frac{\omega_t |\mu_{j0}|^2 |\mu_{i0}|^2 \mathcal{E}_0^3}{16 A c \epsilon_0 \hbar^3} \Theta(T) e^{-i\Omega_{ab}T}$$

$$\times \left(\frac{i}{\omega_t - [\omega_{j0} - i\gamma_{j0}]} \right) \left(\frac{i}{\omega_\tau - [-\omega_{i0} - i\gamma_{i0}]} \right) \quad (3.20)$$

with

$$\Omega_{ab} = \begin{cases} 0 \quad - i\gamma_{00} \quad \text{(for ground-state bleach diagrams) or} \\ \omega_{ji} - i\gamma_{ji} \quad \text{(for the other diagrams)} \end{cases} \quad (3.21)$$

and subscript indices

$$i, j = 1 \text{ or } 1', \quad (3.22)$$

which is a generalization of Eq. (3.5). As described in Section 3.2, the effect of inhomogeneity can be incorporated by smearing out the resonance frequencies ω_{10} and

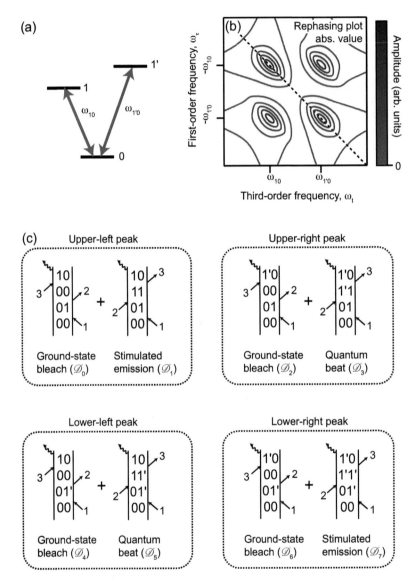

Fig. 3.4 Simulated 2D rephasing plot for an inhomogeneously broadened ensemble of three-state "v" systems. (a) Energy-level diagram. (b) Rephasing one-quantum amplitude spectrum. (c) Double-sided Feynman diagrams.

$\omega_{1'0}$ into finite-width distribution function and summing up ensemble contributions by means of an integral similar to (3.6). To keep the notation from becoming too cumbersome, we gloss over these additional considerations for the time being in the equations here illustrated.

In the case of the upper-left and lower-right diagonal peaks, the double-sided Feynman diagram interactions are identical to the interactions relevant to the two-state system. For both of these peaks, the spectral peak arises from a combination of ground-state bleach and stimulated emission interactions, and the cross-diagonal widths of the peaks give a measure of the homogeneous dephasing rate while the peak height dynamics as a function of the waiting time T give a measure of the population decay rates associated with excited states $|1\rangle$ and $|1'\rangle$.

However, the "v" system's shared ground state offers a qualitatively different set of dynamics, and as such offers a glimpse into the ways in which MDCS can elucidate coherent coupling effects. We may focus, for example, on the upper right cross-peak of Fig. 3.4(b), which arises from the effect that a first-order excitation interaction from the ground state into the lower-energy excited-state $|1\rangle$ can have on the subsequent third-order interaction involving the higher-energy excited state $|1'\rangle$. The lower left peak of Fig. 3.4(b) arises by means of an analogous mechanism. The ground-state bleach interaction for this peak (diagram $\mathscr{D}_2$) operates in a similar manner to the various diagonal-peak ground-state bleach interactions, and can be physically interpreted as follows: the first two pulses drive a fraction of the ground-state population into the excited-state $|1\rangle$, which leaves a reduced ground state population left for the third pulse to drive into excited state $|1'\rangle$ than there would have been in the first two pulses' absence. The result is an emission peak correlating the first-order frequency ω_{10} with the third-order frequency $\omega_{1'0}$. The interaction associated with diagram $\mathscr{D}_3$, on the other hand, is a little bit more subtle. The diagram begins with a first-order interaction resulting in an off-diagonal coherence between the ground state $|0\rangle$ and lower-energy excited state $|1\rangle$. Instead of elevating a fraction of this coherence into an excited-state population, however (which would be the interaction captured by Feynman diagram $\mathscr{D}_1$), the second pulse in diagram $\mathscr{D}_3$ generates *another* coherence, this time between the two excited states $|1\rangle$ and $|1'\rangle$. The third pulse then drives a component of this coherence associated with excited state $|1\rangle$ back down into the ground state, stimulating a third-order coherence between states $|1'\rangle$ and $|0\rangle$ that oscillates at frequency $\omega_{1'0}$.

Among the more interesting aspects of diagram $\mathscr{D}_3$ are that, because the second-order interaction results in an oscillating coherence instead of a steady population, the phase of $\mathscr{D}_3$ evolves as a function of the waiting time T. Thus, the overall amplitude of the cross-diagonal peaks will be seen to correspondingly oscillate with the waiting time T as the amplitudes of diagrams $\mathscr{D}_2$ and $\mathscr{D}_3$ interfere constructively or destructively. These oscillations are synonymous in origin with the intensity oscillations that can be observed in quantum beat spectroscopy, although—as is usually the case—MDCS offers a clearer and less ambiguous spectral signature. Figure 3.5 shows the oscillations of these peaks in the three-state "v" system at various representative waiting time values.

3.4   Incoherent coupling signatures

Coherent coupling effects are not the only means of establishing cross-peak resonances in MDCS. Such peaks can also occur, for example, for a pair of atoms with distinct ground states but in which the excited states are connected by a nonradiative decay

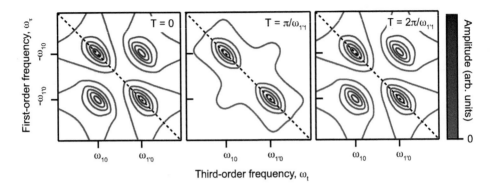

Fig. 3.5 Evolution of the plot from Fig. 3.4(b) as a function of the waiting time T.

channel. An illustration of the energy-level diagram corresponding to such a system and the way in which it compares to the coherently coupled "v" system is depicted in Fig. 3.6.

From a physical standpoint, the lower-left cross peak in Fig. 3.6 arises as a result of the fact that the nonradiative decay channel between states $|1'\rangle$ and $|1\rangle$ amplifies the excited-state electron population of state $|1\rangle$ relative to what it would otherwise be in the case of an isolated two-level system. In consequence, the arrival of the third pulse will necessarily be accompanied by an enhanced degree of stimulated emission as well as a reduced ability of the atom to absorb an electron from the ground state $|0\rangle$ to the excited state due to the fact that the excited state is now more fully occupied than it would have been in the case of an isolated two-level atom. Assuming that state $|1\rangle$ is at lower energy than state $|1'\rangle$ and the temperature is low compared to the energy splitting between them, the process is asymmetrical with respect to excited states $|1'\rangle$ and $|1\rangle$ and so the only cross peak that will appear in this case will be on the spectrum's lower left (lower emission frequency) side. At a temperature comparable to the energy splitting, a peak on the upper left can grow in, but at a slower rate. Beyond this, the incoherent nature of the process will result in a lack of cross peak oscillations with changing T. These differences in the spectral signatures of coherent and incoherent processes are of great help in terms of identifying cross-peak origins.

3.5 Doubly excited states and many-body interactions

One of the more interesting features of MDCS is its ability to reveal the properties of light-matter interactions involving systems with electronic states separated in energy by more than just a single multiple of the photon energy. Examples of such systems in condensed matter include higher-lying Rydberg biexcitons, atomic states with Rydberg indices greater than one, semiconductor systems exhibiting biexcitons and trions, systems including many-body interactions such as excitation-induced shifts (EIS) and excitation-induced dephasing (EID), and systems exhibiting a significant amount of coupling between electronic and vibrational modes.

Figure 3.7 shows an analysis of one such example system of this sort: a three-state "ladder" system in which the energy separation $\hbar\omega_{10}$ between states $|0\rangle$ and $|1\rangle$

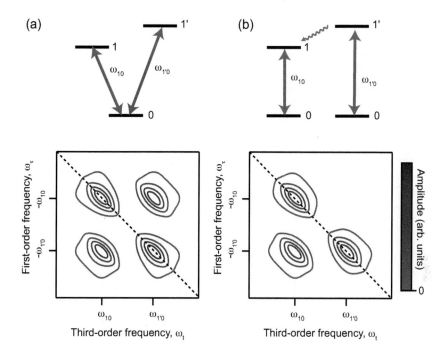

Fig. 3.6 Comparison between (a) the coherent-coupling cross peaks associated with a three--state "v" system, and (b) an incoherent cross peak associated with a pair of two-state systems connected by a nonradiative decay channel from the higher-energy excited state to the lower-energy excited state.

exceeds the separation $\hbar\omega_{21}$ between states $|1\rangle$ and $|2\rangle$, but in which both separations lie within the bandwidth of the excitation laser photon-energy range. States $|0\rangle$ and $|1\rangle$ are coupled by a dipole-allowed transition, as are states $|1\rangle$ and $|2\rangle$. However, state $|0\rangle$ is not coupled to state $|2\rangle$ by a dipole-allowed transition. The double-sided Feynman diagrams comprising the rephasing MDCS plot relevant to this system can be mathematically described by

$$\mathscr{D}_\alpha(\omega_t, T, \omega_\tau) = \pm \frac{\omega_t |\mu_{p,p-1}|^2 |\mu_{10}|^2 \mathcal{E}_0^3}{16 A c \epsilon_0 \hbar^3} \Theta(T) e^{-\Gamma_{10}T}$$

$$\times \left(\frac{i}{\omega_t - [\omega_{p,p-1} - i\gamma_{p,p-1}]} \right) \left(\frac{i}{\omega_\tau - [-\omega_{10} - i\gamma_{10}]} \right) \qquad (3.23)$$

with

$$p = \begin{cases} 1 & \text{for } \alpha = 0, 1 \\ 2 & \text{for } \alpha = 2. \end{cases} \qquad (3.24)$$

Similarly to the two-level system and "v" system described in Sections 3.1 and 3.3, the ladder system exhibits a number of ground-state bleach and stimulated emission interactions associated with Feynman diagrams $\mathscr{D}_0$ and $\mathscr{D}_1$ and resulting in an enhanced

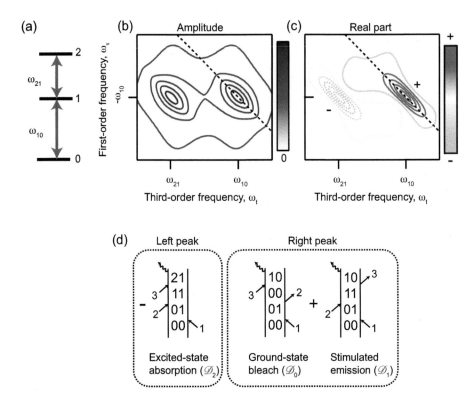

Fig. 3.7 2D rephasing plot corresponding to three-state ladder system. **(a)** Energy-level diagram. **(b)** Absolute value spectrum. **(c)** Real spectrum. **(d)** Relevant double-sided Feynman diagrams.

amount of emission associated with the third-pulse interaction than would have existed in the absence of the first two pulses. As we first mentioned in Section 2.1, however, the existence of the doubly excited state opens up an **excited-state absorption** interaction in which the third pulse experiences more attenuation at some frequencies than would have otherwise been possible in the first and second pulses' absence. The detailed form of this interaction is captured by diagram $\mathscr{D}_2$, and its sign is opposite to the interactions of Feynman diagrams $\mathscr{D}_0$ and $\mathscr{D}_1$, resulting in an overall spectrum with a positive diagonal peak centered at frequency coordinates $(\omega_t, \omega_\tau) = (\omega_{10}, -\omega_{10})$ and a negative off-diagonal peak centered at frequency coordinates $(\omega_t, \omega_\tau) = (\omega_{21}, -\omega_{10})$ as visible in the real part of the spectrum depicted in Fig. 3.7(b). As discussed in more detail in Chapters 7 and 9, this spectrum is closely related to a number of different types of systems, including biexciton signatures in semiconductor quantum dots and exciton and polariton coupling effects semiconductor quantum wells.

Moving beyond one-particle systems, a slightly more complicated but common scenario in the realm of exciton spectroscopy is the situation in which systems comprising multiple interacting atoms experience shifts in their excited-state energies and dephasing rates arising from many-body effects—that is, the mutual interactions that each

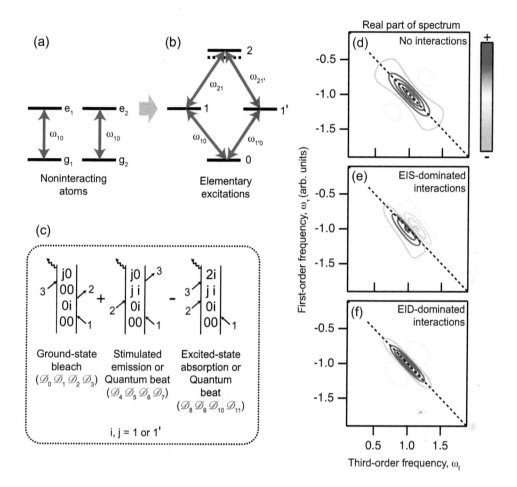

Fig. 3.8 Simulated illustration of many-body effects in single-quantum MDCS. (a)–(b) Energy-level diagrams for a system of coupled two-level systems (a) in the noninteracting-atom picture, and (b) in the elementary-excitation picture. (c) Relevant double-sided Feynman diagrams. (d)–(f) Simulated real part of 2D rephasing spectra (d) in the absence of many-body interactions, (e) in a system dominated by excitation-induced shift (EIS), and (f) in a system dominated by excitation-induced dephasing (EID).

of the different atoms have on each other. Figure 3.8 illustrates the means by which these effects manifest. We imagine for the sake of concreteness a pair of two-level atoms situated in close physical proximity. In the absence of inter-particle interactions, the atoms exert no influence on each other's energy levels, and we can write the system's eigenstates as product states of the form $|n\rangle = |a\rangle|b\rangle$, where the labels $a = (g_1$ or $e_1)$ and $b = (g_2$ or $e_2)$ specify whether or not the first or second atom is in the ground- or excited-state configuration, as pictured in Fig. 3.8(a).

As interactions are turned on, the system can no longer be adequately described within this product-state picture, and a more accurate elementary-excitation picture becomes useful, leading to an accompanying "diamond"-shaped energy-level structure shown in Fig. 3.8(b). State $|0\rangle$ in this picture is the ground state. States $|1\rangle$ and $|1'\rangle$ consist of "singly excited" states residing at energies near (but not necessarily the same as) the excitation energy of the two-level atoms in isolation. State $|2\rangle$ is a "doubly excited" state at roughly the sum of the energies of the two singly excited states.

Double-sided Feynman diagrams associated with the four-state diamond structure are pictured in Fig. 3.8(c). This system may seem complicated to analyze, but it is actually closer to the three-state ladder structure of Fig. 3.7 than might at first be apparent, as illustrated by the parallels in structure between the two different sets of Feynman diagram expansions [compare Figs. 3.8(c) and 3.7(d)]. Moreover, in a variety of different cases, the dipole moments connected to one singly excited state becomes suppressed while the dipole moments of the other state become enhanced, making it possible in light-matter interaction analyses to largely ignore the suppressed state's presence.[3]

In any case, depending on the nature of the interaction between atoms, the energy of the doubly excited state may be larger or smaller than the sum of the energies of the two singly excited states, and we refer to this effect as **excitation-induced shift** (EIS). The dephasing rate associated the single-quantum coherence between singly and doubly excited states can also be larger in the presence of interactions than the dephasing rates for the coherence between singly excited states and the ground state in a phenomenon known as **excitation-induced dephasing** (EID).

For systems in which many-body effects are absent or weak (when $\mu_{2i} \approx \mu_{i0}$, $\gamma_{2i} \approx \gamma_{i0}$, and $\omega_{2i} \approx \omega_{i0}$), the interaction-driven positive and negative spectral peak contributions cancel, leaving behind a signature similar to what would be observed for an isolated two-level atom as depicted in Fig. 3.8(d). In other types of circumstances (e.g., when $\mu_{2i} \approx \sqrt{2}\mu_{i0}$), EIS and EID effects can be the dominant signature in MDCS measurements, resulting in spectral features that bear a strong resemblance to what the spectrum of Fig. 3.7(c) might look like in the limit of $\omega_{21} \to \omega_{10}$. For systems dominated by EIS effects, the positive and negative many-body Feynman diagrams combine to produce a dispersive spectral signature exhibiting a phase-shifted spectral peak maximum as displayed in Fig. 3.8(e) (the typical case is a repulsive interaction such that $\omega_{21} > \omega_{10}$). For systems dominated by EID effects, the positive and negative many-body Feynman diagrams combine to produce a spectral signature with a positive peak but accentuated negative wings as displayed in Fig. 3.8(f) (the typical case is $\gamma_{21} > \gamma_{10}$). Details of these effects as applied to the case of atomic vapors are fleshed out in Chapter 5 and to the case of semiconductors are fleshed out in Chapter 7.

3.6 Double-quantum spectra

The dynamics of doubly excited quantum states are relevant to single-quantum MDCS plots such as the rephasing plots discussed thus far because the employment of mul-

---

[3] As the number of particles becomes large, this suppression-and-enhancement effect forms the basis of superradiance [94].

tiple excitation pulses in single-quantum measurements allows higher-lying states to be interrogated even when single-quantum coherences are the only ones being probed. In such cases, it may be desirable to examine the properties of these double-quantum states in isolation; however, even in the cleanest scenarios, these doubly-excited-state signatures are never fully disentangled from the signatures arising from singly excited states in single-quantum MDCS measurements. This complicating factor is largely removed by employing double-quantum spectroscopy, typically achieved by correlating double-quantum coherences (second-order coherences appearing near twice the resonance frequency) against single-quantum coherences (third-order coherences at the resonance frequency).

A canonical example of double-quantum spectroscopy—as especially relevant to the verification of inter-particle interactions in atomic vapors and the characterization of these interactions in semiconductor quantum dots—is depicted in Fig. 3.9 in the context of two different incarnations of the two-particle diamond structure that was considered in the previous section. In a "symmetric diamond" case, both of the singly excited states have the same energy, and the doubly excited state is slightly shifted relative to twice this energy as depicted in the left side of Fig. 3.9(a). An alternate possibility is an "asymmetric diamond" in which the singly excited energy levels are different as depicted in the right side of Fig. 3.9(a). The measured signal is an S_{III} signal resulting from nonconjugate first- and second-order interactions and a conjugate third-order interaction, and can be experimentally isolated by analyzing a component of the four-wave mixing response that is evolving at carrier frequency $\omega_{III} = \omega_1 + \omega_2 - \omega_3$ and wave vector $\mathbf{k}_{III} = \mathbf{k}_1 + \mathbf{k}_2 - \mathbf{k}_3$.

Regardless of whether a symmetric or asymmetric diamond structure is considered, the perturbative expansion associated with the system's S_{III} signal consists of the sum of eight different double-sided Feynman diagram terms as illustrated in Fig. 3.9(b). In terms of equations, these Feynman diagram contributions can be written in the time domain

$$\mathcal{D}_\alpha(t,T,\tau) = \pm \frac{i\mu_{j0}^*\mu_{2j}^*\mu_{2i}\mu_{i0}\mathcal{E}_0^3}{16Ac\epsilon_0\hbar^3} \frac{\partial}{\partial t} \underbrace{\Theta(t)e^{-i\Omega_{pq}t}}_{3^{rd}\text{-order}} \underbrace{\Theta(T)e^{-i\Omega_{20}T}}_{2^{nd}\text{-order}} \underbrace{\Theta(\tau)e^{-i\Omega_{i0}\tau}}_{1^{st}\text{-order}} \quad (3.25)$$

and in the frequency domain (Fourier transforming the second- and third-order interactions)

$$\mathcal{D}_\alpha(\omega_t, \omega_T, \tau) = \pm \frac{\omega_t \mu_{j0}^*\mu_{2j}^*\mu_{2i}\mu_{i0}\mathcal{E}_0^3}{16Ac\epsilon_0\hbar^3}$$

$$\times \left(\frac{i}{\omega_t - [\omega_{pq} - i\gamma_{pq}]} \right) \left(\frac{i}{\omega_T - [\omega_{20} - i\gamma_{20}]} \right) \Theta(\tau)e^{-i\Omega_{i0}\tau} \quad (3.26)$$

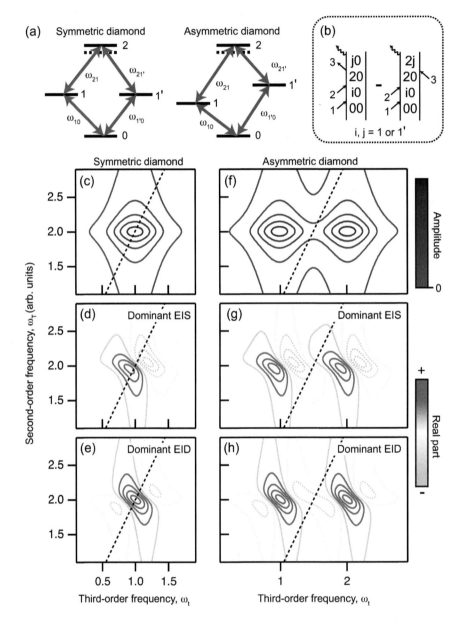

Fig. 3.9 Simulated illustration of double-quantum MDCS. (a)–(b) Example energy-level structures and relevant double-sided Feynman diagrams. (c)–(h) Amplitudes and real parts of the associated double-quantum spectra in the case of dominant EIS (middle panels) and dominant EID (bottom panels). The shape of the amplitude spectrum in these examples is largely insensitive to whether EIS or EID dominates, and so it is shown here only in the case of dominant EID. The dashed lines illustrate where $\omega_T = 2\omega_t$.

with

$$pq = \begin{cases} j0 \text{ for ``positive'' Fig. 3.9(b) diagrams} \\ 2j \text{ for ``negative'' Fig. 3.9(b) diagrams,} \end{cases} \qquad (3.27)$$

and subscript indices $i, j = 1$ or $1'$. We note that although we label the diagrams of a double-quantum spectrum as "positive" or "negative" for the sake of notational convenience, the sign of these diagrams oscillates at the first-order optical resonance frequency as the interaction time τ between pulses 1 and 2 is varied. Thus, this positive and negative association is strictly speaking only true at time $\tau = 0$.

Of key relevance in double-quantum spectroscopy is that—unlike the previous section's description of a rephasing single-quantum experiment on the same system—there can be *no* contribution to the double-quantum signal without the presence of a double-quantum coherence, and perhaps more importantly, the pair of diagrams above will exactly cancel except in the case of a symmetry-breaking feature like interparticle interactions. Simulated spectral signatures corresponding to this model under the different scenarios of EIS-dominated interactions and EID-dominated interactions are depicted below in Figs. 3.9(c)–(h). In the "symmetric diamond" scenario, all eight Feynman diagrams overlap at essentially the same frequency, producing an amplitude spectrum peaked near $(\omega_t, \omega_T) \approx (\omega_{10}, 2\omega_{10})$ as shown in Fig. 3.9(c). In the case of an EIS-dominated spectrum, the real part of the spectrum is characterized by side-by-side positive and negative peaks as illustrated in Fig. 3.9(d). In the case of EID-dominated spectrum, the real part of the spectrum has a positive central peak with negative lobes as illustrated in Fig. 3.9(e). Similar effects occur in the "asymmetric diamond" scenario, except that the resonant lineshapes splits into two peaks straddling the $(\omega_T = 2\omega_t)$ line as illustrated in Figs. 3.9(f)–3.9(h). These latter features reproduce the double-quantum resonance that was introduced at a qualitative level in Fig. 2.4.

3.7 Zero-quantum spectra

Beyond the technique's ability to reveal double-quantum coherences, MDCS enjoys the ability to directly measure zero-quantum coherences, which are the coherences associated with quantum states at nearly (but not quite!) degenerate energies and which are often directly associated with vibrational or electron-phonon coupling phenomena. Such coherences appear as the quantum-beat contributions to single-quantum spectra (as illustrated above in Figs. 3.4 and 3.5) but can be difficult to isolate in these single-quantum spectra from other effects like ground-state bleaching and incoherent coupling. A solution, similarly to the case of double-quantum spectra, can be obtained by plotting correlations between the material's second-order optical response against the third-order four-wave mixing response; however, unlike the case of double-quantum spectroscopy, in zero-quantum spectroscopy the experimental practitioner continues to measure the signal's S_I quantum pathway instead of switching over to the S_{III} pathway.

Figure 3.10 shows an example zero-quantum spectrum, derived from the same three-state "v" system that was used in the generation of Figs. 3.4 and 3.5. The peaks at $\omega_T = \pm(\omega_{1'0} - \omega_{10})$ represent the influence of isolated quantum beats and

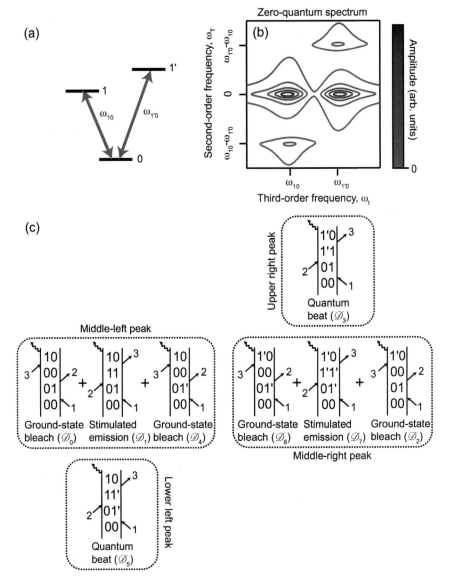

Fig. 3.10 Simulated 2D zero-quantum plot for an inhomogeneously broadened ensemble of three-state "v" systems. (a) Energy-level diagram. (b) Zero-quantum rephasing amplitude spectrum. (c) Double-sided Feynman diagrams.

exhibit linewidths along the vertical direction that are directly associated with the zero-quantum coherences between the $|1\rangle$ and $|1'\rangle$ excited states.

The relevant double-sided Feynman diagrams are pictorially identical to the diagrams used in the generation of Figs. 3.4 and 3.5, but different in terms of their formulaic representation because of the different Fourier transform variables of integration. The mathematical form of the zero-quantum diagrams is

$$\mathscr{D}_\alpha(\omega_t, T, \omega_\tau) = \frac{\omega_t |\mu_{j0}|^2 |\mu_{i0}|^2 \mathcal{E}_0^3}{16Ac\epsilon_0\hbar^3}$$

$$\times \left(\frac{i}{\omega_t - [\omega_{j0} - i\gamma_{j0}]}\right)\left(\frac{i}{\omega_T - [\omega_{ab} - i\gamma_{ab}]}\right)\Theta(\tau)e^{-i\Omega_{0i}\tau} \qquad (3.28)$$

with

$$ab = \begin{cases} 00 & \text{for ground-state bleach diagrams} \\ ji & \text{for the other diagrams} \end{cases} \qquad (3.29)$$

and subscript indices $i, j = 1$ or $1'$.

3.8   Three-dimensional coherent spectroscopy

The advantages of single-quantum spectroscopy and zero-quantum spectroscopy can be combined into a single data set by collecting an S_I spectrum as a function of all three time delays τ, T, and t simultaneously, and then generating a combined frequency-domain plot as a simultaneous function of ω_τ, ω_T, and ω_t. Plots of this sort exemplify three-dimensional coherent spectroscopy (3DCS), and they have been conducted on both atomic vapors and solid-state systems. Details in connection to both of these types of systems are discussed in Chapter 8.

3.9   Nonrephasing pathways and purely absorptive spectra

As noted above in Section 3.1, the interpretation of even the simplest multidimensional spectral signatures can still be complicated by phase twists that are inherent to isolated spectral peaks. In the case of diagonal peaks in which the absolute values of the first- and third-order interaction frequencies are degenerate, these phase twists can be helpfully "unwound" by collecting a sample's nonrephasing spectral signature S_{II} at carrier frequency $\omega_{II} = \omega_1 - \omega_2 + \omega_3$ and wave vector $\mathbf{k}_{II} = \mathbf{k}_1 - \mathbf{k}_2 + \mathbf{k}_3$ in addition to the more traditional rephasing signature S_I acquired at $\omega_I = -\omega_1 + \omega_2 + \omega_3$ and $\mathbf{k}_I = -\mathbf{k}_1 + \mathbf{k}_2 + \mathbf{k}_3$, and then combining these two signatures into the spectrum

$$S_{\text{absorptive}} = S_{II}(\omega_t, T, \omega_\tau) + S_I(\omega_t, T, -\omega_\tau). \qquad (3.30)$$

The real part of the resultant is often termed a **purely absorptive** spectral peak because the phase twists of the constituent spectra cancel, leading to a Lorentzian spectral peak that is strictly positive.

In the case of the two-level system discussed in Section 3.1, the nonrephasing analog of Eq. (3.5) is

$$
\mathcal{D}_{\alpha}(\omega_t, T, \omega_\tau) = \frac{\omega_t |\mu|^4 \mathcal{E}_0^3}{16 A c \epsilon_0 \hbar^3} \Theta(T) e^{-\Gamma_{10} T}
$$
$$
\times \left(\frac{i}{\omega_t - [\omega_{10} - i\gamma_{10}]} \right) \left(\frac{i}{\omega_\tau - [\omega_{10} - i\gamma_{10}]} \right) \qquad (3.31)
$$

leading to the expression,

$$
S_{\text{absorptive}}(\omega_t, T, \omega_\tau) = \frac{\omega_t |\mu|^4 \mathcal{E}_0^3}{16 A c \epsilon_0 \hbar^3} \Theta(T) e^{-\Gamma_{10} T}
$$
$$
\times \left(\frac{i}{\omega_t - [\omega_{10} - i\gamma_{10}]} \right) \left(\frac{2i\gamma_{10}}{[\omega_\tau - \omega_{10}]^2 + \gamma_{10}^2} \right). \qquad (3.32)
$$

The real part of the resulting peak is four-fold symmetric and exhibits horizontal and vertical linewidths $\Delta\omega_{\text{FWHM}} = 2\gamma_{10}$ as shown in Fig. 3.11. This peak is sharper by a factor of $\sqrt{3} \approx 1.73$ than the analogous peak associated with the rephasing-pulse-sequence amplitude spectrum that was depicted previously in Fig. 3.1. It should be noted that although the spectra formed in this way allow the possibility of removing phase twists from diagonal peaks, they have more complicated effects on cross peaks, and thus do not in general remove phase twists from all cases of physical interest.

Beyond their application to the generation of purely absorptive spectra, nonrephasing spectra of the form illustrated by the lower panels of Fig. 3.11(a) are on occasion useful in their own right. Many-body effects appear in slightly different ways in rephasing and nonrephasing spectral forms, and so the simultaneous analysis of both types of forms can help maximally constrain theoretical models of a system, This analysis has been done in the analysis of exciton interactions in GaAs quantum wells [462], as discussed in more detail in Chapter 7. Nonrephasing pulse sequences have been used to effectively isolate individual objects such as quantum dots (which exhibit no inhomogeneity) from the inhomogeneously broadened noise profile associated with other dots that are at the edges of the laser spot [257].

3.10 Finite-pulse effects

The analyses presented up until this point have been based on a model in which the excitation pulses are of the form $\hat{E}_{n,\eta,n}(t - t_n) = \mathcal{E}_0 \delta(t - t_n)$ (refer back to Section 3.1), which is to say that the pulses are assumed to be of infinitesimal duration and therefore of infinite bandwidth. This approximation is good in cases where the laser bandwidth is large in comparison to the range of frequencies associated with the resonance phenomena being observed, but becomes unrealistic as the bandwidth narrows. Moreover, the delta-function pulse approximation disallows the possibility of simulating experiments in which the finite bandwidth of excitation pulses is intentionally utilized to filter out specific MDCS contributions not of interest. An analytically tractable means of capturing the impact of these finite-bandwidth pulses in MDCS is to approximate the excitation pulses as having Gaussian-shaped pulse envelopes as discussed in more detail below and derived in Ref. 362.

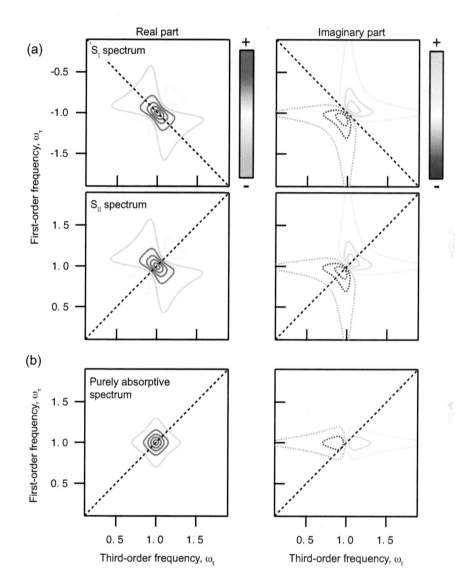

Fig. 3.11 Illustration of the extraction of purely absorptive spectral lineshapes in MDCS. (a) Real and imaginary parts of the single-quantum spectrum from an isolated two-level system under rephasing (S_I) and nonrephasing (S_{II}) pulse sequences. (b) The phase twists of these spectra can be unwound by adding them together according to the formula $S_{\text{absorptive}}(\omega_t, \omega_\tau) = S_{II}(\omega_t, \omega_\tau) + S_I(\omega_t, -\omega_\tau)$. The real part of the resulting spectrum [left side of panel (b)] is strictly positive and exhibits a horizontal linewidth $\Delta\omega_{\text{FWHM}} = 2\gamma_{10}$.

3.10.1   Mathematical formulations

The model can be implemented by defining an envelope function

$$g(t) \equiv \frac{1}{(\delta t_e)\sqrt{2\pi}} e^{-\frac{t^2}{2(\delta t_e)^2}}, \tag{3.33}$$

which is a finite-width Gaussian function with the same units and normalization as the Dirac delta function, and then setting

$$\hat{E}_n(t) = \mathcal{E}_0 \, g(t). \tag{3.34}$$

In the definition here of $g(t)$, we employ the pulse duration parameter δt_e (which is the intensity $1/e$ value half-width, see Section 1.2) to keep the Fourier transform mathematics as notationally simple as possible. The Fourier transform of $g(t)$ is

$$G(\omega) \equiv \int_{-\infty}^{\infty} g(t)e^{i\omega t}dt = e^{\frac{-(\delta t_e)^2 \omega^2}{2}}, \tag{3.35}$$

which yields an angular frequency half-width at $1/e$ intensity bandwidth of

$$(\delta\omega_e) = \frac{1}{(\delta t_e)}. \tag{3.36}$$

Having established these relationships, Eq. (3.34) can be immediately substituted into Eq. (2.23) of the previous chapter. Analytical frequency-domain formulations of the model's double-sided Feynman diagrams in this limit have been obtained and can be written down in forms relevant to 3D spectroscopy as

$$\mathscr{D}_\alpha(\omega_t, \omega_T, \omega_\tau) = \pm \frac{\omega_t \mu_{4,\alpha}\mu_{3,\alpha}\mu_{2,\alpha}\mu_{1,\alpha}\mathcal{E}_0^3}{16 A c \epsilon_0 \hbar^3}$$

$$\times \left(\frac{i}{\omega_t - \Omega_{jk}}\right)\left(\frac{i}{\omega_T - \Omega_{hi}}\right)\left(\frac{i}{\omega_\tau - \Omega_{fg}}\right)$$

$$\times G(\omega_t - \omega_T - \eta_3\omega_3)$$

$$\times G(\omega_T - \omega_\tau - \eta_2\omega_2)$$

$$\times G(\omega_\tau - \eta_1\omega_1), \tag{3.37}$$

in forms relevant to 2D single-quantum spectroscopy as

$$\mathscr{D}_\alpha(\omega_t, T, \omega_\tau) = \pm \frac{\omega_t \mu_{4,\alpha}\mu_{3,\alpha}\mu_{2,\alpha}\mu_{1,\alpha}\mathcal{E}_0^3}{16 A c \epsilon_0 \hbar^3}$$

$$\times \left(\frac{i}{\omega_t - \Omega_{jk}}\right) e^{i\Omega_{hi}T}\left(\frac{i}{\omega_\tau - \Omega_{fg}}\right)$$

$$\times \frac{1}{2}\left[1 + \mathrm{erf}\left(\frac{T + i(\delta t_e)^2(\omega_t + \omega_\tau - \eta_3\omega_3 + \eta_2\omega_2 - 2\Omega_{hi})}{2(\delta t_e)}\right)\right]$$

$$\times G(\omega_t - \eta_3\omega_3 - \Omega_{hi})$$

$$\times G(\omega_\tau + \eta_2\omega_2 - \Omega_{hi})$$

$$\times G(\omega_\tau - \eta_1\omega_1)], \tag{3.38}$$

and in forms relevant to 2D double-quantum or zero-quantum spectroscopy as

$$\mathcal{D}_\alpha(\omega_t, \omega_T, \tau) = \pm \frac{\omega_t \mu_{4,\alpha} \mu_{3,\alpha} \mu_{2,\alpha} \mu_{1,\alpha} \mathcal{E}_0^3}{16 A c \epsilon_0 \hbar^3}$$

$$\times \left(\frac{i}{\omega_t - \Omega_{jk}} \right) \left(\frac{i}{\omega_T - \Omega_{hi}} \right) e^{-i\Omega_{fg}\tau}$$

$$\times \frac{1}{2} \left[1 + \text{erf} \left(\frac{\tau + i(\delta t_e)^2 (\omega_T - \eta_2\omega_2 + \eta_1\omega_1 - 2\Omega_{fg})}{2(\delta t_e)} \right) \right]$$

$$\times G(\omega_t - \omega_T - \eta_3\omega_3)$$

$$\times G^{1/2}(\omega_T - \eta_2\omega_2 + \eta_1\omega_1 - 2\Omega_{fg})$$

$$\times G^{1/2}(\omega_T - \eta_2\omega_2 - \eta_1\omega_1), \tag{3.39}$$

where

$$G^{1/2}(\omega) \equiv \sqrt{G(\omega)}. \tag{3.40}$$

The derivation of these expressions is provided in Ref. [362]. As noted in Eqs. (2.25)–(2.27), the variables $\eta = [\eta_1, \eta_2, \eta_3]$ specify whether the Feynman diagram being considered corresponds to an S_I, S_{II}, or S_{III} phase-matching condition (the output signal will be emitted at wave vector $\mathbf{k}_\eta = \eta_1 \mathbf{k}_1 + \eta_2 \mathbf{k}_2 + \eta_3 \mathbf{k}_3$ and frequency $\omega_\eta = \eta_1\omega_1 + \eta_2\omega_2 + \eta_3\omega_3$).

Among the features worth noting in Eqs. (3.37)–(3.39) are that in each of these expressions, the first two lines are identical to the expressions that would have been obtained in the impulsive-limit case, while finite-pulse effects are captured in the lines below this. Thus, the primary overarching effect of employing this model is a restricted bandwidth compared to the impulsive solution and the replacement of an instantaneous Heaviside theta function $\Theta(x)$ with a more gradual rise function described in terms of error functions. There are a few important subtleties, however. Particularly in the regimes $\tau < \delta t_e$ and $T < \delta t_e$, the resonance frequency Ω_{fg} or Ω_{hi} prominently appears throughout many of the envelope terms in Eqs. (3.38) and (3.39) (note that a Gaussian with complex arguments generates a complex output). The introduction of finite pulses can therefore affect both the amplitude *and* phase of the measured signal in ways that cannot be understood in the impulsive limit. Furthermore, because Ω_{fg} and Ω_{hi} are Feynman-diagram-dependent, so also are finite-pulse effects. It is in general not possible to universally correct a 2D spectrum for bandwidth effects in the absence of knowledge of system resonances.

3.10.2   Example spectra

An application of the finite-pulse effect model to the single-quantum rephasing spectrum of a two-level system that was described in Section 3.1 is depicted in Fig. 3.12, where the figure has been generated in the case of resonant and nearly degenerate excitation such that $\omega_1, \omega_2, \omega_3 \approx \omega_c \approx \omega_{10}$. The two relevant double-sided Feynman diagrams for this scenario can be mathematically described as

$$\mathcal{D}_\alpha(\omega_t, T, \omega_\tau) = \frac{\omega_t |\mu|^4 \mathcal{E}_0^3}{16 A c \epsilon_0 \hbar^3} e^{-\Gamma_{10} T}$$

$$\times \left(\frac{i}{\omega_t - [\omega_{10} - i\gamma_{10}]} \right) \left(\frac{i}{\omega_\tau - [-\omega_{10} - i\gamma_{10}]} \right)$$

$$\times \frac{1}{2} \left[1 + \mathrm{erf} \left(\frac{T + i(\delta t_e)^2 (\omega_t + \omega_\tau + 2i\Gamma_{10})}{2(\delta t_e)} \right) \right]$$

$$\times G(\omega_t - \omega_c + i\Gamma_{10})$$

$$\times G(\omega_\tau + \omega_c + i\Gamma_{10})$$

$$\times G(\omega_\tau + \omega_c), \tag{3.41}$$

which, as noted in the previous subsection, corresponds to a modification of the impulsive limit solution with the most notable addition being the Gaussian bandwidth factors $G(x)$ (compare Eqs. 3.41 and 3.5).

As shown by Figs. 3.12(b)–3.12(d), if the intermediate mixing time is $T > \delta t_e$ and the population decay rate is $\Gamma_{10} \ll \delta\omega_e$, the most important effect of generalizing the impulsive solution to the case of finite pulses is a filtering of visible spectral intensity. Fig. 3.12(b) reproduces the impulsive spectrum of Fig. 3.1 with its characteristic 2D Lorentzian lineshape. Fig. 3.12(c) shows the finite-pulse solution at $T = 4(\delta t_e)$. The most important change is a reduced spectral amplitude at frequencies away from $(\omega_c, -\omega_c)$. Fig. 3.12(d) depicts the quotient of Fig. 3.12(c) and Fig. 3.12(b), and demonstrates that the filtering effect is sharper along the ω_τ direction (with a characteristic width of $(\delta\omega_e)/\sqrt{2}$) than it is along the ω_t direction (where the characteristic width is $\delta\omega_e$). This asymmetry in the finite-pulse effect is frequently compensated by the fact that in coherently detected experiments, the emission signal is most often measured by co-propagating an externally supplied local oscillator pulse with the same intensity profile as the excitation pulses. This local oscillator pulse narrows the observable range of frequencies along the ω_t direction to a comparable width to the range observed along the ω_τ direction. Beyond this, the quotient spectrum is almost entirely real, which can be understood as a consequence of the fact that, for the parameters selected, the imaginary parts of the arguments in the error function and Gaussian filter functions in Eq. (3.41) are small compared to their respective real parts.

As noted at the end of the previous section, however, an important feature of a 2D spectrum in the finite-pulse limit is the appearance of the population decay rate Γ_{10} not just in the impulsive part of the solution given by Eq. (3.41), but also in the finite-pulse factors. Γ_{10} can introduce nontrivial phase effects by two mechanisms. At time $T = 0$, it can introduce phase shifts through its presence in the error function of Eq. (3.41). This effect is shown in Figs. 3.12(e) and 3.12(f). Γ_{10} can also introduce phase shifts, in principle for even $T \gg \delta t_e$, through its presence in the arguments to the Gaussian functions in Eq. (3.41). The results of this in the case where $\Gamma_{10} = 0.4(\delta\omega_e)$ are depicted in Figs. 3.12(g) and 3.12(h). Phase shifts are generally speaking subtle in this simplest case of a two-level system, as demonstrated by the only very slight differences between the finite-pulse spectra depicted in Figs. 3.12(c), 3.12(e), and 3.12(g). It may be important to take such features into account when they play an integral role in spectral interpretation, however [296].

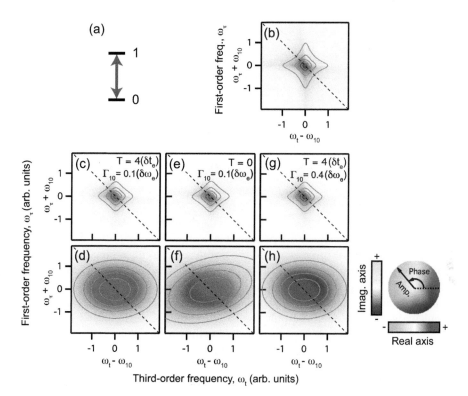

Fig. 3.12 Finite-pulse effects in MDCS, as relevant to a single-quantum rephasing 2D spectrum for a two-level system. The dephasing rate is $\gamma_{01} = \gamma_{10} = 0.2(\delta\omega_e)$. The excitation-pulse carrier frequencies ω_1, ω_2, and ω_3 have been set very close to the resonant frequency $\omega_{10} = -\omega_{01}$. (a) Level structure cartoon. (b) 2D spectrum in the impulsive limit, valid for a generic population decay rate Γ_{10}, and mixing time $T > 0$. (c) 2D spectrum with transform-limited Gaussian pulses, for $\Gamma_{10} = 0.1(\delta\omega_e)$ and $T = 4(\delta t_e)$. (d) Quotient of (c) and (b). (e)–(f) Same as (c) and (d), with $T = 0$. (g)–(h) Same as (c) and (d), with $\Gamma_{10} = 0.4(\delta\omega_e)$. Adapted from Ref. 362.

Finite-pulse effects can become more dramatic in the case of double-quantum spectroscopy, as illustrated in the simulation of Fig. 3.13 corresponding to an asymmetric three-level ladder system. Relevant Feynman diagrams can be described by Eq. (3.39), with $\boldsymbol{\eta} = [1, 1, -1]$, as

$$\mathscr{D}_\alpha(\omega_t, \omega_T, \tau) = \pm \frac{\omega_t |\mu_{21}|^2 |\mu_{10}|^2 \mathcal{E}_0^3}{16 A c \epsilon_0 \hbar^3}$$

$$\times \left(\frac{i}{\omega_t - \Omega_{pq}} \right) \left(\frac{i}{\omega_T - \Omega_{20}} \right) e^{-i\Omega_{10}\tau}$$

$$\times \frac{1}{2} \left[1 + \mathrm{erf} \left(\frac{\tau + i(\delta t_e)^2 (\omega_T - 2\Omega_{10})}{2(\delta t_e)} \right) \right]$$

$$\times G(\omega_t - (\omega_T - \omega_c))$$

$$\times G^{1/2}(\omega_T - 2\Omega_{10})$$

$$\times G^{1/2}(\omega_T - 2\omega_c) \tag{3.42}$$

with $pq = 10$ or 21 [compare to Eq. (3.26)]. The dephasing rates are $\gamma_{10} = 0.2(\delta\omega_e)$ and $\gamma_{20} = 0.4(\delta\omega_e)$. The resonant frequencies are ω_{20}, $\omega_{10} = \omega_{20}/2 - \delta\omega_e$, and $\omega_{21} = \omega_{20}/2 + \delta\omega_e$. Similarly to the example of the two-level system, we examine the case of nearly degenerate excitation pulses, this time such that $\omega_1, \omega_2, \omega_3 \approx \omega_c \approx \omega_{20}/2$.

Notably, finite-pulse effects at $\tau = 0$ can begin to have profound consequences on measured phase. Whereas in the single-quantum case, the relative phase between the impulsive solution and finite-pulse solution is always negligible at the center of the resonance, this phase shift can approach values as large as $\pi/2$ in the double-quantum or zero-quantum case [see Figs. 3.13(c) and 3.13(e), and the argument of the error function in Eq. (3.42)], and values even larger than this when $\tau < 0$. Such effects may be helpful in explaining, for example, the appearance of negative peaks in measurements of the two-quantum spectrum of rhodamine 6G [288, 307]. As is true of one-quantum spectra, the phase effects become less important under well-defined pulse ordering [Figs. 3.13(d) and 3.13(f)].

3.10.3   Further applications

Beyond the above-illustrated case-studies, analytical simulations of finite-pulse effects have been conducted and can have important consequences in cases involving off-resonant pulses, in cases where the rotating-wave approximation begins to break down, in cases involving chirped pulses, and in cases involving inhomogeneous ensembles of noninteracting systems. These situations and more are discussed in Ref. 362. Beyond this, Perlík, et al. have derived a simulation of finite pulses under the assumption of Lorentzian pulse envelopes that includes an analytical time-domain solution in addition to frequency-domain solutions [307]. Complementary work in similar subject areas has also been conducted by Schweigert and Mukamel [337], by Abramavicius, et al. [2], and more recently by Do, et al. [97, 98] and others [196, 292].

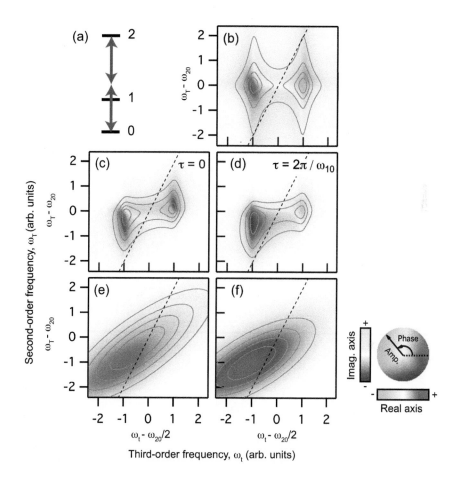

Fig. 3.13 Finite-pulse effects in MDCS, as relevant to a double-quantum 2D spectrum for a 3-level "ladder" system, consisting of a ground state at $E_0 = 0$ and excited states at E_1 and E_2 such that $E_1 = E_2/2 - \hbar(\delta\omega_e)$. The excitation-pulse carrier frequencies are $\omega_1, \omega_2, \omega_3 \approx \omega_{20}/2$. The dephasing rates are $\gamma_{10} = \gamma_{21} = 0.2(\delta\omega_e)$ and $\gamma_{20} = 0.4(\delta\omega_e)$. (a) Level structure cartoon. (b) Delta-function pulse solution, where τ is an integer multiple of $2\pi/\omega_{10}$. (c)–(d) Transform-limited pulse solutions at $\tau = 0$ [for (c)] and $\tau = 2\pi/\omega_{10} = 2\pi(\delta t_e)$ [for (d)]. (e) Quotient of (c) and (a). (f) Quotient of (d) and (a). Adapted from Ref. 362.

4

Experimental implementations

As mentioned in Chapter 2, the challenge of implementing multidimensional coherent spectroscopy in the optical and near-infrared regions of the electromagnetic spectrum has been solved using a number of approaches. In this chapter, we will review two particular implementations—actively stabilized box geometry approaches and phase-modulated collinear geometry approaches—in significant detail. The use of the methods described in this chapter to study specific materials is presented in later chapters. Specifically, Chapter 5 describes the study of atomic vapors, Chapter 7 and Chapter 9 describe the study of semiconductor quantum wells and quantum dots, respectively. The use of MDCS to study atomically thin 2D materials is presented in Chapter 10.

4.1 Experimental requirements and considerations

Experimental implementation of optical 2D coherent spectroscopy is based on measuring the nonlinear response of a sample to multiple excitation fields, which are typically pulses. Most experiments are performed in the time domain, where the nonlinear signal is measured as a function of two or more time delays and numerically Fourier transformed into the frequency domain. The procedure requires precise control of time delays between the pulses to achieve the phase stability and uniform delay steps. A second key challenge is to isolate the weak nonlinear signal from the comparatively strong excitation fields. The third challenge is that the field, not the intensity of the signal, must be measured, which requires some form of heterodyne detection using a reference, also known as a local oscillator, pulse.

4.1.1 Precision and stability of time delays

The measurement of a coherent signal and the discrete Fourier transform require a precise and stable control of time delays between pulses to achieve the phase stability and uniform delay steps. In general, a phase stability is desired such that the fluctuation in path difference between pulses should be under $\lambda/100$, where λ is the wavelength of the laser. For longer wavelengths in the infrared regime, this requirement can be satisfied by using passive stabilization, i.e., mechanical isolation of environmental vibrations. However, this requirement becomes more stringent for short wavelengths in the near-infrared and visible regime. For the typical Ti:sapphire laser wavelength of 800 nm, the $\lambda/100$ stability requires to maintain the path fluctuations at the level of a few nm. At this level, good isolation of vibrations is necessary but not sufficient. A few different strategies are used to stabilize the relative phase. One strategy is the so-called "common optics" approach in which all pulses go through the same optical path so the relative phase between the pulses remain stable. The relative phase can also be actively stabilized by controlling the displacement of a mirror in the optical path according to an error signal generated from the real-time measurement of phase fluctuations. This approach of active stabilization is discussed in detail in Section 4.3. Alternatively, in the phase modulation collinear geometry discussed in Section 4.4, the phase fluctuations are not actively stabilized but are measured, for example by using an auxiliary reference laser, to be accounted for and corrected in the analysis of 2DCS signals.

4.1.2   Isolation of the signal

An important consideration in implementing 2D spectroscopy is isolation of the signal from the strong excitation fields. In addition, the desired signal must be isolated from other nonlinear signals that arise from other possible interactions of the excitation pulses. This separation can be done through wavevector selection, in which the signal of interest is made to propagate in a different direction than the excitation fields and other signals. It can also be done in the frequency domain by generating a signal that is at a unique frequency after heterodyne detection.

As a starting point, consider a time-domain description of the third-order polarization term listed on the right hand side of Eq. (1.1), where the driving field is assumed to be composed of a set of three traveling CW excitation signals of frequencies ω_A, ω_B, and ω_C, and where each of these excitation signals contributes to the polarization signal once. (Although this harmonic-wave excitation scenario does not strictly map in one-to-one fashion onto experiments conducted using pulsed excitation sources, pulsed sources can be constructed out of sums of harmonic signals and so we lose only a small amount of generality in the approximation.) We have

$$P^{(3)}(t)_{\text{signal}} = \epsilon_0 \chi^{(3)}(\omega_A, \omega_B, \omega_C) E_A(t) E_B(t) E_C(t), \tag{4.1}$$

where $P^{(3)}(t)_{\text{signal}}$ is the selected third-order polarization term with frequency ω_S and wave vector $\mathbf{k}_S$, ϵ_0 is the vacuum permittivity, $\chi^{(3)}$ is the third-order susceptibility, and $E_i(i = A, B, C)$ are the electric fields with frequency ω_i and wave vector $\mathbf{k}_i$. Each electric field can be written as the sum of a complex exponential and its complex conjugate,

$$E_i(t) = \frac{1}{2}[\hat{E}_{i0} e^{i(\mathbf{k}_i \cdot \mathbf{r} - \omega_i t)} + \hat{E}_{i0}^* e^{-i(\mathbf{k}_i \cdot \mathbf{r} - \omega_i t)}], \tag{4.2}$$

where $\hat{E}_{i0}$ is the amplitude. The signal wave vector $\mathbf{k}_S$ is the sum of any three wave vectors out of $\pm\mathbf{k}_A$, $\pm\mathbf{k}_B$, and $\pm\mathbf{k}_C$. Assume, now, that $\omega_A = \omega_B = \omega_C = \omega$ and that we restrict our signal detection to be sensitive only to frequencies near ω. The signal wave vector has to be $\mathbf{k}_S = -\mathbf{k}_A + \mathbf{k}_B + \mathbf{k}_C$, $\mathbf{k}_S = \mathbf{k}_A - \mathbf{k}_B + \mathbf{k}_C$, or $\mathbf{k}_S = \mathbf{k}_A + \mathbf{k}_B - \mathbf{k}_C$. Each of these three signals gives different information and can be directly connected back to the S_I, S_{II}, and S_{III} spectra discussed in Chapters 2 and 3.

Having identified the signal as having unique optical properties, we now need to isolate it from a practical standpoint. This isolation can be done by spatial filtering if the excitation fields are arranged in a noncollinear geometry so that the signals emit in different directions according to their wave vectors. A commonly used arrangement for three-pulse excitation is the so-called box geometry [43, 46, 68, 413] shown in

Fig. 1.4(d). In this case, four laser beams propagate in parallel and each beam is located at a corner of a square in the cross-section view with $\mathbf{k}_B$ and $\mathbf{k}_C$ diagonally opposite one another. The four beams are focused by a lens to converge on the sample. Three of the beams excite the sample and generate a third-order signal with wave vector $\mathbf{k}_S = -\mathbf{k}_A + \mathbf{k}_B + \mathbf{k}_C$ which coincides with the propagation direction of the fourth beam. The signal and the excitation beams go through a second lens and are converted back to parallel beams on four corners of square. Although only three pulses are needed to excite the sample, the fourth pulse is useful. It can be used as a guide to align the optics for detecting the nonlinear signal, which is usually too weak to be visible, either by naked eye or a camera/viewer. The fourth beam can also be used as a reference pulse for spectral interferometry. In the box geometry, the detection spatially isolates the signal with wave vector $\mathbf{k}_S = -\mathbf{k}_A + \mathbf{k}_B + \mathbf{k}_C$ and rejects the signal with other wavevectors and most of the isotropic background emission such as fluorescence and scatter. The box geometry can be implemented with the "common optics" approach. Alternatively, the relative phases between pulses can be actively stabilized [43, 423, 461] by using interferometric error signals as a feedback. This approach is discussed in detail in Section 4.3.

In general, the wavevector of the polarization does not match the wavevector of the emitted electric field, which is determined by its frequency and the index of refraction of the material. In certain geometries, such as the box geometry described in the previous paragraph and shown in Fig. 1.4(d), the wavevectors of the polarization and emitted field are matched, so these geometries are known as "phase-matched." The phase conjugate geometry, shown in Fig. 1.4(c) is also phase matched. The two-pulse and three-pulse planar geometries, shown in Figs. 1.4(a) and (b) are not phase matched, however, for small angles between the incident beams and thin samples, such as a semiconductor epilayer or monolayer of a van der Waals material, the phase mismatch is negligible.

The spatial filtering to select a specific signal does not work in the collinear geometry in which all excitation beams co-propagate. In this case, the wave vectors of all signals at different orders are either aligned along the excitation beam or zero (isotropic emission). A detector aligned to the laser beam receives all signals as well as the excitation pulses. One way to select the desired signal is to perform a phase cycling procedure [350, 391, 424], in which the phase of each excitation pulse is toggled to change the phase of the signal from certain pathways but not others. The signal from specific pathways can be isolated by a proper combination of subtraction and addition of signals acquired with different phase configurations of excitation pulses. In some cases, this procedure can be complicated and take as many as 16 steps of phase cycling [394] to isolate a certain signal. Alternatively, a specific signal can be selected in the frequency domain by frequency tagging excitation pulses [35, 144, 285, 391]. In this approach, the phase of each excitation pulse is modulated by an acousto-optic modulator (AOM) at a frequency Ω_i such that $E_i = \frac{1}{2}(\hat{E}_i(t)e^{i(\mathbf{k}_i \cdot \mathbf{r} - \omega_i t - \Omega_i t)} + \hat{E}_i^*(t)e^{-i(\mathbf{k}_i \cdot \mathbf{r} - \omega_i t - \Omega_i t)})$. Each pulse is tagged with a slightly different frequency. Instead of using wave vectors, the signals from different processes are modulated at a specific frequency resulting from the mixing of AOM frequencies. The desired signal can then be selectively detected at the corresponding frequency using a lock-in amplifier. This approach is discussed in detail in Section 4.4.

The need to isolate a weak nonlinear signal from the strong excitation light is a need that is ubiquitous in nonlinear spectroscopy, not just MDCS. Thus many of the methods described here find application in other modalities of nonlinear spectroscopy.

4.1.3   Detection of the signal

A key ingredient of MDCS is that the field of the signal, not simply its intensity, is detected. Since optical detectors, such as photodiodes, charge-coupled devices (CCDs), or photo-multiplier tubes, detect the intensity of light, measuring the signal requires the use of heterodyne methods. In heterodyne detection, the interference between the signal-to-be-measured and an known reference field, often called the "local oscillator," is measured using an intensity detector. The resulting fringes allow the phase of the signal to be determined, however, it is always the case that multiple measurements are needed to determine the phase. These measurements can be done either in the time domain, where the timing of the local oscillator with respect to the signal is varied, or in the frequency domain, where there is a fixed delay between the signal and the local oscillator and there spectral interference pattern is measured [212].

In MDCS, the change in the phase of the signal as the delay between excitation pulses is varied provides critical information. Thus it not just the change in phase with respect to time or frequency, which corresponds to chirp, but the overall phase that must be measured. This creates more stringent requirements on the field measurement: namely the phase of the local oscillator must be stable over a long enough period of time such that the signal for many different excitation delays can be measured.

4.2   Overview of experimental approaches

There are multiple approaches to achieving the requirements described in Section 4.1. In this section we will give an overview of the most common ones. In the following sections we go into more depth on two of them.

One approach is to use "common optics." The idea is to have excitation pulses go through the same optical path so that the pulses are subject to the same beam path fluctuations. This way, the relative phase between the pulses remains stable even with fluctuations in the optical path. Reflection from mirrors is the most likely to change the phase of a beam due to motion of the mirror. There is a one-to-one mapping between the displacement of a mirror and change in path length, and, consequently vibrations causing micron-scale motion can randomize the phase by 2π or more. Transmissive optics, such as lenses, have a much smaller, even negligible effect.

The "common optics" approach has been realized in several schemes. Usually, two pulses can be generated by splitting a pulse at a beam splitter and their time delay is controlled by inserting a delay stage in the path; however, the two pulses do not share the same optical path and they are not phase-locked without additional stabilization. The challenge of the "common optics" approach is how to generate multiple pulses and control their delays while maintaining a common path. Instead of using a beam splitter, a beam can be diffracted into two directions by diffractive optics such as a grating. Implementations [46, 68] based on diffractive optics pass two laser beams through a transmission grating with low groove density (30 grooves per mm). The positive and negative first-order diffractions of each beam generate two beams. The resulting four beams provide three excitation pulses and a reference pulse in the box geometry. The delay of each pulse can be controlled with in-beam glass slides or prisms. In these schemes, the phases are only locked pairwise while the second time delay is not stabilized. Therefore, it does not allow 2D spectroscopy that requires scanning T with interferometric precision. Moreover, diffraction optics and slides/prisms introduce significant dispersion for ultrashort pulses. The time delays controlled by glass slides or prisms are limited to a few picoseconds.

Femtosecond pulse shapers can be used to split a pulse into multiple replicas and control the time delays, in which case all pulses propagate through the same optical path and are phase locked [141, 394]. Pulse-shaper approaches enable various types of 2D spectroscopy and provide opportunities to vary temporal and spectral pulse shapes; however, some disadvantages include limited time delays, satellite pulses, and a reduced throughput. By using a 2D spatial-light modulator and diffractive optics, MDCS using pulse-shaping techniques has also been implemented using a non-colinear geometry [413]. There is also an elegant approach [342] that does not use diffraction optics or pulse shapers but only conventional optics to achieve pairwise phase stabilization. The limitations are that the second time delay is not stabilized and it lacks the flexibility in controlling time delays.

Besides the hardware realization, the data acquisition and processing is also an important part of optical 2D coherent spectroscopy implementations. In the time domain approaches, the data are recorded in the time domain as a function of time delays. The time-domain data are usually converted into the frequency domain numerically by using fast Fourier transform (FFT). In some cases, the inverse Laplace transform is performed instead of Fourier transform to highlight certain features in the spectra. Several technical issues regarding FFTs are discussed in Section 4.6. The Fourier transform requires high-quality data with a large number of equally spaced data points. A novel signal processing method known as compressed sensing has been proposed and demonstrated [326] in optical 2D coherent spectroscopy. This method allows for random undersampling of the experimental data, down to a few percent of the data set required for FFT, without loss in spectral resolution. Providing the same data set as for FFT, compressed sensing can improve the spectral resolution by an order of magnitude.

Although most experiments are implemented in the time domain with femtosecond lasers, optical 2D coherent spectroscopy can also be realized in the frequency domain with narrow-band lasers [55–57, 443] or in a mix of frequency and time domains [295].

In the remainder of this chapter, we focus on two specific implementations, one in the box geometry and another in the collinear geometry, as examples to discuss the basic principles in implementing optical 2D coherent spectroscopy. We note that there are many different implementations [123] that are tailored to specific applications.

4.3   Actively stabilized box geometry

In this section, we describe an experimental implementation of optical 2D coherent spectroscopy with actively stabilized optical pulses arranged in the box geometry [43].

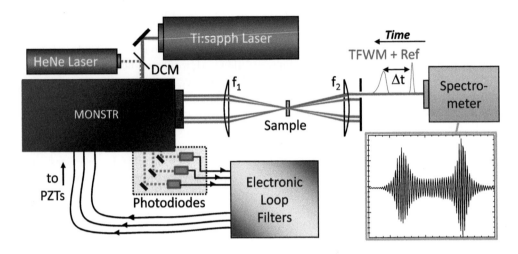

Fig. 4.1 Schematic of the entire experimental setup for performing optical 2D coherent spectroscopy with actively stabilized pulses in the box geometry. Adapted from Ref. 43.

The schematic of the entire experimental setup is shown in Fig. 4.1. The central piece of the setup is a stable platform dubbed the "Multidimensional Optical Nonlinear Spectrometer" (MONSTR). Fed by a femtosecond (fs) mode-locked Ti:sapphire oscillator and a continuous-wave (CW) helium-neon laser operating at 632.8 nm, the MONSTR splits a femtosecond pulse into four pulses that are phase-stabilized by three feedback loops using the interference of co-propagating CW beams as error signals. The four pulses are focused by a lens to converge on the sample. Three of the pulses excite the sample and generate a TFWM signal. The fourth pulse is attenuated and used as a reference for heterodyne detection. The fourth pulse can either go through or around the sample. Subsequently, the TFWM signal and the reference pulse are combined and their interferogram is detected by a spectrometer with a CCD camera. A series of interferograms is recorded as one time delay is incrementally stepped. A 2D spectrum can be generated by Fourier transforming the 2D array of recorded data.

The technical challenge of this implementation is to control all pulses with subwavelength precision over a delay range up to about one nanosecond. This is accomplished in MONSTR by using precision translation stages, sturdy mechanical construction, and active stabilization with feedback loops. The apparatus is based on a nested Michelson interferometer, as shown schematically in Fig. 4.2(a). Specifically, an input femtosecond pulse is split into four copies by three 50/50 beam splitters. To introduce delay lines, "roof mirrors," or possibly corner-cube retroreflectors mounted on four translation stages, are used. The delay lines control the path length and hence the delay via the time-of-flight. Stage "U" changes the delay of pulses "C" and "Ref" simultaneously while other stages each control the timing of one pulse. At the same time, the CW laser beam co-propagates with the pulses along the same path. At the exit, a dichroic mirror (DCM) reflects the CW laser and transmits the pulses. The physical realization uses only one DCM at the exit even though four DCMs are shown in Fig. 4.2. After exiting the DCM, the pulses go through "common optics" so only the paths prior to the DCM requires the active stabilization.

For the CW laser, the beam splitters and the DCM form a nested Michelson interferometer where the two arms of a "master" interferometer are themselves Michelson interferometers. The interferometric outputs of the CW laser beams at the three beam splitters are monitored by photodetectors and used as error signals for the beam path length fluctuations. Each error signal is fed to an electronic servo loop whose output drives a piezo-electric transducer that moves a mirror to compensate fluctuations in the beam path. There are three feedback loops to stabilize the relative phases between the four pulses. Two loops stabilize the pairs of pulses A^*/B and C/Ref. The third loop locks the phase between the two pairs. This active stabilization is able to compensate small fluctuations ($< 1/4$ of the CW laser wavelength) in the beam path at frequencies up to about 10 kHz. Therefore, the whole platform has to be passively stable to a sufficiently low level of fluctuations. To do so, each arm (which is a Michelson interferometer itself) of the main interferometer is constructed on a separate aluminum block. One of the blocks is flipped upside down and placed on the top of the other to form a double deck configuration whose output beams are aligned in the box geometry. The entire setup is situated on an optical table with vibration isolators.

With both the passive and active stabilization, the platform can achieve a stability of greater than 1/100 of the CW laser wavelength.

This scheme is implemented in the MONSTR, which consists of a lower deck and an upper deck, as shown in Figs. 4.2(b) and 4.2(c), respectively, as CAD drawings. The lower deck houses the interferometer for pulses C and Ref. The transmission of the first beam splitter (BS) is routed to a long delay stage (U) which moves pulses C and Ref simultaneously with respect to the other two pulses. The beam is then split into two by the second BS. One of them is directed to the exit DCM and the other to a short delay stage (Z) that changes the delay between pulses C and Ref. A mirror is mounted on a piezo (PZT) for active stabilization. The reflection from the first BS is directed to the top deck when the MONSTR is fully assembled. The top deck houses the interferometer for pulses A^* and B. The input beam is split into two. Each beam goes through a short delays stage (X and Y) and exits at the DCM. Two mirrors are mounted PZTs. One of them is used to stabilize the phase between pulses A^* and B and the other for the stabilization between the top and bottom decks. The delay stages are direct-drive linear translation stages with nanometer resolution and stability. The short stages have 5 cm of travel and the long stage has 20 cm. All mirrors are metallic mirrors coated with protected silver. The BSs are thin, broadband, low group-velocity dispersion with coating centered at 800 nm. To balance the dispersion in the reflection arm of each BS, a compensation plate (CP) made from the same substrate as BS is inserted in the beam path. All BSs and CPs are also anti-reflection coated and mounted in strain-free mounts to prevent strain-induced birefringence and beam spatial mode deformation. Each deck is made from a cast aluminum block with a mass of 19 kg to provide a sturdy and thermal-stable platform.

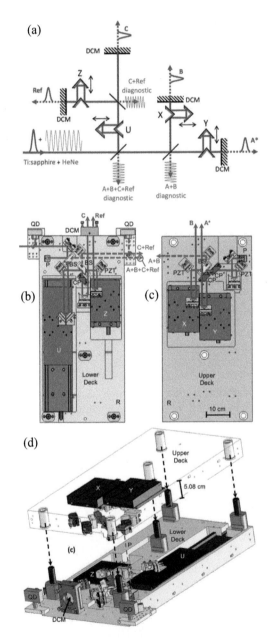

Fig. 4.2 (a) Schematic of optics, (b) CAD drawing of the bottom deck, (c) CAD drawing of the top deck, and (d) assembled top and bottoms decks of the MONSTR. Adapted from Ref. 43.

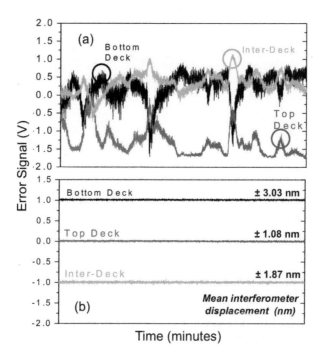

Fig. 4.3 (a) The interferometric error signals from the top, bottom, and inter decks when the servo loop is disabled. (b) The interferometric error signals when the servo loop is enabled. The traces are shifted vertically for clarity. Adapted from Ref. 43.

The alignment of the MONSTR starts from each individual deck. With all components installed, the CW laser beam is used to guide the alignment. The beam is aligned parallel to the motion of stages such that the output beams do not deviate when the stages are scanned. The two output beams from each deck are parallel and 25.4 mm apart. Once the decks are aligned, the top deck is flipped over and mounted on the bottom deck by five mounting posts, as shown in Fig. 4.2(d). Four parallel beams from the two decks form a square cross section that is 25.4 mm along each side. A 50 mm diameter DCM is attached to the exit, reflecting the CW beams. The assembly is then completed with side panels to reduce the air flow. Besides forming the box geometry, the folded design also ensures that the two interferometers are close to each other so they are subject to similar fluctuations due to changes in temperature and air pressure. The passive stabilization provided by this design is critical for the active stabilization.

The active stabilization utilizes servo loops to realize the negative feedback. The servo loop can be realized with either an analog circuit or a field-programmable gate array (FPGA). The servo loops use the interferometric error signals as the inputs. The error signals with the feedback loops disengaged are shown in Fig. 4.3(a). The error signals show slow drifts of phase. When the feedback stabilization is engaged, the error signals are locked at 0 V to maintain the phase. The variations in the locked error signals have a normal distribution with a standard deviation corresponding to

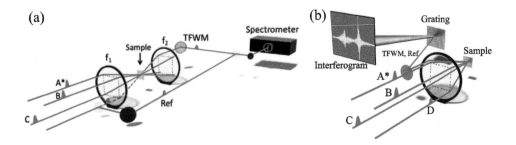

Fig. 4.4 Schematic showing the box-geometry 2DCS experiment in (a) the transmission mode and (b) the reflection mode. The detection of the TFWM signal is done by the spectral interferometry with a reference pulse. Figure (b) is adapted from Ref. 215.

motion of 2 nm. For the excitation laser pulses of $\lambda = 800$ nm, this variation corresponds to the phase stability of $\lambda/400$. There are several specific issues that should be considered in the feedback loop. First, the servo locks to the error signal of 0 V but cannot tell if the interferometric output changes by a half fringe or more. Second, the PZT-driven mirrors in the feedback loops have slow response, limiting the bandwidth of the feedback loops to a few kHz. Therefore, the feedback loop can only correct slow fluctuations within a half wavelength. The proper operation of the feedback loop requires a sufficiently good passive stabilization by having a sturdy mechanical design, stable translation stages, well-isolated optical tables, stable ambient temperature and pressure, etc. Third, the servo can lock to 0 V at either the positive or negative slope of an interference fringe, which requires a proper sign ($+$ or $-$) in the gain setting to achieve a negative feedback. This sign needs to be checked and corrected if necessary every time when a servo loop is engaged.

After exiting the MONSTR, the four phase-locked pulses arranged in the box geometry are converged by a lens to the same spot on the sample, as shown in Fig. 4.4(a), in the transmission mode. One of the phase-matched directions for the generated TFWM signal is $\mathbf{k}_S = -\mathbf{k}_A + \mathbf{k}_B + \mathbf{k}_C$, which is the propagation direction of the fourth pulse. The generated TFWM signal and transmitted excitation beams are collimated by another lens. The TFWM signal is directed to a spectrometer while the three transmitted beams are blocked. The experiment can also be carried out in the reflection mode, as shown in Fig. 4.4(b), for optically thick or opaque samples. In this case, the TFWM signal is generated and reflected at the surface of the sample, along the same direction of pulse D. The TFWM signal is collimated by the same lens and picked up by an edge mirror or a beam splitter. The TFWM signal can be slightly shifted away from the path of pulse A^* by tilting the sample.

The TFWM signal field is measured by spectral interferometry [212] between the signal and reference pulse. The measurement requires that the reference pulse remains phase locked with the excitation pulses and the signal. In the transmission mode, the reference pulse can either go through the common optics or be routed around the sample and recombined with the TFWM signal at a beam splitter. In the later case, the reference pulse needs to be re-locked with pulse C to maintain the phase stability. In the reflection mode, the mechanical vibration of the sample causes phase fluctuations in the TFWM signal. The reference pulse is reflected off the sample and co-propagates with the TFWM signal so that they remain phase locked even in the presence of sample vibration. Before arriving at the sample, the reference pulse is attenuated and delayed relative to the excitation pulses so the excitation due to the reference pulse is negligible; however, the reference pulse can be reshaped by the resonant interaction with the sample. The induced phase distortion of the reference pulse needs to be characterized by an auxiliary spectral interferometry experiment and to be accounted for in data analysis [212].

The interferogram of the TFWM signal and the reference pulse recorded by the spectrometer is $I_{SI} = |E_S + E_R|^2$. The TFWM signal field is $E_S = \hat{E}_S e^{-i\omega\tau_S} e^{i\phi_S}$ and the reference field is $E_R = \hat{E}_R e^{-i\omega\tau_R} e^{i\phi_R}$, where $\hat{E}_{S,R}$, $\tau_{S,R}$, and $\phi_{S,R}$ are the electric field magnitude, delay, and phase, respectively. The intensity of recorded interferogram is

$$I_{SI} = |E_R + E_S|^2 = |E_R|^2 + |E_S|^2 + E_S E_R^* + E_S^* E_R. \tag{4.3}$$

To extract the TFWM signal field, the reference pulse spectrum $I_{Ref} = |E_R|^2$ and the TFWM signal spectrum $I_{Sig} = |E_S|^2$ are also independently measured by the spectrometer and subtracted from the interferogram $(I_{SI} - I_{Sig} - I_{Ref})$. The remaining interference terms are separated in the frequency domain and either one can be isolated via Fourier filtering. We choose the term

$$E_S E_R^* = \hat{E}_S \hat{E}_R^* e^{-i\omega(\tau_S - \tau_R)} e^{i(\phi_S - \phi_R)}. \tag{4.4}$$

The TFWM signal is a result of the nonlinear response to the excitation fields

$$E_S \propto iP^{(3)} \propto \chi^{(3)} \hat{E}_A^* \hat{E}_B \hat{E}_C e^{-i\omega(-\tau_A + \tau_B + \tau_C)} e^{i(-\phi_A + \phi_B + \phi_C)}, \tag{4.5}$$

so the interference term is

$$E_S E_R^* \propto \chi^{(3)} \hat{E}_A^* \hat{E}_B \hat{E}_C \hat{E}_R^* e^{-i\omega(-\tau_A + \tau_B + \tau_C - \tau_R)} e^{i(-\phi_A + \phi_B + \phi_C - \phi_R)}. \tag{4.6}$$

Comparing Eqs. 4.4 and 4.6, the third-order susceptibility $\chi^{(3)}$ can be obtained and its phase $\phi_{\chi^{(3)}}$ depends on the phase of the recorded interferogram as well as the excitation and reference pulses

$$\phi_{\chi^{(3)}} = \phi_{SR} - (-\phi_A + \phi_B + \phi_C - \phi_R), \tag{4.7}$$

where $\phi_{SR} = \phi_S - \phi_R$. The pairwise phase differences can be measured from the respective pairwise interference following the procedure described in Ref. 41. Therefore, the global phase of the third-order nonlinear susceptibility can be determined in optical 2D coherent spectroscopic measurement. The resulting 2D spectra include both the amplitude and phase information (or real and imaginary parts) of the sample's nonlinear optical response. The phase information in 2D spectra is an important feature and sometime is essential in applications such as revealing many-body interactions in a semiconductor quantum well [220].

An important technical issue in optical 2D coherent spectroscopy is how to improve the signal-to-noise ratio (SNR) and thus the detection sensitivity. In the experiment, only the nonlinear signal corresponding to specific excitation quantum pathways is supposed to be measured. The nonlinear signals from other excitation pathways, fluorescence, and non-resonance scatter of excitation laser beams are background. The presence of a fluctuating background can diminish the SNR and ultimately the sensitivity of the technique. It is important to realize that even if the fractional fluctuations of the background are small, if the background itself is large, the absolute fluctuations can be large compared to the signal itself. New emerging applications of 2D spectroscopy may pose challenges in the detection sensitivity. For instance, measurements on atomically thin two-dimensional materials are complicated by strong laser scatter from the substrate in comparison to a weak nonlinear signal from the sample itself. To implement optical 2D coherent spectroscopy in samples involving strong background noise or weak signal or both, it is essential to have an effective method to reduce the background noise and increase SNR.

The phase cycling methods discussed previously are one effective and common approach to increasing the SNR [284, 350, 391, 424]. The phase of each excitation pulse is toggled to change the phase of the signal from certain pathways but not others. The desired signals from specific pathways can be isolated by a proper combination of subtracting and adding spectra acquired with different phase configurations of excitation pulses. The TFWM signal depends on the phases of all three excitation pulses as in Eq. 4.5. If any of the excitation pulses, for example, pulse B, has a π phase shift, both the TFWM signal E_{TFWM} and the scatter of pulse E_B have a sign change ($e^{i\pi} = -1$) while the fields of other pulses and their scatter remain the same. If two pulses B and C each have a π phase shift, the TFWM signal does not change the sign but the two pulses do. By toggling the phase shift between 0 and π of two excitation pulses, a phase cycling procedure can be implemented to eliminate the background.

Table 4.1 Phase cycling operations by toggling the phases of two pulses.

Operation	E_S	E_{TFWM}	E_A	E_B	E_C
$\Delta\phi_B = 0, \ \Delta\phi_C = 0$	S_1	$+$	$+$	$+$	$+$
$\Delta\phi_B = \pi, \ \Delta\phi_C = 0$	S_2	$-$	$+$	$-$	$+$
$\Delta\phi_B = \pi, \ \Delta\phi_C = \pi$	S_3	$+$	$+$	$-$	$-$
$\Delta\phi_B = 0, \ \Delta\phi_C = \pi$	S_4	$-$	$+$	$+$	$-$

Considering the background, the total signal E_S measured in the experiment includes the TFWM signal and the scatter of each pulse, that is

$$E_S = E_{TFWM} + E_A + E_B + E_C. \tag{4.8}$$

To eliminate the background and extract the TFWM signal, we perform a phase cycling procedure with four operations, as shown in Table 4.1. In the first step, the phases of pulses are left unchanged. As a reference, the phases of pulses B and C are considered zero ($\Delta\phi_B = 0, \ \Delta\phi_C = 0$) and the signs of signal fields are $+$. The measured total signal is denoted as S_1. In the second step, a phase shift $\Delta\phi_B = \pi$ is introduced to pulse B. The signs of E_{TFWM} and E_B flip to $-$ and the resulting total signal is S_2. Subtracting the signals measured in these two operations gives $S_1 - S_2 = 2(E_{TFWM} + E_B)$. That is, the background noise due to scattering of pulses

A and C can already be eliminated by changing the phase once. This two-step phase cycling might be sufficient in some cases where pulses A and C are the main sources of background noise. To further eliminate the noise due to pulse B, the phase shift of another pulse and two more operations are needed. In step 3, both pulses B and C have phase shifts $\Delta\phi_B = \pi$, $\Delta\phi_C = \pi$. The sign of E_{TFWM} changes back to $+$ while E_B and E_C have a $-$ sign. The signal measured in step 3 is denoted as S_3. In the last step, the phase shifts are set to $\Delta\phi_B = 0$, $\Delta\phi_C = \pi$ and the operation flips the signs for E_{TFWM} and E_B compared to that in the previous step. The resulting signal is S_4. Using the signals obtained in these four operations, we can calculate $S_1 - S_2 + S_3 - S_4 = 4E_{TFWM}$. Therefore, the TFWM signal can be extracted without the background of all excitation pulses after the four-step phase cycling as

$$E_{TFWM} = \frac{1}{4}(S_1 - S_2 + S_3 - S_4). \tag{4.9}$$

The phase modulation of a pulse can be achieved by varying its time delay [43]. In MONSTR, the time delay can only increment a distance commensurate with an integer number of helium-neon (HeNe) reference laser (632.8 nm) fringes to ensure the phase stabilization. Since the excitation pulses and the HeNe reference laser have different wavelengths, the pulse delay needs to be varied more than a half cycle to achieve a phase shift close to π. For example, when the excitation laser is tuned to 768 nm for measurements on K atoms, the shortest increment of the delay stage is 6 HeNe fringes corresponding to approximately 5π at the excitation laser wavelength. This approach affects the precision of the pulse time delay and may not provide an exact π phase shift, especially when the laser wavelength is not "convenient." Moreover, the data acquisition time is significantly longer since moving the delay stage is the most time-consuming operation in the 2D scan. Alternatively, the phase of a pulse can be modulated by a liquid crystal phase retarder inserted in the beam path [284]. The liquid crystal phase retarder can be calibrated to shift the phase exactly by π. The phase retarder has a fast switching speed with a typical rise and fall time of 34 ms and 360 μs, respectively. The extra dispersion due to the liquid crystal phase retarder could be a concern for ultrashort sub-10-fs pulses. In this case, the dispersion can be pre-compensated by negatively chirping the input excitation pulses.

4.4 Phase modulated collinear geometry

In this section, we describe a collinear optical 2D coherent spectroscopy implementation [285, 391] with excitation pulses being phase modulated by acousto-optic modulators (AOMs). The basic idea is to tag the excitation pulses with different phase modulation frequencies so that the nonlinear signal corresponding to the excitation pulses can be selectively detected by referencing a proper mixing frequency of the pulse modulation frequencies. The reference frequency can be obtained by measuring the interferometric output of excitation pulses [391] or an auxiliary continuous-wave (CW) laser [285]. The setup does not require active phase stabilization. The phase fluctuations are monitored in real time by the interferometric output and taken into account in the reference for lock-in detection.

The schematic of a collinear setup is shown in Fig. 4.5(a). An input pulse is split into four pulses and combined back into one beam by two Mach-Zehnder interferometers nested within a larger Mach-Zehnder interferometer. The four pules are denoted A, B, C, and D and the corresponding time delays between them are τ, T, and t. The three delays are controlled by three translation stages. In each arm of the interferometer, the pulse passes through an AOM driven by a radio frequency Ω_i where $i = A, B, C, D$. The first-order diffraction is used for experiment while the zeroth-order beam is blocked. The frequencies for AOMs are slightly different so that each pulse is uniquely tagged by the modulation frequency. Moreover, the nonlinear signals resulting from two or more pulses oscillate at the difference of the AOM frequencies and can be isolated the frequency domain by lock-in detection. For instance, the signal due to the excitation of pulses A and B oscillates at the difference frequency $\Omega_{AB} = \Omega_A - \Omega_B$ and can be measured by a lock-in amplifier referencing to Ω_{AB}. The reference signal is obtained from an auxiliary CW laser that goes through the same optics as the excitation laser. The excitation laser beam goes through the center of every optics along the optical axis, while the CW laser beam is offset from the optical axis so the CW and excitation laser beams are spatially separated at the outputs. The beating signal of the CW laser can be measured at the interferometric outputs by photo detectors labeled REF 1, REF 2, and REF 3 in Fig. 4.5(a). These beating signals and their mixing can be used as reference for lock-in detection.

In the configuration shown in Fig. 4.5(a), all four pulses are incident on the sample. The first three pulses generate a third-order coherence in the sample. The role of the fourth pulse is to convert the third-order coherence to a fourth-order population that can be detected as a fluorescence [391] or photocurrent [285] signal. Since the fluorescence emission is not directional, the detection of fluorescence can be done in reflection, transmission, or from the side, offering flexibility in the setup. In many cases, measuring the signal resulting from the fourth-order population is equivalent to measuring the FWM signal due to the third-order coherence; however, the two are not entirely equivalent especially when the experiment involves doubly excited states. Their similarities and differences have been discussed in [254, 306, 336]. For a system with both singly and doubly excited states, the third-order coherence can be converted into a population in both singly and doubly excited states. If the quantum yields of singly and doubly excited states to emit fluorescence are different, the fluorescence detection is not equivalent to the direct measurement of third-order coherence with three pulses. The difference in quantum yields can be due to different non-radiative relaxation pathways for singly and doubly excited states, which is common in liquid and solid states samples. The collinear geometry also allows the direct measurement of third-order signals by using the configuration shown in Fig. 4.5(b). In this case, only three pulses are incident on the sample to generate a third-order FWM signal. The fourth pulse, D, is not incident on the sample but combined with the emitted signal to perform heterodyne detection. Pulse D serves as a reference pulse that samples the signal at time t. The time-resolved four-wave mixing signal can be measured by scanning the time delay between pulses C and D. One disadvantage of this approach is that the phase of the signal depends on the position of the sample, which can fluctuate due to vibrations, temperature variations or mechanical instabilities. This problem can

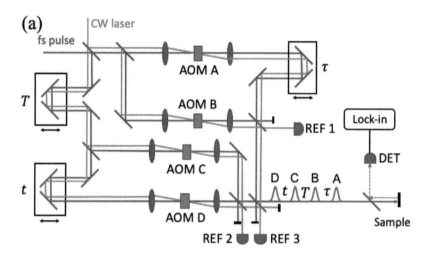

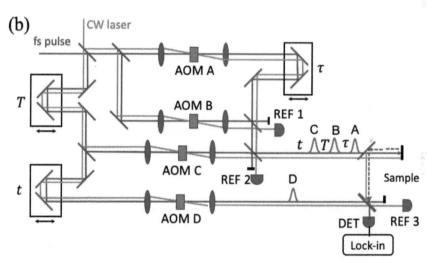

Fig. 4.5 Schematic showing the collinear 2DCS experiment. (a) The configuration uses four excitation pulses and detects fluorescence as the signal. (b) The configuration uses three excitation pulses and the nonlinear four-wave mixing signal is measuring by heterodyne detection using the fourth pulse as reference.

be overcome by reflecting the reference beam off of the sample, however, care is needed to account for any modification in the complex spectrum of the reference, and also the possibility that the reference beam itself excites the sample and thus modifies the response to the excitation beams.

Measurement with photocurrent detection is especially useful when examining photovoltaic materials and other photo-devices, since it provides direct access to ultrafast dynamics of the device under typical operating conditions. Because only current-

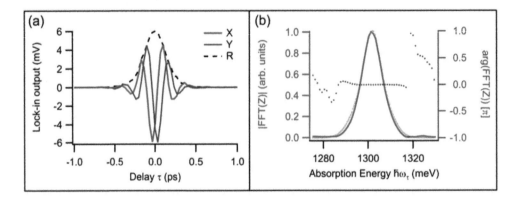

Fig. 4.6 (a) Field auto-correlation measurement of the pulse. X and Y are the in-phase and in-quadrature outputs of the lock-in amplifier and $R = \sqrt{X^2 + Y^2}$ is the field amplitude. (b) The power spectrum (red) of the laser pulse obtained by Fourier transforming the auto-correlation spectrum. The retrieved phase is shown in blue dots. A normalized laser spectrum recorded with a spectrometer is shown in green circles for comparison. Adapted from Ref. 285 with permission.

generating processes can contribute to the detected signal, many background responses such as scattering of the excitation pulses or coherent artifacts are suppressed from detection, resulting in exceptional noise suppression and dynamic range. Detecting different "action variables" (fluorescence versus photocurrent for these experiments) can provide different physical insight into the mechanisms governing ultrafast carrier dynamics in semiconductor nanostructures. In a study investigating carrier multiplication processes in colloidal quantum dot solar cells [177], the signatures of multi-exciton generation were not observed in the fluorescence spectrum for any excitation wavelength due to non-radiative Auger recombination on a timescale faster than the luminescence lifetime but slower than charge carrier separation and photocurrent generation.

The principle of lock-in detection can be demonstrated by measuring the wave mixing of two pulses A and B to retrieve the field auto-correlation spectrum of the laser pulse. To do so, the photodetector DET in Fig. 4.5(a) replaces the sample to detect the laser pulses directly. Alternatively, the sample can be replaced with a mirror or a glass plate to reflect the laser pulses. Pulses C and D are blocked so that only pulses A and B are incident on the detector. The output of DET is fed to a lock-in amplifier and demodulated by using the beating signal measured by REF 1 as the reference. The lock-in amplifier uses a bandpass filter around the beat note frequency to isolate the signal resulting from a wave-mixing process where pulses A and B each contribute once. Other contributions and noise are filtered out by the lock-in. The lock-in output provides a direct access to the complex signal $Z = X + iY$ and the absolute phase (global phase) can be calibrated [95] to obtain the phase of sample's optical response. The time delay τ is scanned during the measurement. The signal phase at the lock-in output does not evolve at the optical frequency of the laser pulse, but at a reduced frequency given by the difference between the signal and CW laser frequencies as $\nu^* = |\nu_{sig} - \nu_{cw}|$. This is essentially a physical undersampling of the optical signal. The stepping of the stage is only required to sample the signal at the reduced frequency instead of the optical frequency. This undersampling also improves the signal-to-noise ratio by reducing the impact of delay fluctuation $\delta\tau$, which scales with the reduced frequency as $\delta\tau \cdot \nu^*$. Effectively, the CW laser monitors and compensates for phase noise due to optical path fluctuation in the setup. The signal measured by the lock-in amplifier is shown in Fig. 4.6(a). Since pulses A and B are copies of the same pulse, the result is a field auto-correlation spectrum of the laser pulse. The y-axis is the output voltage of the lock-in amplifier and the x-axis is the delay τ converted from the stage movement. The red and blue curves are the in-phase output X and the in-quadrature output Y of the lock-in amplifier. The black dash line is the field amplitude $R = \sqrt{X^2 + Y^2}$. Note that X and Y oscillate with τ at the reduced frequency $\nu^* = 3.5$ THz instead of the optical frequency. The time-domain signal $Z = X + iY$ can be Fourier transformed into the frequency domain by using fast Fourier transform (FFT). The power spectrum of the laser is provided by $|\text{FFT}(Z)|^2$ and shown as the red curve in Fig. 4.6(b). The spectral phase is given by $\arg(\text{FFT}(Z))$ and plotted as blue dots. The retrieved laser spectrum matches well with the spectrum (green circles) measured directly by using a spectrometer.

The nonlinear signal due to wave mixing of four pulses can be measured in a similar way. The signal is recorded as two or three delays are stepped and then Fourier transformed in the frequency domain with respect to each delay to produce a 2D or 3D spectrum. To detect the signal due to a particular excitation pulse sequence, the corresponding reference signal needs to be generated from the CW laser beating signals. For instance, the reference frequency $\Omega_{SI} = -\Omega_A + \Omega_B + \Omega_C - \Omega_D$ is required for the rephasing excitation pulse sequence shown in Fig. 4.5(a), where pulses A and D are considered conjugated. The beating frequency $\Omega_{AB} = \Omega_A - \Omega_B$ and $\Omega_{CD} = \Omega_C - \Omega_D$ are detected by photo detectors REF 1 and REF 2 in Fig. 4.5(a). The beat notes $\cos(\Omega_{AB}t^*)$ and $\cos(\Omega_{CD}t^*)$ are fed to a digital signal processor (DSP) based wave mixer. The digital signal processor (DSP) applies the Hilbert transform to each beat note to shift the phase of positive frequency components by $\pi/2$ to get

$$H[\cos(\Omega_{AB}t^*)] = \sin(\Omega_{AB}t^*); \quad H[\cos(\Omega_{CD}t^*)] = \sin(\Omega_{CD}t^*). \qquad (4.10)$$

With both cosine and sine components, the DSP can calculate the sum and difference of Ω_{AB} and Ω_{CD} by using the identity

$$\cos(\Omega_{AB}t^* \pm \Omega_{CD}t^*) = \cos(\Omega_{AB}t^*)\cos(\Omega_{CD}t^*) \mp \sin(\Omega_{AB}t^*)\sin(\Omega_{CD}t^*). \qquad (4.11)$$

This operation gives the reference frequency

$$\Omega_{SI} = \Omega_{CD} - \Omega_{AB} = -\Omega_A + \Omega_B + \Omega_C - \Omega_D, \qquad (4.12)$$

which can be used to detect the rephasing signal. Instead of REF 1 and REF 2, all beat notes of four AOM frequencies can be extracted from photodetector REF 3 and used by the DSP for frequency mixing. Using a similar process, the reference frequency $\Omega_{SII} = \Omega_A - \Omega_B + \Omega_C - \Omega_D$ can be obtained for detecting the non-rephasing signal, and the reference frequency $\Omega_{SIII} = \Omega_A + \Omega_B - \Omega_C - \Omega_D$ for the two-quantum signal.

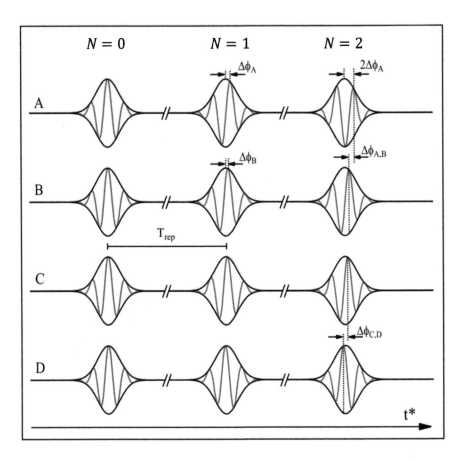

Fig. 4.7 Illustration of the carrier-envelope phase cycling in four pulse trains. The difference in the carrier-envelope phase between the beams is due to different phase modulation frequencies by the AOMs. The time delays are set to 0 here for simplicity. Adapted from Ref. 285 with permission.

This detection scheme provides great flexibility in experimentation. The photodetector registers all possible signals, which are post-processed to select a specific signal in the lock-in amplifier by demodulating the detector output at the corresponding reference frequency. By using multiple demodulating channels with each referencing to a different frequency, it is possible to retrieve different types of 2D spectra in one measurement [13]. One can also measure nonlinear signals with higher orders than the fourth order. The higher-order signals are generated by one or more excitation pulses acting multiple times. The signal can be isolated by using the reference frequency corresponding to the pulse sequence. Nonlinear signals up to the 14th order have been measured to generate multi-quantum 2D spectra [460].

The frequency modulation and signal detection scheme in this approach can be considered as a dynamic pulse-to-pulse phase cycling. The pulse train output from a mode-locked laser carries an initial carrier-envelope offset f_{CE}. The shift of the carrier-envelope phase from pulse to pulse is $\Delta\phi_{CE} = 2\pi f_{CE}/f_{rep}$, where f_{rep} is the laser repetition rate. For each arm of the nested interferometer, the pulse train acquires an additional carrier-envelope offset phase due to the AOM phase modulation at the respective frequency Ω_i. This additional shift leads to a different pulse-to-pulse carrier-envelope phase shift for each pulse train if Ω_i is different for each AOM. Explicitly, the pulse-to-pulse carrier-envelope phase shift for each beam is

$$\Delta\phi_i = 2\pi(\Omega_i + f_{CE})T_{rep}, \tag{4.13}$$

where $T_{rep} = 1/f_{rep}$. Therefore, comparing the pulses in two pulse trains, the phase difference is cycled from pulse to pulse. This is illustrated in Fig. 4.7. Assume the carrier-envelope offset is zero for the first pulse ($N = 0$), for the Nth pulses in the pulse train i and j, the phase difference is

$$\Delta\phi_{i,j} = 2\pi N T_{rep}(\Omega_i - \Omega_j). \tag{4.14}$$

The phase difference between beams A and B is $\Delta\phi_{A,B} = 2\pi N T_{rep}\Omega_{AB}$ and it is $\Delta\phi_{C,D} = 2\pi N T_{rep}\Omega_{C,D}$ between beams C and D. The phase difference is cycled at the respective frequency Ω_{AB} and Ω_{CD}. Therefore, the phase and amplitude of the resulting nonlinear signal oscillates at the corresponding reference frequencies Ω_{SI}, Ω_{SII}, or Ω_{SIII} and can be selected by the lock-in detection. This dynamic phase-cycling scheme does not require a stable f_{CE} since any fluctuations in f_{CE} are the same in all four pulse trains and cancel out in the phase difference between two pulse trains.

In this approach, the auxiliary CW laser is essential to monitoring phase fluctuations and providing reference frequencies for lock-in detection. The wavelength of the CW laser is required to be near (a few nm) to that of the signal so that the reduced frequency ν^* is low. This requires the CW laser wavelength to be changed every time we change to a sample that has a different resonant wavelength, which might be inconvenient. To solve this problem, an alternative method can use a reference laser having an arbitrary wavelength with respect to the signal wavelength [256].

In this improved reference technique, a field-programmable gate array (FPGA) is used to calculate the path length difference in real time. The intensity modulation of the interference between the reference laser beams from two arms of the interferometer is measured and used to calculate the optical phase difference between the two paths. The phase changes as the delay between the two paths is scanned or there is a path length fluctuation. The wrapped phases for different reference laser wavelengths are shown in Fig. 4.8(a). The wrapped phase varies between $-\pi$ and π. The phase can be unwrapped in real time by the FPGA. The unwrapped phases are plotted in Fig. 4.8(b) for different wavelengths. The curves illustrate that, in terms of the unwrapped phase, the phase fluctuations measured with one reference laser wavelength can be scaled to the phase difference for any other wavelengths. The scaling can be accomplished by multiplying the measured phase with the ratio of the actual reference laser wavelength and the desired reference wavelength. The calculated phase for the new reference wavelength is then applied to the demodulation frequency for the lock-in detection. This process is performed in real time by the FPGA and provides a demodulation reference frequency for any arbitrary reference laser wavelength. By

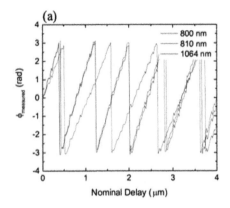 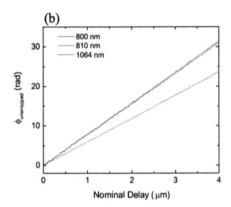

Fig. 4.8 (a) The wrapped optical phases for different reference CW laser wavelengths. (b) The unwrapped phases for difference wavelengths. The phase difference for one wavelength can be used to calculated the corresponding phase difference for another wavelength by multiplying the reference phase by the ratio of wavelengths. Adapted from Ref. 256.

using this technique, we can use a reference CW laser with one fixed wavelength for any signal wavelengths. The reference laser wavelength can be chosen to be longer than all possible resonant wavelengths so the reference laser beam does not excite the sample in case the reference laser is incident on the sample.

To obtain a 2D spectrum, two time delays (for example, τ and t) need to be scanned to measure a 2D time-domain signal $S(\tau, t)$. The time delays are typically scanned using discrete steps. The scanned range of the delays is determined by the time scale of measured dynamics. The delays are scanned by using equidistant steps which are determined by the required measurement bandwidth. For each step of the measurement, the stages are moved to achieve pre-defined delays in τ and t. We then wait for the stages to settle and for the lock-in amplifier to converge on the correct values before recording the measurement for this step. Each step can take from hundreds of ms to a few s depending on the lock-in time constant and averaging setting. This process is repeated for all combinations of delays τ and t. For a 2D spectrum, the total acquisition time scales with the number of steps along one dimension squared. Therefore, the acquisition time can be very long (hours) for high-resolution or large-bandwidth 2D spectra.

The acquisition time can be drastically reduced by continuous scanning. For continuous scanning, the delay stage controlling one time delay continuously moves at a constant speed while the measurements from the lock-in amplifier are recorded continuously. A two-channel lock-in amplifier is needed to record both the 2D signal generated by the pulses and the interferometric signal of the CW laser. The recorded interferometric signal is used to measure the unwrapped phase which gives the time delays. For continuous scanning, the measurement bandwidth is no longer determined by the stepping of delays but the stage velocity and the lock-in time constant TC. The final stage of a lock-in amplifier integrates the signal over the time constant TC and works as a bandpass filter. The effective step size is set by TC. Considering the stage velocity v_s, the effective optical bandwidth of the measurement is $(\frac{2v_s}{c} \cdot TC)^{-1}$, where c is the speed of light. For example, a time constant of 10 ms and a stage velocity of 1 mm/s result in an optical bandwidth of about 15 THz.

4.5   Comparison of different approaches

The approaches in both the box and collinear geometry are appropriate for the study of the coherent nonlinear optical response of many different samples. Both apparatuses can generate sequences of excitation pulses with precisely controlled and long (up to ns) inter-pulse delays, providing the capabilities to obtain different types of 2D spectra.

The box geometry has an advantage of the spatial filtering of the nonlinear signal. Since the TFWM signal emits in a specific direction different from the excitation beams, the TFWM signal can be selectively detected while most background signals of the excitation beams are spatially filtered, resulting in a good signal-to-noise ratio. However, the crossing beams cannot be easily focused to a small spot, limiting the spatial resolution to about 10 μm. The excitation beams are only overlapped around the focal point within the range of the Rayleigh length. The effective interaction range for the experiment is small and the sample needs to be placed in the Rayleigh range. In comparison, the co-propagating beams in the collinear geometry have better spatial overlap and longer effective range in the sample, providing a higher detection sensitivity in dilute samples. The collinear beams can be coupled into an objective lens to achieve a tight focus for a higher spatial resolution, close to the diffraction limit. A disadvantage of the collinear approach is that the signals from processes at all orders and the excitation laser beams are all detected along with the desired nonlinear signal. This results in a strong background and the desired nonlinear signal can only be post-selected at a specific demodulation reference frequency. In addition, care must be taken that nonlinearity in the photodetector can result in spurious signals at the same frequency as the TFWM signal. On the other hand, this feature can be an advantage to allow convenient detection of higher-order (higher than third-order) nonlinear signals [460]. The higher-order signals can be generated if some excitation pulses are considered acting multiple times. These signals can be detected in the same direction by referencing to proper mixing frequencies that account for multiple actions of excitation pulses. Another advantage of the collinear approach is its compatibility with imaging since it uses a single beam that can be raster scanned over the sample.

4.6   Data analysis

To generate 2D spectra presented in the frequency domain, the experimental data acquired in the time domain need to be converted into the frequency domain by using a fast Fourier transform (FFT). In the collinear approach, the 2D spectral data are recorded as a function of two time delays and a 2D spectrum is generated by a 2D FFT of the time-domain data. In the box-geometry approach, the data are recorded in the frequency domain for one dimension and in the time domain for the other. In principle, it requires only a 1D FFT to generate a 2D spectrum; however, the raw spectra in the frequency domain can be first transformed into the time domain so that a Fourier filter can be applied to select only the signal in a proper time range. The processed time-domain data can then be 2D Fourier transformed into the frequency domain to generate a 2D spectrum.

Since FFTs are essential in 2DCS data analysis, we first review the basic concepts of FFTs. An FFT algorithm computes the discrete Fourier transform (DFT) of a data sequence. Consider a 1D time-domain data sequence $x_0(t = 0), x_1(t = \Delta t), x_2(t = 2\Delta t), ..., x_{N-1}(t = (N-1)\Delta t)$, there are N data points with an equal spacing Δt in the time delay. The total time delay that is scanned is $T = \Delta t N$. The corresponding frequency-domain data sequence is given by the DFT as

$$S_k = \sum_{l=0}^{N-1} x_n e^{i2\pi kl/N}, \quad k = 0, ..., N-1. \tag{4.15}$$

This definition is evaluated by an FFT algorithm to generate a frequency-domain data sequence $S_0(f = 0), S_1(f = \Delta f), S_2(f = 2\Delta f), ..., S_{(N-1)}(f = (N-1)\Delta f)$. The sequence has N data points with an equal frequency spacing of Δf. The total frequency range or the bandwidth is $F = \Delta f N$. The time-domain parameters can be related to the parameters in the frequency domain to determine the step size and the total scan time needed in the experiment. The frequency resolution is determined by the total range of the time delay as $\Delta f = 1/T$. The bandwidth is determined by the step size of the time delay as $F = 1/\Delta t$.

Ideally, the scan in each dimension should be long enough until the signal disappears completely, i.e., goes to zero. In practice, the experiment might not be able to provide a sufficiently long delay. Sometimes the signal does not go to zero due to a background even for a sufficiently long delay. In this case, a Fourier transform of the sudden change from a nonzero signal to zero can lead to artificial side bands in the frequency domain. A window function can be applied to the time-domain data to ensure a smooth transition to zero at the end of data sequence. Some examples of smoothing window functions include Hanning window function

$$w(l) = 0.5 - 0.5\cos(2\pi l/N), \quad 0 \le l \le N, \tag{4.16}$$

and Hamming window function

$$w(l) = 0.54 - 0.46\cos(2\pi l/N), \quad 0 \le l \le N. \tag{4.17}$$

After applying a window function, additional zeros can be appended to the time-domain data sequence to artificially increase the resolution in the frequency domain.

Another important issue is undersampling. The Nyquist criterion states that the sampling rate F_s has to be at least twice the bandwidth of the signal to measure the real frequency of the signal. When the signal is undersampled, the measured frequency is distorted in a phenomenon known as aliasing. The measured signal appears to have a frequency below $F_s/2$ which is known as the Nyquist frequency or folding frequency. This effect can be explained by using the "fan-fold" method, as illustrated in Fig. 4.9. The frequency domain is divided into zones of $F_s/2$. Zone 1 has a frequency range from 0 to $F_s/2$, zone 2 has a range from $F_s/2$ to F_s, and zone i has a range from

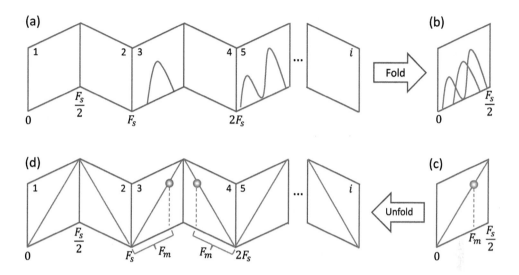

Fig. 4.9 Undersampling explained by the "fan-fold" method. (a) Frequency zones defined by the Nyquist frequency $F_s/2$. (b) Higher frequency signals appear to have frequencies below $F_s/2$ as the fan paper is folded. (c) An undersampled signal appears to have a measured frequency F_m below the Nyquist frequency $F_s/2$. (d) The real frequency of an undersampled signal can be calculated from the measured frequency F_m by unfolding the fan paper.

$(i-1)F_s/2$ to $iF_s/2$. If the signals to be measured have higher frequencies than the Nyquist frequency $F_s/2$, as shown in Fig. 4.9(a), the signals appear to be in zone 1, as shown in Fig. 4.9(b), as if the zones are fan-folded and overlapped onto zone 1. For a signal with a measured frequency F_m, as shown in Fig. 4.9(c), the real frequency of the signal can be retrieved if the signal frequency is known to be within a specific zone, say zone i. As shown in 4.9(d), the algorithm to calculate the real frequency is different, depending on whether the real frequency is in an odd or even numbered zone. For the signal in an odd numbered zone (i is odd), the real frequency can be retrieved from the measured frequency F_m as

$$(i-1)F_s/2 + F_m. \tag{4.18}$$

If the signal is in an even number zone (i is even), the real frequency is

$$iF_s/2 - F_m. \tag{4.19}$$

In the optical region, the experiment usually undersamples the optical carrier frequency so the apparent measured frequency needs to be "unfolded" to retrieve the real frequency. The sampling rate should be carefully chosen such that the signal of interest falls entirely into only one the Nyquist frequency zones. Besides using Fourier transform, other alternative methods are also used, such as compressive sensing [326] and linear prediction from singular value decomposition [383].

5

Multidimensional coherent spectroscopy of atomic ensembles

Alkali metal atomic vapors, particularly potassium (K) and rubidium (Rb), were used initially as a model system to demonstrate the techniques of multidimensional coherent spectroscopy. Both K and Rb atoms are extensively studied in atomic physics. Their energy structures and other physical and optical parameters have been well characterized. The standard K and Rb vapor cells are commercially available and relatively easy to use. The D-lines of K and Rb can be covered by the bandwidth of a typical Ti:sapphire femtosecond oscillator. The two D-lines resemble a three-level V system, which is simple to interpret yet sufficiently complex to demonstrate many advantages of multidimensional coherent spectroscopy. The resulting 2D spectra can be calculated from the optical Bloch equations and confirm the basic principles of implementing and understanding 2D spectroscopy. An early demonstration [379] of two-dimensional spectroscopy of atomic vapor was performed in a sodium (Na) vapor where the ground-state Zeeman levels were excited by a radio-frequency pulse sequence. Optical two-dimensional spectroscopy of the electronic transitions in Rb was first performed by using a 16-step phase cycling with a collinear three-pulse sequence [394]. Other approaches of 2D spectroscopy were also implemented in Rb [391, 417] and K vapors [78]. Subsequently, atomic vapors were often used as a model system to demonstrate, for instance, optical 3D spectroscopy [214], 2D spectroscopy in reflection geometry [215], and pulse propagation effects in 2D spectra [217, 370].

Despite being studied initially as a model system, later experiments on atomic vapors revealed surprisingly interesting physics of many-body interaction and correlation in atoms. Experimentally confirmed understanding of many-body interaction and correlation in atoms is fundamental in many-body physics and essential for various problems in cold atoms and molecules, optical atomic clocks, semiconductors, photosynthesis, etc. Advances in cooling and trapping atoms have provided a well-controlled environment for studying many-body physics. On the other hand, a thermalized atomic vapor provides a broader range of mean interatomic separation and more atoms for experiments and thermal motion introduces effects that might be shared by many natural processes in chemical and biologic systems. Experiments in atomic vapors can complement studies of many-body physics in cold atoms and molecules. Various spectroscopic techniques have been used to study many-body effects in atomic vapors. However, most techniques are limited to highly dense atomic vapors (the atomic density $\mathcal{N} > 10^{16}$ cm^{-3}) and not sufficiently sensitive to many-body effects in more dilute vapors. Optical two-dimensional coherent spectroscopy provides an extremely

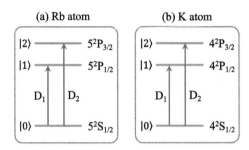

Fig. 5.1 Energy states related to D-lines for (a) Rb atom and (b) K atom.

sensitive detection of many-body interaction and correlation effects in atomic vapors. Double-quantum 2D spectra revealed two-atom collective resonances due to dipole-dipole interaction in both K and Rb vapors [79, 129, 234] even at a density as low as 10^8 cm^{-3} [459]. The technique was also extended to multi-quantum two-dimensional coherent spectroscopy to probe multi-atom Dicke states with up to seven atoms [460].

The potential of 2D spectroscopy in studying many-body physics and other advantages motivated efforts to implement optical 2D coherent spectroscopy in cold atoms and molecules. Optical 2D coherent spectroscopy has been performed with Rb$_2$ and Rb$_3$ molecules prepared by helium nanodroplet isolation in a molecular beam. However, realizing the full potential of the technique requires a significantly improved frequency resolution to resolve detail energy structures of cold atoms/molecules. The required frequency resolution was made possible by incorporating femtosecond frequency combs in 2D spectroscopy [232, 234], which will be discussed separately in the next chapter.

5.1 Single- and zero-quantum 2D spectra of atomic vapors

The energy states related to D-lines of Rb and K atoms form a typical three-level V system, as shown in Fig. 5.1. For Rb, the three states include the ground state $|0\rangle = |5^2S_{1/2}\rangle$ and two excited states $|1\rangle = |5^2P_{1/2}\rangle$ and $|2\rangle = |5^2P_{3/2}\rangle$. For K, the three states are $|0\rangle = |4^2S_{1/2}\rangle$, $|1\rangle = |4^2P_{1/2}\rangle$, and $|2\rangle = |4^2P_{3/2}\rangle$. Both Rb and K each have two dominant isotopes: ^{85}Rb (72.2%) and ^{87}Rb (27.8%) for Rb and ^{39}K (93.26%) and ^{41}K (6.73%) for K. Some important optical parameters of the D-lines are listed in Table 5.1 and Table 5.2 for Rb and K isotopes, respectively. The experiments discussed in this chapter do not have sufficient frequency resolution to resolve either the hyperfine levels or the energy differences between the isotopes, which are not considered here. In some cases, an argon buffer gas was introduced to broaden the transitions such that the hyperfine and Zeeman sublevels are not resolvable or relevant in the measurement. The D-lines of K atom are separated by 3.4 nm and thus can be easily covered by the bandwidth of a typical Ti:sapphire femtosecond oscillator. The D-lines of Rb atom are separated by 14.7 nm which requires a broader bandwidth corresponding to sub-100-fs pulses. If both D-lines are excited, the results can be interpreted with the three-level system.

Table 5.1 Important optical parameters of D-lines for Rb isotopes ^{85}Rb and ^{87}Rb.

		D_1 line $(5^2S_{1/2} \rightarrow 5^2P_{1/2})$	D_2 line $(5^2S_{1/2} \rightarrow 5^2P_{3/2})$
Frequency	^{85}Rb	$2\pi \cdot 377.107\ 385\ 690(46)$ THz	$2\pi \cdot 384.230\ 406\ 373(14)$ THz
	^{87}Rb	$2\pi \cdot 377.107\ 463\ 5(4)$ THz	$2\pi \cdot 384.230\ 484\ 468\ 5(62)$ THz
Wavelength	^{85}Rb	$794.979\ 014\ 933(96)$ nm	$780.241\ 368\ 271(27)$ nm
(Vacuum)	^{87}Rb	$794.978\ 850\ 9(8)$ nm	$780.241\ 209\ 686(13)$ nm
Wave Number	^{85}Rb	$12\ 578.948\ 390\ 0(15)$ cm^{-1}	$12\ 816.546\ 784\ 96(45)$ cm^{-1}
	^{87}Rb	$12\ 578.950\ 985(13)$ cm^{-1}	$12\ 816.549\ 389\ 93(21)$ cm^{-1}
Lifetime	^{85}Rb	$27.679(27)$ ns	$26.2348(77)$ ns
	^{87}Rb	$27.70(4)$ ns	$26.24(4)$ ns
Natural Linewidth	^{85}Rb	$2\pi \cdot 5.7500(56)$ MHz	$2\pi \cdot 6.0666(18)$ MHz
(FWHM)	^{87}Rb	$2\pi \cdot 5.746(8)$ MHz	$2\pi \cdot 6.065(9)$ MHz
Transition Dipole	^{85}Rb	$2.5377(12) \times 10^{-29}$ C·m	$3.584\ 25(52) \times 10^{-29}$ C·m
	^{87}Rb	$2.537(3) \times 10^{-29}$ C·m	$3.584(4) \times 10^{-29}$ C·m

Table 5.2 Important optical parameters of D-lines for K isotopes ^{39}K and ^{41}K.

		D$_1$ line (4^2S$_{1/2} \rightarrow 4^2$P$_{1/2}$)	D$_2$ line (4^2S$_{1/2} \rightarrow 4^2$P$_{3/2}$)
Frequency	^{39}K	$2\pi \cdot 389.286\ 058\ 716(62)$ THz	$2\pi \cdot 391.016\ 170\ 03(12)$ THz
	^{41}K	$2\pi \cdot 389.286\ 294\ 205(62)$ THz	$2\pi \cdot 391.016\ 406\ 21(12)$ THz
Wavelength (Vacuum)	^{39}K	$770.108\ 385\ 049(123)$ nm	$766.700\ 921\ 822(24)$ nm
	^{41}K	$770.107\ 919\ 192(123)$ nm	$766.700\ 458\ 70(2)$ nm
Wave Number	^{39}K	$12\ 985.185\ 192\ 8(21)$ cm^{-1}	$13\ 042.895\ 496\ 4(4)$ cm^{-1}
	^{41}K	$12\ 985.193\ 050\ 0(21)$ cm^{-1}	$13\ 042.903\ 375(1)$ cm^{-1}
Lifetime	^{39}K	$26.72(5)$ ns	$26.37(5)$ ns
	^{41}K	$26.72(5)$ ns	$26.37(5)$ ns
Natural Linewidth (FWHM)	^{39}K	$2\pi \cdot 5.956(11)$ MHz	$2\pi \cdot 6.035(11)$ MHz
	^{41}K	$2\pi \cdot 5.956(11)$ MHz	$2\pi \cdot 6.035(11)$ MHz
Transition Dipole		$3.485(0) \times 10^{-29}$ C·m	$4.928(1) \times 10^{-29}$ C·m

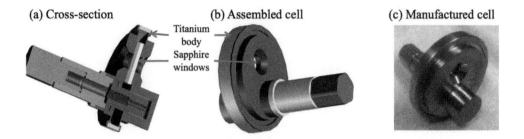

(a) Cross-section   (b) Assembled cell   (c) Manufactured cell

Titanium body
Sapphire windows

Fig. 5.2 High-temperature vapor cell made of a titanium body and sapphire windows. (a) Cross-section and (b) assembled cell shown in a 3D model. (c) A manufactured high-temperature cell.

Atomic vapor samples are commercially available as standard vapor reference cells. These are usually cylindrical glass cells with optical quality windows made of glass or quartz. From a stem on the cell body, the cell is evacuated to 10^{-8} torr and back filled with alkali metal before the stem is sealed. The alkali metal represents as a small amount of solid which evaporates into vapor in the cell. The vapor saturation pressure is determined by the temperature of the coldest point in the cell and can be calculated according to an empirical relationship [289]. The number density can then be calculated by using the ideal gas law. Therefore, the atomic number density in an atomic vapor cell can be conveniently controlled by varying the cell temperature. However, the glass cell can only operate at relatively low temperatures ($< 200\ ^{\circ}\text{C}$) since alkali metals react quickly with glass at higher temperatures. When a standard glass cell is used, the experiment needs to be performed in a way to avoid pulse propagation effects in an optically thick sample.

Alternatively, some experiments used special custom-made cells. One example is a high-temperature vapor cell that consists of a titanium body and two sapphire windows, as shown in Fig. 5.2, similar to the high-temperature reflection vapor cell [243]. The cell can operate at temperatures up to $800\ ^{\circ}\text{C}$. The two sapphire windows form a narrow gap that can be adjusted from $\sim 10\ \mu\text{m}$ to a few mm for experiments. A short optical path of the cell is critical to ensure that the optical density is smaller than 1 to avoid the spectral distortion in 2D spectra when the experiment is performed in the transmission geometry.

Optical two-dimensional coherent spectroscopy of Rb vapor was first reported in [394]. The experiment used an acousto-optic pulse shaper to create collinear three-pulse sequence. The time delays and relative phases between pulses can be precisely controlled by the pulse shaper. The three-pulse sequence excites a Rb atomic vapor and the fluorescence is detected. As shown in Fig. 5.3, the first two pulses are equivalent to a three-pulse sequence with the second time delay $T = 0$, i.e., the second pulse acts twice. The third pulse subsequently converts the third-order polarization into a fourth-order population, which is detected as fluorescence. Unlike the box-geometry, this collinear setup cannot spatially separate signals according to different phase-matching conditions. The detected fluorescence signal includes contributions from all possible excitation pathways at all orders. To select the desired photon echo

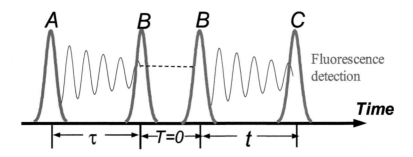

Fig. 5.3 Three-pulse sequence excites a fourth-order population that emits a fluorescence signal. The second pulse B is applied twice, equivalent to two pulses with $T = 0$. The phase of each pulse can be controlled independently by the pulse shaper for phase cycling.

signal, a 16-step phase cycling procedure was used in this experiment. The idea is to toggle the phase of each excitation pulse to change the phase of the signal from certain pathways but not others. The signal from specific pathways can then be isolated by subtracting or adding a proper combination of signals acquired with different phase configurations of excitation pulses. The photon echo signal can be selected by combining 16 measurements as

$$\rho_{16} = (\rho_{xxx} - \rho_{xx\bar{x}} - \rho_{\bar{x}xx} + \rho_{\bar{x}x\bar{x}} - \rho_{yxy} + \rho_{\bar{y}xy} + \rho_{yx\bar{y}} - \rho_{\bar{y}x\bar{y}})$$
$$+ i(\rho_{\bar{x}x\bar{y}} - \rho_{\bar{x}xy} + \rho_{xxy} - \rho_{xx\bar{y}} + \rho_{yxx} - \rho_{yx\bar{x}} - \rho_{\bar{y}xx} + \rho_{\bar{y}x\bar{x}}), \qquad (5.1)$$

where ρ corresponds to the excited state population, subscript symbols $x, y, \bar{x}$, and $\bar{y}$ represent a phase of 0, $\pi/2$, π, and $3\pi/2$, respectively. The resulting 2D spectrum consists of two diagonal peaks corresponding to the two D-lines and two cross-diagonal peaks due to the coupling between the two transitions. Since the second time delay is fixed at $T = 0$, the setup cannot perform 2D measurements that require scanning T. The multi-step phase cycling also limits the acquisition speed.

Optical two-dimensional coherent spectroscopy of K vapor was also demonstrated [78], in which single-quantum rephasing, non-rephasing, and zero-quantum 2D spectra were reported. The experiment was performed in the box geometry with four phase-locked pulses. Three pulses are incident on the sample cell to generate a TFWM signal at the fourth corner of the box. The fourth pulse is routed around the sample and combined with the TFWM signal for spectral interferometry. Pulse A is considered as a conjugated pulse for the phase matched direction of the measured signal. The excitation sequence can be either rephasing or non-rephasing depending on pulse A arriving first or second, respectively. The K vapor, along with an argon buffer gas, is contained in a high-temperature thin cell that has an optical path of $\sim 20~\mu$m. The vapor cell can be heated to over 200 °C to achieve a high density while maintaining an optical density smaller than 1. As shown in Fig. 5.4(a), the laser spectrum (red dotted line) covers both D-lines as indicated by the absorption spectrum (blue line).

Single-quantum spectra were obtained by scanning the time delay τ, between the first and second excitation pulses, in either a rephasing or a non-rephasing excitation

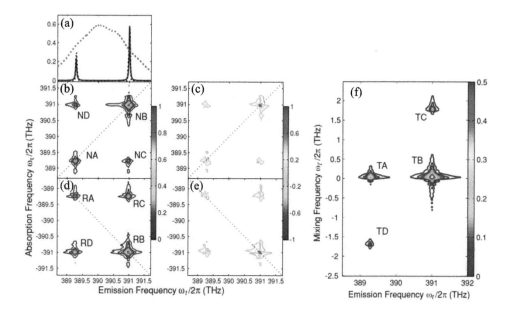

Fig. 5.4 Experimental 2D spectra of a K vapor. (a) Absorption (blue) and laser spectra (red). (b) Non-rephasing single-quantum 2D amplitude spectrum. (c) The real part of the spectrum in (b). (d) Rephasing single-quantum 2D amplitude spectrum. (e) The real part of the spectrum in (d). (e) Zero-quantum 2D amplitude spectrum. Adapted from Ref. 78.

pulse sequence. Figures 5.4(b) and (d) show the amplitude of a non-rephasing and a rephasing, respectively, one-quantum 2D spectra. Both spectra have two resonances (NA, NB, RA, and RB) on the diagonal line corresponding to the D_1 and D_2 transitions, and two cross resonances (NC, ND, RC, and RD) due to the coupling of the two transitions. The experiment also measured the phase so that the real part of the spectra can be plotted as shown in Figs. 5.4(c) and (e) for the non-rephasing and rephasing spectrum, respectively. The real part of the resonances has an absorptive lineshape, which is expected for this simple three-level system. The strength of the cross peaks is the geometric mean of the associated diagonal peaks, as expected from a Bloch model. Zero-quantum spectra were obtained by scanning the time delay T, between the second and third pulses, in either a rephasing or non-rephasing excitation pulse sequence. A zero-quantum 2D spectrum is shown in Fig. 5.4(f). The mixing frequency dimension is associated with the dynamics during the time period T. The two resonances (TC and TD) with nonzero mixing frequencies involve coherences oscillating at these frequencies, while the resonances (TA and TB) with a zero mixing frequency include a population decay during T.

A more rigorous interpretation of 2D spectra requires that each resonance be associated with one or more excitation quantum pathways by using doubled-sided Feynman diagrams (see section 2.3.2). By tracking the evolution of the density matrix during each time period, the expected peak locations can be determined in a 2D spectrum.

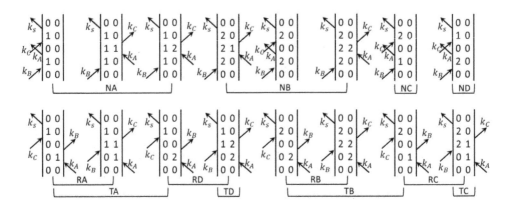

Fig. 5.5 Double-sided Feynman diagrams representing all excitation quantum pathways that contribute to the 2D spectra in Fig. 5.4. The diagrams are grouped and labeled by the corresponding 2D peaks.

For the 2D spectra in Fig. 5.4, all relevant excitation quantum pathways are represented by the double-sided Feynman diagrams in Fig. 5.5. The three states are $|0\rangle = |4^2S_{1/2}\rangle$, $|1\rangle = |4^2P_{1/2}\rangle$, and $|2\rangle = |4^2P_{3/2}\rangle$, as shown in Fig. 5.1. We only consider the third-order nonlinear signal in the chosen signal direction under the excitation of rephasing and non-rephasing pulse sequences. The diagrams are labeled by their corresponding resonances in the 2D spectra. In the non-rephasing 2D spectrum, each diagonal peak has contributions from three pathways and one of them involves the coherence (ρ_{12}) between the two excited states. Each cross peak corresponds to only one pathway that does not include ρ_{12}. In the rephasing 2D spectrum, each diagonal peak has contributions from two pathways that describe similar processes. The cross peaks each include two pathways that have different dynamics during the second time delay T, i.e., a population decay and an oscillating coherence ρ_{12}. Therefore, the pathways with an oscillating term can be separated in a zero-quantum spectrum by scanning T. The resulting spectral resonances are peaks TC and TD with a mixing frequency of ω_{12} or ω_{21}. However, the peaks with a zero mixing frequency each have contributions from three pathways.

Double-sided Feynman diagrams help us understand the signal origin and the underlying processes of each spectral peak in 2D spectra. The pathway analysis reveals both advantages and limitations of two-dimensional coherent spectroscopy. Compared to 1D spectra, a 2D spectrum clearly isolates and identifies the couplings between two transitions and the quantum coherence between two states. However, the excitation quantum pathways are not completely separated in 2D spectra.

5.2   MDCS in optically thick samples

Besides validating basic experimental techniques and data interpretations, atomic vapors can be used to explore more complex effects in 2D spectroscopy. One example is to investigate how the line shape of 2D spectra is affected by an optically thick sample [217, 370]. The line shape of 2D spectra provides important information such as

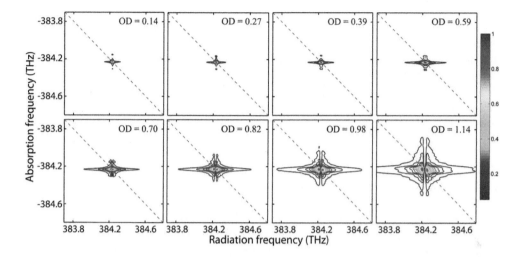

Fig. 5.6 Experimental 2D spectra of Rb D_2 transition at different optical densities (OD). The spectra were taken with the reference pulse going around the sample. Adapted from Ref. 217.

inhomogeneity and temporal dynamics. However, the spectral line shape can be significantly distorted in optically thick samples where pulse propagation can reshape both the excitation pulses and the signal going through the sample. Accurate interpretation of 2D spectra taken in transmission must account for the effects of pulse propagation to retrieve the microscopic nonlinear response. In particular, the possibility to measure and interpret 2D spectra on optical thick sample has the advantage of improving signal strengths and signal-to-noise ratio. The eased constraint on the optical thickness may allow easier sample preparation and experiments on more different samples.

The pulse propagation effects on 2D spectra were mainly studied in theoretical models [181, 182, 457]. An atomic vapor provides an ideal model to experimentally study the effects of pulse propagation. The optical density in a vapor cell can be continuously varied by changing the temperature without changing most of other parameters such as the physical thickness. The experiment was performed with a Rb vapor cell with a thickness of about 0.5 mm. The laser spectrum was tuned to excite only the D_2 transition so the resulting spectra have only one spectral resonance. In the experiment, the optical density (OD), as defined by OD$= -\log_{10}(T)$ with T being the transmission, was varied from 0.14 to 1.14 and confirmed by the linear absorption measurement. Rephasing single-quantum 2D spectra were taken in the transmission box geometry with the reference pulse routed around the sample so the reference pulse does not experience propagation effects. The obtained 2D spectra, as shown in Fig. 5.6 for different values of OD, reflect the spectral effects of pulse propagation. As the OD increases, the standard star-shaped peak becomes broader, asymmetric, and distorted. The spectral broadening is caused partially by the faster dephasing rate due to elevated temperatures. However, the asymmetry is due to the propagation effect, which has been shown [181] to cause more severe spectral distortion along the radiation frequency axis than the absorption frequency axis. The spectrum starts splitting at OD = 0.59 and forms a clear gap at OD = 1.14. Such a splitting was not predicted by previous theoretical models [181, 182, 457] at similar optical densities.

The observed spectral distortion can be understood as the resonant reabsorption of the TFWM signal. The reabsorption depletes the spectral amplitude near the line center, resulting in a broader line shape in the radiation frequency direction at modest OD and a spectral gap at high OD. The spectrum can also be severely distorted in the absorption frequency direction if the excitation pulses experience strong propagation effects. The spectral distortions can experimentally be corrected by routing the reference pulse through the same sample. The reference pulse characterizes the propagation effects and can partially correct the distortion by using the transmitted reference pulse in the spectral interferometry. More extensive discussion about this topic can be found in [217].

A theoretical model has been developed [370] to quantitatively reproduce the spectral distortion in 2D spectra due to pulse propagation. The model calculates the generation and spatial propagation of the third-order nonlinear signal based on an exact, three-dimensional solution of Maxwell's equations [27]. The algorithm generates a complex-valued 3D spectrum in the frequency domain, from which a 2D spectrum at a specific waiting time T can be extracted. An advantage of this approach is that the third-order nonlinear response is calculated only once for each point on a 3D grid instead of being evaluated at small time intervals as in convolution algorithms used to model pulse propagation effects [112, 172, 421]. In the time domain, calculating the third-order nonlinear polarization usually involves a 3D convolution of the third-order nonlinear susceptibility with the three excitation pulses. This requires $O(N^2)$ operations where N is the total number of points in the 3D grid. Alternatively, the third-order signal can be calculated in the frequency domain by multiplication following a 3D fast Fourier transform of the third-order nonlinear susceptibility, requiring $O(N \log_2 N)$ operations. The algorithm is optimized for calculating the entire 3D spectrum, which includes the complete set of 2D spectra. The pulse propagation can be applied to the 3D signal by a single multiplication in the 3D frequency domain. These advantages reduce the computation time. The tradeoff of this approach is the need for more random access memory. Overall, the computation can be performed on personal computers for 3D grids of 1024^3 points to match available resolution in most experiments.

The simulated rephasing 2D spectra at different OD values are shown in Fig. 5.7 to compare with the experimental 2D spectra in Fig. 5.6. The undistorted 2D spectrum at OD = 0 exhibits a sysmetric star-shaped 2D peak. As the OD increases, the line shape broadens in both ω_τ and ω_t dimensions but the width in ω_t increaes faster than that in ω_τ. The 2D peak becomes asymmetric between the ω_τ and ω_t dimensions. The peak starts to split in the ω_t dimension at higher OD and the splitting forms a gap down to the 70% contour at OD= 1.14. The simulated results are in a good agreement with the experimental spectra.

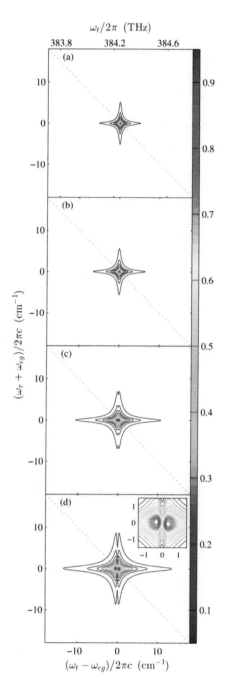

Fig. 5.7 Simulated rephasing 2D spectra of an atomic vapor with OD values of (a) 0, (b) 0.14, (c) 0.59, and (d) 1.14. The amplitude of the spectra is plotted. Adapted from Ref. 370.

5.3   Probing many-body interactions with double-quantum 2D spectroscopy

The dipole-dipole interaction plays an important role in many systems such as cold atoms and molecules [171, 437], optical atomic clock [247, 384], semiconductors [52], and photosynthesis [106]. Its long-range nature extends the interaction beyond nearest neighbors, resulting in collective and emergent phenomena [11] that cannot be understood by a simple extrapolation of the microscopic laws of a few particles. Experiments [103, 348] have shown results that cannot be quantitatively explained by accounting for only two-atom interactions. Experimentally confirmed understanding of many-body dipole-dipole interactions is essential. The ability to generate and trap cold atoms and molecules has advanced the study of atom-atom and molecule-molecule interactions at long ranges. Photoassociation spectroscopy [171] involving cold atoms has successfully isolated the dipole-dipole interaction between two atoms, and precision measurements of the London dispersion coefficient C_3, excited state lifetimes, s-wave scattering lengths, and retardation corrections pose rigorous tests for theory. On the other hand, rich structural and dynamic information about dipole-dipole interactions can be found in atomic vapors at room or higher temperatures. Experiments with thermalized atoms complement studies in cold atoms for a full understanding of dipole-dipole interactions. Although a few measurements [72, 103, 216, 218, 242, 244, 253, 329, 348, 416] have been done in hot atomic vapors, the fundamental understanding of dipole-dipole interaction strength, effective interaction range, number of interacting atoms, interaction dynamics, excitation-induced many-body states, effects of a buffer gas, and effects of thermal motion is still incomplete.

The potential energy of the long-range interaction between two atoms can be written as

$$V(R) = \pm\frac{C_3}{R^3} - \frac{C_6}{R^6} - \frac{C_8}{R^8} - \cdots, \tag{5.2}$$

where R is the interatomic separation and C_3, C_6, C_8, $\cdots$ are known as London dispersion coefficients. The dipole-dipole interaction potential is determined by C_3. The expression of the dipole-dipole interaction between two identical atoms was derived by King and van Vleck [189]. Briefly, the interaction term of the Hamiltonian is

$$V_{AB} = \frac{1}{4\pi\epsilon_0 R^3}[\vec{\mu}_A \cdot \vec{\mu}_B - \frac{3(\vec{\mu}_A \cdot \vec{R})(\vec{\mu}_B \cdot \vec{R})}{R^2}], \tag{5.3}$$

where $\vec{\mu}_A$ and $\vec{\mu}_B$ are the dipole moments of atoms A and B. Note that the atoms have no permanent dipole moments and they are coupled by the dipole-allowed transitions. This description can be generalized for the dipole-dipole interactions among three and more atoms and the interaction term is given by $V = \sum_{m,n,m\neq n} V_{mn}$, where V_{mn} is the interaction term of atoms m and n in the ensemble.

Extensive theoretical work [437] has been done to study the dipole-dipole interactions in cold atoms. The theory has been successful in calculating excited state lifetimes, s-wave scattering lengths, and retardation corrections that can be extracted from experiments [107, 170]. For hot vapors, theories [7, 10, 213, 317, 366, 385] have been developed to calculate the spectral broadening and shift due to the interactions in the presence of thermal motion. However, the effects of dipole-dipole interactions on the lineshape is subtle in the mix of other effects and the theoretical modeling is difficult. These theoretical analyses are all limited to the two-particle (binary) interactions at short ranges (Weisskopf radius) in the fast-collision (impact) limit or in the opposite, static limit. To overcome these limitations, Leegwater and Mukamel [208] developed a theoretical method based on the exciton picture to calculate the spectral broadening and shift. The theory accounts for the many-body dipole-dipole interactions at long ranges and suggests that the interactions affect the lineshape at any density.

Various experimental techniques have been used to study the dipole-dipole interactions in atomic ensembles. Photoassociation spectroscopy [437] in cold atoms has been extremely successful in mapping out the potential energy curve, which yields information about the atomic interactions. For hot vapors, early studies focused on the spectral broadening and shift at very high atomic densities ($\mathscr{N} > 10^{16}$ cm^{-3}), where selective reflection spectroscopy [216, 218, 253, 329, 416] has revealed effects of the local field and dipole-dipole interactions. Time-domain techniques have also been developed using ultrafast lasers. Quantum beating experiments [103, 348] on alkali atomic vapors measured the energy shift and the results can only be quantitatively explained with the interactions involving at least five atoms. Transient four-wave mixing experiments [72, 243, 244] on potassium vapor enable one to probe the transient dynamics of the interactions and reveal the non-Markovian behaviors.

Although the experimental approaches mentioned above have provided valuable insights into the dipole-dipole interaction, their applications are limited in several respects. Photoassociation spectroscopy can achieve sufficient frequency resolution only in cold atoms that have a limited range of interatomic separations. Techniques for hot atomic vapors are not sensitive at low atomic densities. Selective reflection and quantum beating cannot probe interaction dynamics. Moreover, the effects of dipole-dipole interactions are always masked by other more pronounced effects in hot vapors such as the Doppler broadening. These methods have difficulties in isolating effects of the dipole-dipole interaction from manifestation of other interaction types. Thus many aspects of the dipole-dipole interaction have to be inferred.

Alternatively, a powerful technique to study the dipole-dipole interaction is double-quantum 2D spectroscopy, in which quantum pathways that include a double-quantum coherence can be observed. They have been shown to be particularly sensitive to the presence of many-body interactions [453] and provide background-free detection to weak dipole-dipole interactions in dilute atomic vapors [79, 129, 459].

Double-quantum 2D spectroscopy requires an excitation pulse sequence with a particular time ordering as shown in Fig. 5.8(a). A double-quantum coherence can be excited by this pulse sequence if the sample is a three-level ladder system consisting of a ground state, $|0\rangle$; a singly excited state, $|1\rangle$; and a doubly excited state, $|2\rangle$. The first pulse with wave vector $\mathbf{k}_B$ prepares the system in a superposition between states $|0\rangle$ and $|1\rangle$, i.e., in a single-quantum coherence. The second pulse ($\mathbf{k}_C$) converts the single-quantum coherence to a double-quantum coherence between states $|0\rangle$ and $|2\rangle$. The third pulse ($\mathbf{k}_A$) converts the double-quantum coherence back to a single-quantum coherence either between states $|0\rangle$ and $|1\rangle$ or between states $|1\rangle$ and $|2\rangle$,

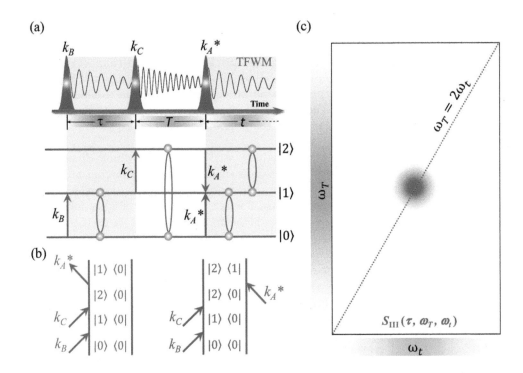

Fig. 5.8 (a) The time ordering of excitation pulses that can generate double-quantum coherence in a three-level ladder system. (b) Double-sided Feynman diagrams representing the
two pathways involving double-quantum coherences. (c) A schematic double-quantum 2D
spectrum with a resonance on the diagonal line $\omega_T = 2\omega_t$. Part of this figure is adapted from
Ref. 79.

which radiates a signal field. The corresponding quantum pathways can be described
using the double-sided Feynman diagrams shown in Fig. 5.8(b). To obtain a double-
quantum 2D spectrum, the TFWM signal is recorded while scanning the second time
delay T. Fourier transforming the signal with respect to T and t generates a double-
quantum 2D spectrum. The spectrum correlates the coherence dynamics during T
with the signal emission during t. Assuming the energy difference between $|0\rangle$ and
$|2\rangle$ is exactly twice the energy difference between $|0\rangle$ and $|1\rangle$, the double-quantum
coherence during T oscillates at twice the frequency of the single-quantum coherence.
In this case, the spectrum features a peak at a double-quantum frequency that is twice
of its emission frequency, i.e., the peak is on the diagonal line $\omega_T = 2\omega_t$, as shown in
Fig. 5.8(c). For individual atoms, the presence of a double-quantum signal relies on
the existence of a doubly excited state.

However, for K atoms, within the laser bandwidth tuned to excite the D-lines, there
are no atomic energy levels at twice the frequency of the laser. As shown in Fig. 5.9(a),
the D_1 and D_2 lines have transition frequencies of 389.29 and 391.02 THz, respectively.
The closest higher-lying states 5P and 4D, at frequencies 740.81 and 821.36 THz, are

both well outside the spectral range at twice the frequency of the laser. Thus no double-quantum signal is expected due to the absence of a doubly excited state. Surprisingly, the observed double-quantum spectrum in Fig. 5.9(d) shows clear resonances, even though there are no atomic states at these energies. The observed resonances are at exactly twice the frequencies of the D_1 and D_2 lines and their sum frequency in the double-quantum frequency axis and at the D_1 and D_2 lines in the emission frequency axis. A rephasing single-quantum spectrum is also shown in Fig. 5.9(e) for reference. The real part is plotted in these spectra.

This observation can be explained using a simple picture by combining the Hilbert spaces of two individual atoms into a Hilbert describing both simultaneously, as shown in Fig. 5.9(b), and working in the joint Hilbert space of both atoms. There are no doubly excited states for two isolated non-interacting two-level atoms. However, the joint space of two atoms has four levels consisting of a ground state, where both atoms are in the ground states; two singly excited states, where one of the two atoms is excited; and a doubly excited state, where both atoms are excited (two-atom Dicke states). In this four-level system, the double-quantum coherences can occur between the ground state and the doubly excited state. For K atoms (a three-level V system), the two-atom collective states are shown in Fig. 5.9(c). However, the change in description should not alter the underlying physics. It seems contrary that the nonlinear susceptibility calculated for the four-level system has terms that are not present for the two isolated two-level systems. In fact, the double-quantum signals associated with the two pathways shown in Fig. 5.8(b) interfere destructively and exactly cancel each other if the ground state to the single excited state transitions are identical to the singly excited state to doubly excited state transitions. The situation changes if there is a dipole-dipole interaction between the two atoms. As shown in Fig. 5.9(b), the interaction between the two atoms leads to an energy splitting between the singly excited states $|10\rangle$ and $|01'\rangle$, an energy shift to the doubly excited state $|11'\rangle$, and a different dephasing rate for the upper transitions. In this case, the symmetry in the four-level system is broken and the interference cancellation of the available pathways will be incomplete, resulting in a nonzero double-quantum signal. Since a slight mismatch gives rise to incomplete cancellation, the double-quantum signal provides an extremely sensitive and background-free detection to weak atom-atom interactions. The fact that the double-quantum resonances in the spectrum show no observable energy shift from the D_1 and D_2 lines indicates that the interaction is weak.

The joint Hilbert space description is useful in explaining why interactions result in the double-quantum resonance. This description only considers the interaction of a pair of atoms. A more complete theoretical treatment based on the exciton picture considers an ensemble of interacting atoms. The exciton formalism will be used to calculate the nonlinear optical response of a collective resonance due to the many-body dipole-dipole interactions in an atomic ensemble. The exciton formalism is commonly used in condensed matter such as semiconductors and molecular crystals. An exciton is a bound electron-hole pair in semiconductors, while in molecules it is an excitation distributed across molecular constituents. Similar to the molecular case, N_p exciton states in an atomic ensemble describe states where N_p atoms are excited and all other atoms are in the ground state.

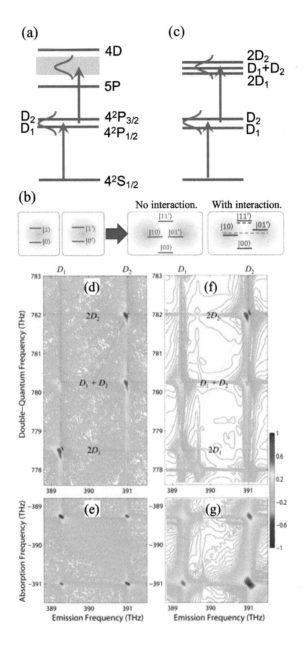

Fig. 5.9 (a) Relevant energy levels of individual K atoms. (b) Two two-level atoms can be considered as a four-level system in their joint Hilbert space. (c) Energy levels of two-atom collective states. Experimental (d) double-quantum and (e) single-quantum 2D spectra of a K vapor. Simulated (f) double-quantum and (g) single-quantum 2D spectrum based on the exciton formalism. The real part is plotted in these spectra. Part of this figure is adapted from Ref. 79.

The third-order nonlinear optical response can be calculated based on the nonlinear exciton equations [281]. In this excitonic picture, the effective Hamiltonian describing the electronic excitation of interacting atoms is

$$H(t) = \sum_m \Omega_m(t) \hat{B}_m^\dagger \hat{B}_m + \sum_{m,n,m \neq n} [J_{mn}(t) \hat{B}_m^\dagger \hat{B}_n + K_{mn}(t) \hat{B}_m^\dagger \hat{B}_n^\dagger \hat{B}_m \hat{B}_n], \quad (5.4)$$

where $\hat{B}_m^\dagger$ and $\hat{B}_m$ are the exciton creation and annihilation operators on the state m and m is a multi-index consisting of the atom number and the intra-atom exciton state (e.g., the D_1 or D_2 line); Ω_m is the excitation energy of exciton m; J_{mn} is the one-exciton coupling strength between excitons m and n; and K_{mn} is the two-exciton coupling parameter. The atoms interact through the dipole-dipole coupling

$$J_{mn}(t) = \frac{1}{4\pi\epsilon} \left(\frac{\vec{\mu}_m \cdot \vec{\mu}_n}{R_{mn}(t)^3} - \frac{3[\vec{\mu}_m \cdot \vec{R}_{mn}(t)][\vec{\mu}_n \cdot \vec{R}_{mn}(t)]}{R_{mn}(t)^5} \right), \quad (5.5)$$

where $\vec{\mu}_{m,n}$ are the transition dipole moments, R_{mn} is the distance between atoms m and n, and ϵ is the dielectric constant. The interaction Hamiltonian for the field-atom interaction is

$$H_{int}(t) = -E(\vec{r}, t) \cdot V(t) = -E(\vec{r}, t) \cdot \sum_m \vec{\mu}_m(t)[\hat{B}_m^\dagger(t) + \hat{B}_m(t)], \quad (5.6)$$

where $E(\vec{r}, t)$ is the electric field. All elements of the Hamiltonian are time-dependent through the changing configuration of the atoms.

Simulation of the 2D spectra requires three steps: first, the dynamic trajectories of an ensemble of atoms are generated using the molecular dynamics package GROMACS; second, the trajectories are used to calculate the interatomic distances between atom pairs and the coupling parameter J_{mn} for each snapshot of the trajectory; and third, the 2D spectra are generated using the calculated interatomic couplings. At the last step, 2D spectra are generated using the SPECTRON package [109]. The double-quantum coherence spectrum can be calculated in the time domain as

$$S_{III}(\tau, T, t) = 2(\frac{i}{\hbar})^4 \sum_{t_s} \sum_{n_4} \vec{\mu}_{n_4}(\tau + T + t + t_s)$$

$$\times R_{n_4}^{S_{III}}(\tau + T + t + t_s, T + t + t_s, t + t_s, t_s). \quad (5.7)$$

The signal is averaged over several start times, t_s, along the trajectory, with

$$R_{n_4}^{S_{III}}(\tau_4, \tau_3, \tau_2, \tau_1)$$

$$= \sum_{m_4 m_1} \int_{\tau_3}^{\tau_4} ds G_{n_4, m_4}(\tau_4, s) K_{m_4 m_1}(s) \times \psi_{m_4 m_1}^{(2)}(s; \tau_2; \tau_1) \psi_{m_1}^{(1)*}(s; \tau_3) \quad (5.8)$$

for the one-exciton Green function $G_{n_2, n_1}(\tau_2, \tau_1) = \langle g | \hat{B}_{n_2} U(\tau_2, \tau_1) \hat{B}_{n_1}^\dagger | g \rangle$, where $U(\tau_2, \tau_1)$ is the time evolution operator connected to $H(t)$. The excitonic functions are calculated by direct integration of the Schrödinger equation

$$i\hbar\frac{\partial}{\partial t}|\psi^{(1)}(t;\tau_3)\rangle = H(t)|\psi^{(1)}(t;\tau_3)\rangle \tag{5.9}$$

$$i\hbar\frac{\partial}{\partial t}|\psi^{(2)}(t;\tau_2;\tau_1)\rangle = H(t)|\psi^{(2)}(t;\tau_2;\tau_1)\rangle \tag{5.10}$$

with the proper initial conditions. Double-quantum 2D spectra can be obtained by Fourier transforming $S_{III}(\tau, T, t)$ with respect to τ and t.

In the simulation, the time-dependent interatomic couplings are calculated based on the simulated trajectories of 20 potassium atoms. The 2D spectra are generated using the calculated interatomic couplings. The simulated double- and single-quantum 2D spectra are shown in Figs. 5.9(f) and (g), respectively. They are in agreement with the experimental spectra, matching the double-quantum resonance positions and the dispersive lineshape. The double-quantum signal disappears when the dipole-dipole interaction is turned off. This study shows that the observed double-quantum signal is attributed to a collective resonance induced by weak interatomic dipole-dipole interactions, which cannot be ignored even at a low atomic density.

Double-quantum 2D spectroscopy has also been used to study the dipole-dipole interaction in a Rb vapor [129]. Different from K, the energy level structure of Rb atoms provides a real doubly excited state for individual atoms. As shown in Fig. 5.10(a), the relevant energy levels are $|0\rangle = |5^2S_{1/2}\rangle$, $|1\rangle = |5^2P_{3/2}\rangle$ at the frequency of 384.2 THz, and $|2\rangle = |5^2D\rangle$ at the frequency of 770.5 THz. The laser spectrum is centered at half the energy difference between states $|2\rangle$ and $|0\rangle$. The laser has a sufficient bandwidth such that the single excitation covers the transition $|0\rangle \rightarrow |1\rangle$ (D2 line), while the double excitation covers the single-atom state $|2\rangle$ and the two-atom doubly excited state (both atoms in $|1\rangle$). The system can simultaneously produce the double-quantum signals due to the doubly excited state of individual atoms and many-body interactions between atoms. An experimental double-quantum 2D spectrum of Rb is shown in Fig. 5.10(b). For Peaks 2 and 3, the double-quantum frequency is 770.5 THz and the emission frequency is either 384.2 or 386.3 THz. These two resonances are due to the doubly excited state of individual atoms. However, Peak 1 has a double-quantum frequency of 768.4 THz, which is exactly twice the emission frequency 384.2 THz. This resonance is contributed by the many-body interaction between atoms. The spectrum can be reproduced by a theoretical simulation including the dipole-dipole interaction between atoms, as shown in Fig. 5.10(c).

The simultaneous detection of the double-quantum signals from both single atoms and atom-atom interactions provide an opportunity to quantitatively study the interactions by using the single-atom signal as a reference, since the double-quantum signal from individual atoms is well understood. For example, we can study the density dependence of the signal due to the dipole-dipole interaction. It is problematic to accurately measure the absolute magnitude of individual spectral features as the density varies, however, the relative magnitudes between different features can be extracted with high precision. Double-quantum 2D spectra of an Rb vapor can be obtained at different densities. The relative strength of Peak 1 compared to Peaks 2 and 3 shows how the double-quantum signals due to the dipole-dipole interactions and individual atoms scale with the density. This information is useful in comparing

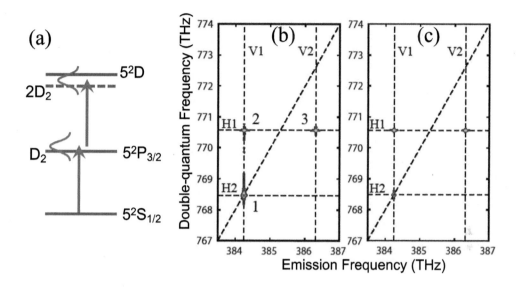

Fig. 5.10 (a) Relevant energy levels of individual Rb atoms and two-atom collective state (dash line). (b) Experimental and (c) simulated two-quantum 2D spectrum of a Rb vapor. Adapted from Ref. 129.

experiments with the calculations from theoretical models of many-body dipole-dipole interactions.

Double-quantum 2D spectroscopy of K and Rb vapors was first performed in the box geometry [79, 129]. The measurement can also be done in the collinear setup [459] which provides an improved detection sensitivity primarily due to the better overlap of copropagating excitation laser beams and the use of lock-in amplifier. The improved sensitivity enables the detection of long-range dipole-dipole interaction in atomic vapors at extremely low densities. Double-quantum 2D spectra indicating dipole-dipole interactions were obtained in K and Rb atomic vapors with atomic densities as low as 4.81×10^8 cm^{-3} and 8.40×10^9 cm^{-3}, respectively. At these densities, the mean atomic separation $\langle R \rangle = 2(\frac{4\pi}{3}\mathcal{N})^{-1/3}$ is 15.8 μm (or $3.0 \times 10^5 a_0$) for K and 6.1 μm (or $1.2 \times 10^5 a_0$) for Rb, where a_0 is the Bohr radius.

The effects of dipole-dipole interactions in K and Rb atomic vapors can be probed by several other spectroscopic techniques [47, 72, 79, 103, 129, 216, 218, 234, 242, 244, 253, 329, 348, 416]. Due to the difference in detection sensitivity of these methods, the effective range of dipole-dipole interactions confirmed by these experiments varies substantially. The detected interaction range reported for different methods is illustrated in Fig. 5.11. The atomic density is plotted in solid lines as a function of temperature for K (blue) and Rb (orange) vapors. The corresponding mean interatomic separation is plotted as dashed lines. The reported lowest densities at which the dipole-dipole interaction was detected in K and Rb vapors are marked with diamond for TFWM [72, 242, 244], triangle for selective reflection [216, 218, 253, 329, 416], pentagon for

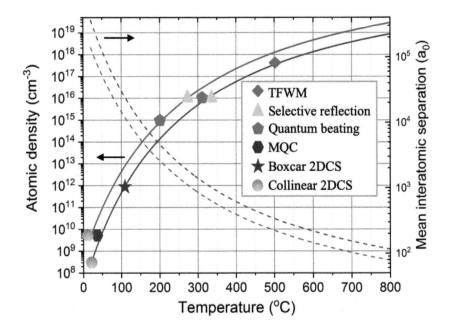

Fig. 5.11 Sensitivity of detecting dipole-dipole interaction in atomic vapors by various techniques. Atomic density (solid lines) and mean interatomic separation (dashed lines) at different temperatures are shown for K (blue) and Rb (orange) vapors. The reported lowest density is marked with diamond for TFWM, triangle for selective reflection, pentagon for quantum beating, hexagon for MQC, star for boxcar 2DCS, and circle for collinear 2DCS. Adapted from Ref. 459.

quantum beating [103, 348], hexagon for multi-quantum coherence (MQC) detection [47], and star for 2D coherent spectroscopy in the box geometry [79, 129]. The techniques of TFWM, selective reflection, and quantum beating are effective for densities higher than 10^{15} cm^{-3} which require cell temperatures higher than 200 °C. Conventional glass alkali metal vapor cells cannot work at such temperatures since alkali metals react quickly with glass above 200 °C. The box geometry 2D spectroscopy can detect dipole-dipole interactions at a density of 10^{12} cm^{-3}. The technique measuring multi-quantum coherence (MQC) can detect interactions in a Rb vapor at room temperature with a density of 8×10^9 cm^{-3} and in an atom beam with a density of 8×10^6 cm^{-3} [47], the lowest density at which the dipole-dipole interaction was reported. Using the collinear 2D spectroscopy technique, dipole-dipole interactions can be detected at a density of 4.81×10^8 cm^{-3} for K and 8.40×10^9 cm^{-3} for Rb.

These measurements in atomic vapors at various densities indicate that dipole-dipole interactions should be considered at all densities ranging from 10^8 to 10^{19} cm^{-3}. These account for a large range across 11 orders of magnitude in density for atomic vapors at temperatures from 25 to 800 °C. Moreover, at the lowest density, the mean interatomic separation $\langle R \rangle = 2(\frac{4\pi}{3}\mathscr{N})^{-1/3}$ is 15.8 μm or $3.0 \times 10^5 a_0$ for K and 6.1

μm or $1.2 \times 10^5 a_0$ for Rb. The experiment confirms the long-range nature of dipole-dipole interaction with an effective interaction range up to at least 15.8 μm. This 10-μm range interaction might have implications in experiments with optical lattices since the interaction is not just limited to the same or nearest lattice sites but also can extend to the sites further away. The long-range interaction is crucial for atom-based quantum simulators [30, 261, 409] to perform many-body simulation and the dipole-dipole interaction might provide an alternative approach. Two major technical concerns of implementing 2D coherent spectroscopy in cold atoms are sensitivity and frequency resolution. The experiment in low-density atomic vapors shows that collinear approach has a sufficient sensitivity for the number of atoms and the density in a typical magneto-optical trap. The required frequency resolution can be achieved by using femtosecond frequency combs [232, 234], which will be discussed in the next chapter.

5.4   Probing many-body correlations with multi-quantum 2D spectroscopy

Double-quantum 2D spectroscopy can be extended to multi-quantum 2D spectroscopy in which the multi-quantum coherence between the ground state and a high-lying energy state can be created and detected by a proper excitation pulse sequence. The multi-quantum coherence can be associated with a collective state of multiple particles, providing a tool to probe many-body correlations.

A fundamentally important many-body state, known as Dicke state, is of great interest and responsible for many intriguing collective behaviors of many-body systems. Introduced by Dicke in 1954 [94], a coherent collection of N_p atoms can act collectively as one big dipole and form Dicke states. A signature phenomenon is the cooperative spontaneous emission from a Dicke state known as superradiance [361] whose radiation rate scales with N_p but intensity is proportional to N_p^2 instead of N_p. Some other interesting phenomena associated with Dicke states include single-photon superradiance [339, 340], collective Lamb shift [322, 338], Dicke quantum phase transition [22, 148, 287], multi-particle entanglement [18, 187, 309, 325, 440]. As a many-body model system with exact solutions [390], Dicke states provide a unique opportunity to study how many-body properties scale with the number of atoms. Experimentally, it requires the preparation of scalable and deterministic Dicke states. The preparation and manipulation of such states have been demonstrated with trapped ions [325], entangled photons [187, 309, 440], and superconducting qubits [18]. For neutral atoms and molecules, however, the collective excitation has been demonstrated mainly for either two particles [79, 126, 129, 155, 234] or a large ensemble [22, 148, 361], as opposed to a few particles. Multi-quantum 2D spectroscopy enables the creation and detection of Dicke states with a scalable and deterministic number of atoms in an atomic vapor [460].

A triple-quantum coherence in semiconductor quantum wells was measured in triple-quantum 2D spectroscopy performed in the box geometry [411]. However, it is difficult to perform higher-order multi-quantum 2D spectroscopy in the box geometry due to the spatial phase-matching conditions. It turns out that the collinear

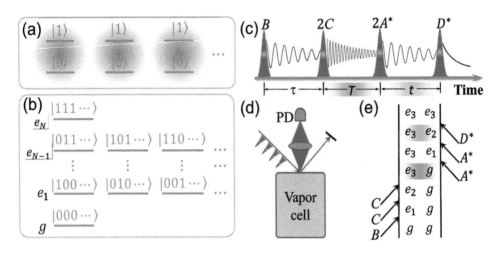

Fig. 5.12 (a) Energy level diagrams of individual two-level atoms with ground state $|0\rangle$ and excited state $|1\rangle$. (b) Energy level diagrams of Dicke states of N_p two-level atoms in the symmetric basis. (c) Excitation pulse sequence for three-quantum 2D spectroscopy. (d) Experimental schematic showing excitation pulses and fluorescence detection by a photo detector (PD). (e) One of the possible excitation pathways represented by a double-sided Feynman diagram. Adapted from Ref. 460.

2D spectroscopy setup provides a more flexible platform to implement higher-order multi-quantum 2D spectroscopy.

To explain how multi-quantum 2D spectroscopy detects multi-quantum coherence associated with Dicke states, we consider a system consisting of N_p atoms with two energy levels, as shown in Fig. 5.12(a). Each individual atom has two states: ground state $|0\rangle$ and excited $|1\rangle$. The whole system of N_p atoms can be described in the joint basis of N_p correlated atoms by using Dicke states [94] in the symmetric basis, as shown in Fig. 5.12(b). The ground state g has all atoms in state $|0\rangle$. The first excited state e_1 has one atom excited and others in state $|0\rangle$. Each next higher-energy excited state has one more atom excited and all N_p atoms are excited in state e_N. For N_p correlated atoms, it is possible to generate a coherence ρ_{ge_N} between states $g = |000\cdots\rangle$ and $e_N = |111\cdots\rangle$, which is known as N_p-quantum coherence. To observe the N_p-atom Dicke state e_N, one measures the corresponding N_p-quantum coherence ρ_{ge_N}, which oscillates at N_p times the transition frequency ω_{10} of a single atom. This approach eliminates the ambiguity in detecting spontaneous emission where signals from all manifolds of the ladder have the same frequency ω_{10}. It also ensures that the involved state is a coherent collection of atoms which is required for generating N_p-quantum coherence.

Multi-quantum 2D spectroscopy can be implemented in a collinear geometry. The experiment is described here by using triple-quantum 2D spectroscopy as an example. The pulse sequence is shown in Fig. 5.12(c). Four copropagating pulses B, C, A^*, and D^* are incident on the vapor in that order, as shown in Fig. 5.12(d). Pulses C and A^* each act twice and pulses A^* and D^* are considered conjugated. The time delays between pulse pairs BC, CA^*, and A^*D^* are defined as τ, T, and t, respectively. One possible excitation pathway involving triple-quantum coherence ρ_{ge_3} is illustrated by the double-sided Feynman diagram in Fig. 5.12(e). In this process, pulse B generates a single-quantum coherence between g and e_1; pulse C acts twice converting it to a triple-quantum coherence between g and e_3, which evolves during T; applying pulse A^* twice subsequently converts it into a single-quantum coherence between e_3 and e_2, which evolves during t; pulse D^* then turns the single-quantum coherence into a population in e_3 emitting a fluorescence signal. This pulse sequence can excite eight similar processes that involve the triple-quantum coherence ρ_{ge_3}.

In the experiment, each pulse is phase-modulated by an AOM at a specific frequency ω_i $(i = A, B, C, D)$. The fluorescence signal due to the excitation of this particular pulse sequence is modulated at the frequency $\omega_{3Q} = \omega_B - \omega_D + 2(\omega_C - \omega_A)$, thus it can be isolated by lock-in detection using ω_{3Q} as the reference frequency. The signal is recorded as a function of T and t. Fourier transforming the time-domain signal generates a 2D spectrum with two frequency dimensions ω_T and ω_t corresponding to T and t, respectively. The 2D spectrum displays the correlation between the dynamics during T and t as a spectral resonance on the diagonal line $\omega_T = 3\omega_t$. The signal indicates the existence of triple-quantum coherence and interatomic interactions that lead to an incomplete cancellation of contributions from the possible processes [79, 129]. This experiment can be generalized to measure N_p-quantum coherence for N_p-atom Dicke states by applying C and A^* pulses $(N_p - 1)$ times each and selecting the signal at the frequency $\omega_{NQ} = \omega_B - \omega_D + (N-1)(\omega_C - \omega_A)$. In this case, N_p-quantum coherence ρ_{ge_N} is created and evolves in the time period T. The resulting 2D spectrum correlates the N_p-quantum coherence in T with the single-quantum coherence in t. The spectral peak should have a N_p-quantum frequency that is N_p times the single-quantum frequency. Therefore, N_p-quantum 2DCS provides a specific detection of N_p-atom state.

The experiment was performed on a potassium (K) atomic vapor contained in a glass reference cell. A femtosecond (fs) oscillator provides excitation pulses with a duration of ~150 fs at a 76-MHz repetition rate. The laser wavelength is tuned to cover both D_1 and D_2 transitions of K atoms. Considering both D_1 and D_2 singly excited states, the Dicke states of K atoms have multiple possible energies that are combinations of D_1 and D_2. The hyperfine structures are not resolved here due to limited frequency resolution. For two-atom states, the doubly excited state energies can be $2D_1$ with two atoms in D_1, $2D_2$ with two atoms in D_2, and D_1+D_2 with one atom in D_1 and other in D_2. All these two-atom states are within the spectrum of the double laser frequency and can give rise to signals in two-quantum 2D spectra. As shown in Fig. 5.13(a), a double-quantum 2D spectrum of a K vapor was acquired at a number density of $\mathcal{N} = 5.32 \times 10^{13}$ cm^{-3}. The spectral amplitude is plotted. The vertical axis ω_T represents the double-quantum frequency that is associated with double-quantum coherences during the time period T. The horizontal axis ω_t is the emission frequency of single-quantum coherences during t. The spectrum features four spectral peaks. Their double-quantum frequencies are $2D_1$, D_1+D_2, and $2D_2$, matching the energies of two-atom doubly excited states. The emission frequency is D_1 (D_2) if the double-quantum frequency is $2D_1$ ($2D_2$) so the two corresponding peaks are located on the $\omega_T = 2\omega_t$

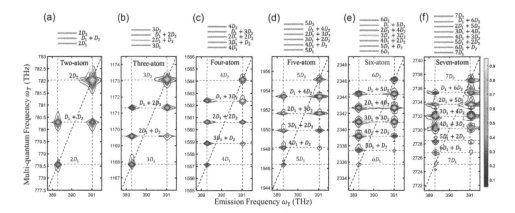

Fig. 5.13 Experimental multi-quantum 2D spectra reveal Dicke states of (a) two atoms, (b) three atoms, (c) four atoms, (d) five atoms, (e) six atoms, and (f) seven atoms. In each case, possible energy levels of multi-atom excited states are shown on the top of 2D spectrum. Adapted from Ref. 460.

diagonal line. If the double-quantum frequency is D_1+D_2, the emission frequency can be either D_1 or D_2, resulting in two off-diagonal peaks. The double-quantum coherence between the ground state and two-atom doubly excited states relies on the existence of two-atom Dicke states. The signal in 2D spectrum also requires interaction between the two atoms that breaks the symmetry to avoid a complete cancellation of possible excitation pathways [79]. However, the interatomic interaction is weak and no energy shift is observed within the frequency resolution in our experiment. Therefore, the double-quantum 2D spectrum of K is a direct consequence of the double-quantum coherence due to the collective state of two correlated, weakly interacting atoms, providing direct evidence of two-atom Dicke states.

Similarly, N_p-atom Dicke states of K can be detected by using N_p-quantum 2D spectroscopy. For the case of three K atoms, the triply excited states have four energies $3D_1$, $2D_1+D_2$, $D_1 + 2D_2$, and $3D_2$, as shown in Fig. 5.13(b). The resulting triple-quantum 2D spectrum includes six peaks with three-quantum frequencies that match three-atom triply excited states, as indicated by the horizontal dashed lines with corresponding labels. The peaks with a triple-quantum frequency of $3D_1$ or $3D_2$ are located on the $\omega_T = 3\omega_t$ diagonal line and have a emission frequency of D_1 or D_2, respectively. When the triple-quantum frequency has mixed contributions from D_1 and D_2, the emission frequency can be both D_1 and D_2, leading to four off-diagonal peaks. For the cases of more K atoms, the corresponding multi-quantum 2D spectra are shown in Fig. 5.13(c), (d), (e), and (f) for Dicke states of four, five, six, and seven atoms, respectively. The N_p-atom excited states have $N_p + 1$ energies due to possible combinations of N_p atoms each being in either D_1 or D_2 states. The spectra display a similar spectral pattern as the spectra for two- and three-atom Dicke states. The multi-quantum frequencies in the vertical direction match energies of the corresponding multi-atom excited states, while the emission frequencies are D_1 and D_2. The two peaks involving only D_1 or D_2 are on the corresponding diagonal line $\omega_T = N_p\omega_t$ in

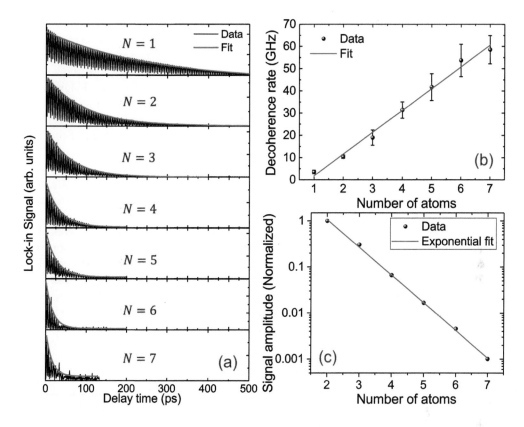

Fig. 5.14 (a) The decoherence dynamics of N_p-quantum coherences for different N_p. (b) The extracted decoherence rate plotted as a function of the number of atoms. The red line is a linear fit. (c) The signal amplitude of N_p-quantum 2D spectra plotted as a function of the number of atoms. The red line is an exponential fit. Adapted from Ref. 460.

each spectrum, while other peaks are off-diagonal. The spectral pattern in the N_p-quantum spectrum is a direct result of N_p-atom Dicke states. In each 2D spectrum, the maximum signal amplitude is normalized to one. The relative strength of spectral peaks is determined by the dipole moments of D_1 and D_2 transitions and the laser spectral shape. The multi-quantum 2D spectra in Fig. 5.13 are the observation of Dicke states consisting of a scalable and deterministic number of atoms up to seven. The Dicke states with a specific number of atoms can be deterministically selected by using proper multi-quantum 2DCS, allowing possibilities to study the dependence of many-body properties on the number of atoms.

The experimental access to multi-atom Dicke states allows the dependence study of a property on the number of atoms. As an example, the decoherence rate of N_p-quantum coherence associated with N_p-atom Dicke states and its scaling with the number of atoms can be measured. In this experiment, the N_p-quantum ($N_p > 1$) coherence evolves during the time period T. To measure the decoherence dynamics of N_p-quantum coherence, T is scanned while keeping τ and t fixed. The signal originated from N_p-quantum coherence is selected by lock-in detection at the corresponding reference frequency, similar to the detection method in multi-quantum 2D spectroscopy. The experiment is sightly different for the single-quantum coherence which is measured by scanning τ under the excitation of the single-quantum rephasing pulse sequence. The measured dynamics of N_p-quantum coherence are shown in Fig. 5.14(a) for $N_p = 1$ to 7, where the lock-in signal is plotted as a function of the scanned delay time. The signal should oscillate at a period of 578 fs due to the beating between D_1 and D_2, which is under sampled at a step size of 667 fs in these measurements. The peak amplitude of the signal is fit to an exponential decay function $Ae^{-\Gamma T}$, where A is the amplitude and Γ is the decoherence rate. The extracted decoherence rates are plotted as black dots in Fig. 5.14(b) for different number of atoms. The error bars represent the standard deviations of multiple measurements. The decoherence rate increases with N_p and can be fit to a linear function (red line), suggesting that the decoherence rate scales linearly with the number of atoms N_p. This scaling dependence can be understood by considering the decoherence processes in atomic vapor. The primary contributions to the coherence loss is the population decay and the interatomic collision. The population decay rate of N_p-atom state scales with N_p. The collision probability of N_p atoms is N_p times the probability of a single atom colliding with others under the same temperature and number density. So the decoherence rate due to interatomic collision is $N_p\Gamma^*$ with Γ^* being the collision decoherence rate of a single atom, which is determined by the mean interatomic separation and mean thermal velocity. Considering the two decoherence processes, the overall decoherence rate is $\Gamma = N_p(\gamma/2 + \Gamma^*)$. For K atom the D-lines have a lifetime of 26 ns, we can extract $\Gamma^* = 9.78$ GHz from the fitting. The dependence of decoherence rate on N_p is measured and further confirms that the observed N_p-quantum coherence is indeed a cooperative property of N_p-atom Dicke states opposed to individual atoms.

6

Frequency comb-based multidimensional coherent spectroscopy

Time-domain implementations [43, 101, 121, 124, 152, 286, 332, 369, 391, 392, 411] of optical MDCS based on femtosecond lasers have some intrinsic limitations. The spectral resolution is relatively low, limited by a combination of the longest achievable time delay and the spectrometer resolution, if one used. When the time delays are scanned by mechanically stepping translation stages, the measurement requires a long data acquisition time. Due to these limitations, most time-domain methods have only been used to probe systems that have broad resonances, which correspond to fast dephasing dynamics. They have not been able to fully resolve energy levels in atomic systems (either cold and Doppler broadened) where the relaxation dynamics are measured in nanoseconds and energy level splittings are of the order of tens to hundreds of MHz [167, 230, 238, 239, 255]. In addition, most time-domain methods struggle to distinguish very congested ro-vibrational levels in complex molecules and identify the energy transfer between these states. We note that optical MDCS has been used to measure single and double-quantum spectra of atomic systems [78, 79, 129, 214, 215, 217, 370, 391, 394, 417, 459, 460] (Rb and K vapors) with the goals of demonstrating optical MDCS techniques and investigating long-range dipole-dipole interactions in dilute atomic vapors, as discussed in the previous chapter. However, due to resolution limitations, these measurements did not fully resolve the energy levels in the system. In some cases, an argon (Ar) buffer gas was introduced into the vapor cell to artificially broaden the resonances and match them to the spectral resolution. The experiments were able to demonstrate various optical MDCS techniques and give insight about basic properties and underline physics of dipole-dipole interactions. However, the broadening led to the modification of the resonance lineshapes [354] and hence the measurements did not provide a complete picture of the interactions and dynamics in an atomic vapor. In addition, the measurements could not differentiate the interactions between the same and different isotope atoms, which is critical for understating the formation of homo and hetero-nuclear molecules.

To overcome the spectral resolution and acquisition speed limitations, a novel approach to optical MDCS was developed by utilizing frequency combs and implementing a dual-comb spectroscopy (DCS) detection technique [64, 65].

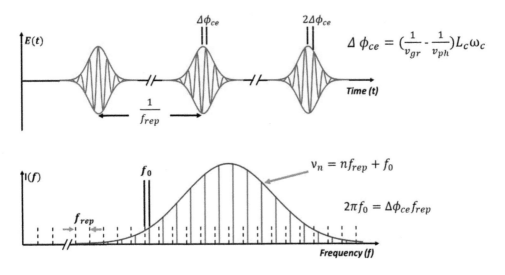

Fig. 6.1 Frequency comb in the time (top) and frequency domains (bottom).

6.1 Introduction to frequency combs and dual-comb spectroscopy

A frequency comb is typically generated using a mode-locked laser that outputs a highly periodic train of ultrashort pulses. The time-domain representation of a comb is shown in Fig. 6.1 (top figure) where the pulses are separated by $1/f_{\rm rep}$ (repetition frequency). Each pulse is a copy of the previous one but differs by a phase increment of $\Delta\phi_{ce}$ (carrier-envelope phase slip). This phase change occurs because in a laser cavity the pulse carrier and its envelope travel with phase and group velocities respectively. The phase slip is

$$\Delta\phi_{ce} = \left(\frac{1}{v_{gr}} - \frac{1}{v_{ph}}\right) L_c \omega_c \tag{6.1}$$

where v_{gr} and v_{ph} are the group and phase velocities, respectively, L_c is the laser cavity length and ω_c is the carrier frequency [69].

After stabilizing $f_{\rm rep}$ and $\Delta\phi_{ce}$ (either actively or passively) a power spectrum of a mode-locked laser can be obtained by taking a Fourier transform of the pulse train, which produces a comb of regularly spaced frequencies, as shown in Fig. 6.1 (bottom figure). The frequency of each tooth is $f_n = nf_{\rm rep} + f_0$, where n is an integer, f_0 would be the frequency of the very first tooth ($n = 0$), which is related to the carrier-envelope phase slip by $2\pi f_0 = \Delta\phi_{ce}f_{\rm rep}$.

Since the development of the frequency comb technology [70, 75, 456] a method know as dual-comb spectroscopy has emerged as a very powerful optical method [64, 65, 180]. In DCS, one frequency comb (typically a mode-locked laser) is used to excite the sample and the response is sampled in time with another comb (local oscillator LO) that has a slightly different repetition rate. The resulting interferogram is captured by a single photodetector, as shown in Fig. 6.2 (a). In the frequency domain, the DCS arrangement produces a radio frequency (RF) comb spectrum that

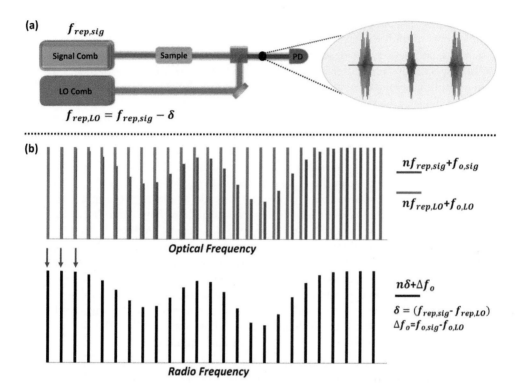

Fig. 6.2 (a) Dual-comb spectroscopy in the time domain. The repetition rates for the signal comb and the local oscillator comb are slightly different by δ. The signal comb probes the sample and the local oscillator (LO) comb reads out the response. (b) Dual-comb spectroscopy in the frequency domain. The radio frequency comb spectrum directly maps to the optical absorption spectrum. $f_{0,\text{sig}}$ and $f_{0,LO}$ correspond to the offset frequencies of the signal and LO combs.

results from these two optical combs beating against each other on the photodetector [365]. The RF spectrum directly maps to the optical absorption spectrum of the sample, as shown in Fig. 6.2 (b). DCS is similar to Fourier transform infrared (FTIR) spectroscopy but contains no moving elements and hence enables the measurement of a broad absorption spectrum rapidly and with high spectral resolution. In addition dual-comb spectroscopy is becoming compact with the development of micro-resonators [191, 246, 347, 377, 378, 408] and diode combs [84, 99] for applications outside the laboratory (LIDAR, chemical detection, etc.) [39, 64, 135].

Dual-comb methods have been extended to include nonlinear effects to implement spectroscopy and imaging. Dual-comb spectroscopy was demonstrated that used the Raman-induced Kerr-effect [163]. The Raman effect was also employed to perform dual-comb spectro-imaging [164, 269]. Two-photon and stepwise excitation spectroscopy at the doppler limit was demonstrated [157] followed by the demonstration of doppler free spectroscopy [262]. Time resolved dual-comb spectroscopy was demonstrated using a separate pump pulse [12]. While these results demonstrate that dual-comb spectroscopy can be successfully employed in nonlinear spectroscopy, they lack the full power of multidimensional techniques.

In the next section, we describe the combination of DCS and four-wave-mixing (FWM) (specifically photon echo) spectroscopy to set the stage for frequency comb based multidimensional coherent spectroscopy. We will also show how linear and FWM signals can be separated without using complex static phase cycling schemes. In the following section we introduce the concept of comb-based MDCS [232, 233] which enables the measurement of a multidimensional coherent spectrum rapidly and with high resolution. These improvements make optical MDCS more relevant for systems with slow dephasing rates, for example, atomic and molecular systems. To show the improved resolution and acquisition speed, the comb-based technique is applied to Doppler broadened rubidium atoms to measure a single-quantum two-dimensional coherent spectrum. In the subsequent section, we also show how comb-based MDCS can be used to measure a double-quantum two-dimensional spectrum that probes extremely weak many-body interactions (dipole-dipole interactions) in Rb atomic vapor and gives insight into the effects of thermal motion on dipole-dipole interactions [234]. In the last section of the chapter, we introduce a novel approach to comb-based multidimensional coherent spectroscopy, tri-comb spectroscopy (TCS) [241] that contains no mechanical moving parts and enables the measurement of a comb-resolution two-dimensional spectrum in under half a second. This approach has the potential to become a field deployable device for chemical sensing applications.

6.2 Frequency comb-based four-wave-mixing spectroscopy

As described in the earlier chapters, multidimensional coherent spectroscopy (MDCS) is based on a four-wave-mixing (FWM) process where the coherent interaction of two or three waves with a nonlinear medium produces a new signal wave. Before we introduce the concept of frequency comb-based MDCS, we show how frequency combs can be used to generate, isolate, and detect FWM signals rapidly and with high resolution. We note that the experiment described below uses collinear excitation frequency combs and the separation of FWM and linear signals is performed in the RF domain. However, frequency combs can also be used in the noncollinear excitation geometry [168] where the FWM signal is detected in the phase matched direction (see Section 4.1).

A schematic diagram of frequency comb-based four-wave-mixing spectroscopy is shown in Fig. 6.3, where two home-built Kerr-lens mode-locked Ti:Sapphire lasers centered at 800 nm were used. The repetition frequencies of the combs (f_{rep_1}=93.543954 MHz for Comb 1 and $f_{\text{rep}_{LO}}$=93.543954 MHz+200.2 Hz for the LO comb), were phase locked to a direct-digital synthesizer using a feedback loop. The comb offset frequencies were not stabilized. The output of Comb 1 was split into two parts using a half wave plate (HWP) and a polarizing beam splitter (PBS 1). The offset frequency of the first part was shifted by an acousto-optical modulator (AOM) and recombined with the second part on PBS 2. Optical paths for the two arms were adjusted to overlap the two pulse trains in time. Before interacting with the sample, the beams were projected on the same linear polarization state using a polarizer. The sample used for this proof-of-concept experiment was 10 layers of 10 nm GaAs quantum wells (QW) separated

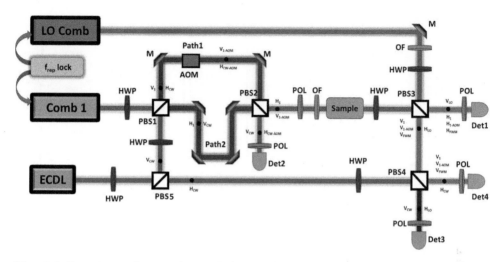

Fig. 6.3 Experimental setup for comb-based FWM spectroscopy: ECDL—External cavity diode laser, LO—Local oscillator, FWM—Four-wave-mixing, HWP—Half wave plate, PBS—Polarizing beam splitter, AOM—Acousto-optical modulator, M—Mirror, POL—Polarizer, OF—Optical filter, BPF—RF Band pass filter, Det—detector. H—Horizontal and V—Vertical indicate the linear polarization states of the beams. Subscripts 1 and 2 indicate the path of the beams. Adapted from Ref. 231

by 10 nm thick $Al_{0.3}Ga_{0.7}As$ barriers (see Fig. 7.2). The sample was cooled down to 7 K. The laser beams were optically filtered to excite only the heavy hole (HH) excitonic resonance, attenuated and focused to a 30 μm spot on the sample. Average powers for beams traveling through path 1 and 2 were 300 and 600 μW respectively. The FWM signals emitted from the sample, along with the incident beams, were combined with the LO comb (optically filtered to match filtered Comb 1) on PBS 3. Half wave plates were adjusted such that the most of the light from each beam was sent to a photodetector (Det 1) to obtain a RF comb spectrum whereas only small fractions were sent to PBS 4 for phase correction (described below).

The output of Det 1 contains both linear and FWM RF signals, which are spectrally separated in the RF domain. Figure 6.4 shows the generation and separation of the FWM signal in the frequency domain. The blue lines correspond to the original (signal) Comb 1 lines ($f_{n,1} = n f_{rep,1} + f_{0,1}$), the black lines correspond to AOM shifted comb lines ($f_{k,2} = k f_{rep,1} + f_{0,1} + f_{AOM}$), and the red and magenta lines correspond to the generated FWM signals ($-f_{n,1} + f_{m,2} + f_{k,2}$ and $-f_{n,2} + f_{m,1} + f_{k,1}$). The green lines correspond to the LO comb teeth ($f_{s,LO} = s f_{rep,LO} + f_{0,LO}$). In these expressions $f_{rep,1}$, $f_{0,1}$, $f_{rep,LO}$, and $f_{0,LO}$ correspond to repetition and offset frequencies of the Comb 1 and LO combs, f_{AOM} is the AOM drive frequency, and n, k, m, and s are integer numbers. The detector (Det 1) in Fig. 6.3 measures the superposition of all of these lines, which due to interference between them produces multi-heterodyne beat signals in the RF domain where the linear and FWM signals are spectrally separated (Fig. 6.4, bottom plot). Note that each FWM comb tooth represents a coherent sum of waves with different integers (m, n, k); $(m-1, n-1, k)$; $(m, n+1, k-1)$; $(m-2, n-2, k)$; etc.

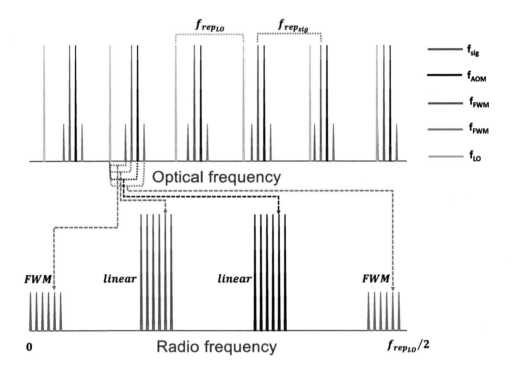

Fig. 6.4 Schematic diagram showing how the linear and FWM comb lines are separated after down converting them from the optical to RF domain. Adapted from Ref. 231.

For zero delay between pulses, both red and magenta waves interfere constructively, whereas only one color survives for a finite delay. In the experiment, one of the FWM signals was isolated using a RF bandpass filter before digitizing.

It is important to note that to obtain a comb structure, the phase fluctuations (described below) of the FWM RF comb teeth have to be monitored and corrected. The phase of a FWM RF comb tooth (e.g., the red comb in Fig. 6.4) is

$$\phi_{FWM}^{RF} = (\phi_1 - \phi_2) + (\phi_{LO} - \phi_2) \tag{6.2}$$

where ϕ_2, ϕ_1 and ϕ_{LO} are the phases of the original, AOM frequency shifted Comb 1 and LO comb teeth respectively. Any uncorrelated fluctuations of these phases will broaden the FWM comb teeth in the RF spectrum. In this co-linear experimental arrangement, the broadening could be due to the fluctuating offset frequencies, residual relative repetition frequency fluctuations, and optical path fluctuations between path 1 and path 2. These fluctuations will affect the $(\phi_{LO}-\phi_2)$ and $(\phi_1-\phi_2)$ terms in Eq. (6.2) respectively.

In the experiment, these fluctuations were measured using a CW external cavity diode laser (shown in Fig. 6.4) that had a wavelength tuned near the HH resonance. The CW laser beam was split into two parts using a HWP and PBS 5. One part of the beam was used to monitor optical path fluctuations between path 1 and path 2.

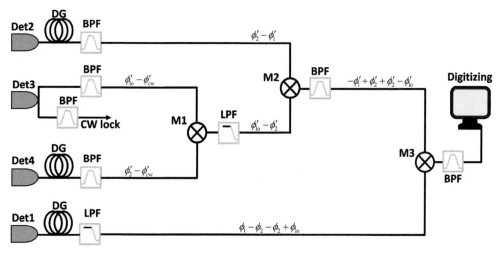

Fig. 6.5 Schematic diagram of the phase correction scheme: ϕ_2', ϕ_1' and ϕ_{LO}' correspond to phases of comb 1, AOM shifted comb 1 and LO comb teeth nearest to CW laser frequency. ϕ_{CW}' is the phase of CW laser. Det—Detector, LPF—Low pass filter, BPF—Band pass filter, DG—Delay generator, M—Mixer. Adapted from Ref. 231.

It was split again using a HWP and PBS 1. These beams propagated through path 1 and 2, combined on PBS 2, and a beat signal was measured on a photodetector (Det 2). The second part of the CW laser beam was used to determine the relative offset frequencies, including fluctuations, of Comb 1 and LO comb. It was split on PBS 4, and interfered with the LO comb and Comb 1 on Det 3 and Det 4 photodetectors, respectively. The beat signal on Det 2 that measures the relative path fluctuations between path 1 and path 2 and is also equivalent to measuring the relative phase $(\phi_2' - \phi_1')$, where ϕ_2' and ϕ_1' correspond to the phases of the original and the AOM shifted Comb 1 teeth nearest to the CW laser frequency. The beat signals measured by Det 3 and Det 4 reflect the optical frequency fluctuations (due to offset and residual repetition frequency fluctuations) of Comb 1 and LO comb teeth nearest to the CW laser frequency: $(\phi_2' - \phi_{CW}')$, $(\phi_1' - \phi_{CW}')$, and $(\phi_{LO}' - \phi_{CW}')$, where (ϕ_{CW}') is the phase of the CW laser and (ϕ_{LO}') is the phase of the LO comb tooth nearest to the CW laser frequency.

These beat signals were used to generate a correction signal in real time. Figure 6.5 shows a diagram of the phase correction scheme. The beat signal from Det 3 was split into two parts. One part of the signal was bandpass filtered and used for stabilizing the CW laser frequency using a slow loop filter. The second part of the Det 3 signal was mixed with the beat signal between the CW laser and a tooth from Comb 1 (traveling through path 2) from Det 4 that was isolated using a RF band pass filter. The output of the mixer 1 was filtered in the frequency domain to isolate the signal that had the common CW laser noise canceled and mixed with the signal from Det 2. The output of mixer 2 was band pass filtered again to isolate the signal corresponding to a tooth of the FWM RF comb. Finally, this reference was mixed with the FWM RF comb, filtered, and the phase noise free signal was digitized.

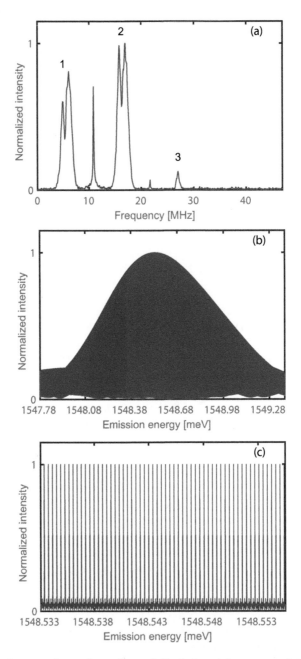

Fig. 6.6 (a) Fourier transform of one "burst." Peak 1 and 2 correspond to linear signals, 3 corresponds to the FWM signal. (b) Fourier transform of filtered FWM time-domain signal (25 "bursts"). (c) Portion of figure (b) on an expanded scale. Adapted from Ref. 231.

Figure 6.6 presents the experimental results. The Fourier transform of one "burst" (corresponding to the temporal overlap of the linear and FWM signals with LO pulses) is shown in Fig. 6.6 (a). Peaks 1 and 2 correspond to the linear dual-comb signals, which are the result of the heterodyne detection of the AOM shifted Comb 1 and the original comb 1 with the LO comb, respectively. The dips are due to the linear absorption of the GaAs quantum well. Peak 3 corresponds to the heterodyne beat between the FWM signal generated by the sample and the LO comb. The figure clearly shows the separation of the linear and FWM contributions. In this figure, only one FWM signal is shown. The absence of the second FWM signal is because the pulses traveling through path 2 were slightly delayed in time with respect to the pulses traveling through path 1 (Fig. 6.3). In this case only one FWM signal is generated due to causality. The spike around 11 MHz corresponds to a beat signal between the original and the AOM frequency shifted combs.

After filtering the FWM RF signal, the Fourier transform of 25 bursts is shown in Fig. 6.6 (b) (frequency remapped to the optical domain) and a portion is shown in Fig. 6.6 (c) on an expanded scale. The measured spectrum clearly shows the comb structure. The measured linewidth of the FWM signal, corresponding to inhomogeneous broadening of QW, is in good agreement with the literature value.

6.3 Frequency comb-based single-quantum 2D spectroscopy

In the previous section we showed the generation of a FWM signal (particularly a photon echo) using frequency combs in a collinear excitation geometry. We also showed how linear and FWM signals can be separated in the RF domain and how to correct the phase fluctuations to obtain a comb structure and determine the inhomogeneous linewidth. However, the signal is one dimensional and it contains limited spectroscopic information. For example the signal cannot provide homogeneous linewidth of the resonance since it is only for a single, fixed delay between the excitation pulses. Also, in general, one-dimensional spectra cannot provide information such as coupling between the excited states (e.g., distinguish a V-type three-level system (with ω_1 and ω_2 resonance frequencies) from the sample that contains two independent two-level systems that have the same resonant frequencies). To obtain these measurements, single-quantum multidimensional coherent spectra are required. To obtain a single-quantum 2D spectrum we need to measure the FWM signal (described in the previous section) as a function of the time delay between Comb 1 and the AOM shifted Comb 1 and then take a Fourier transform of the recorded signal with respect to the delay time. In this section we will describe in detail how to perform this measurement.

Following the approach described in Ref. 232, 235, a schematic diagram of frequency comb-based single-quantum multidimensional coherent spectroscopy (M-DCS^2) is shown in Fig. 6.7 (a). Two Kerr-lens mode-locked Ti:Sapphire lasers are used and their repetition frequencies are locked with a small offset. The comb offset frequencies are not stabilized but the phase fluctuations due to the relative repetition frequency between the two combs, offset frequencies, and path length fluctuations are measured and corrected [231]. The output of the signal comb is split into two parts. One part is frequency shifted using an AOM and then combined with the other part, whose delay is controlled using a retro-reflector mounted on a mechanical stage. The combined beams

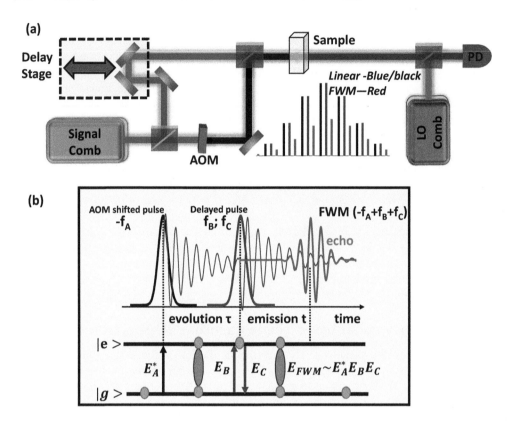

Fig. 6.7 (a) Experimental setup for M-DCS². Two combs with different offset frequencies are generated using a signal comb and an accousto-optical modulator. These combs then interact with the sample and generate a FWM signal at a different time delay between combs. The FWM signal is sampled using a LO comb. (b) Photon echo excitation scheme where $|g\rangle$ is the ground state and $|e\rangle$ is the excited state.

are then focused to a 5 μm spot in a 0.5 mm thick vapor cell. The cell contained ^{87}Rb and ^{85}Rb atoms at 100 °C. The beams are optically filtered to excite only the D_1 lines of both Rb isotopes. For this experiment a photon echo excitation scheme is used to generate a four-wave-mixing (FWM) signal, as shown in Fig. 6.7 (b). In this scheme the electric field of the signal is proportional to the complex conjugate of the first pulse (E_A^*), which excites the coherence between the ground state and the excited state. The second pulse (E_B) then converts this coherence into the population of the excited (or ground) state and then it (E_C) converts this population into the third-order coherence that radiates the FWM signal (photon echo shown in red). The FWM signal is then interfered with the LO comb, which has a slightly different repetition frequency, on a photodetector and isolated from the linear signals in the RF domain. The separation of linear and FWM signals is described in the previous section. To generate the second axis for the two-dimensional spectrum, the delay between the excitation pulses is varied from 0 to 3.3 ns with 10 ps steps using the stage. A two-dimensional coherent

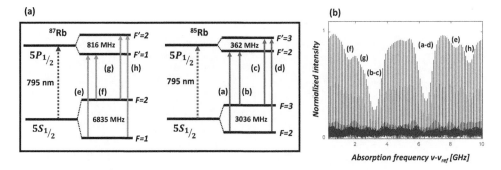

Fig. 6.8 (a) Energy level diagram of ^{87}Rb and ^{85}Rb atoms D_1 hyperfine lines. (b) Measured transmission spectrum for ^{87}Rb and ^{85}Rb. ν_{ref}=377.103258084 THz.

spectrum is then generated by calculating Fourier transforms of the FWM signal with respect to the emission and evolution times.

The results are shown in Figs. 6.8 and 6.9. Figure 6.8(a) and (b) show the energy level diagram and the measured linear absorption spectrum of the D_1 lines of ^{87}Rb and ^{85}Rb atoms, respectively. The absorption axis is plotted with respect to an arbitrary reference ν_{ref}=377.103258084 THz. The natural linewidths of the hyperfine lines (a-h) for Rb atoms are about 6 MHz, however, the absorption profile at 100 °C is Doppler broadened up to $\sim$ 600 MHz and hence the hyperfine lines are overlapped. In Fig. 6.9 (a) and (b) two-dimensional energy spectra measured by co-linearly (HHH-H) and cross-linearly (HVV-H) polarized excitation pulses are shown. The negative values on the absorption/evolution axis reflect the negative phase evolution, due to the phase conjugation of the first pulses contribution to the signal field, during the evolution period in the photon echo excitation sequence, as illustrated in Fig. 6.7 (b).

The diagonal peaks (along the (0,0) and (10,-10) line) correspond to absorption (evolution) at the same resonance frequency (a-h) as the emission. The elongation in the diagonal direction is due to Doppler broadening. Along the cross-diagonal direction the inhomogeneity is removed and the line shapes reflect the homogeneous linewidth. Although the resolution in this direction is limited by the scan range achievable with the mechanical stage, the spectra resolve the hyperfine structure and all possible couplings between the resonances appearing at unique locations [232]. It is also clear that the two-dimensional energy spectra do not show the coupling peaks between the two isotopes ^{87}Rb and ^{85}Rb, indicating that they behave as two independent atoms (the details are described at the end of the chapter).

It is also noticeable that some of the diagonal peaks are suppressed for HVVH case compared to HHHH case. In Fig. 6.10 (a) and (b) the diagonal slices of Fig. 6.9 (a) and (b) are plotted. The slices show that peaks at (g,c,a) frequencies are suppressed and the peak at (e) frequency is absent for HVVH case. This behavior can be explained by calculating all possible double-sided Feynman diagram for each state (see section 2.3.1), including the magnetic sub-levels, in HHHH and HVVH cases respectively. The simulation shows that for the HVVH case the FWM signals for F to $F' = F$ transitions have the opposite sign compared to the F to $F' = F \pm 1$ transitions, which causes the

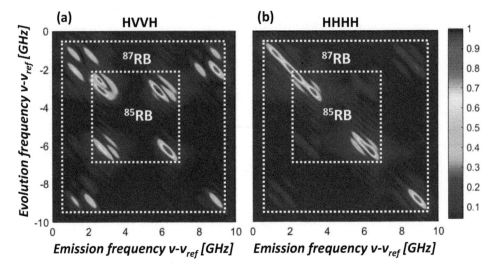

Fig. 6.9 Measured two-dimensional spectra generated by (a) cross-linearly (HVVH) and (b) co-linearly (HHHH) polarized excitations pulses. H—Horizontal, V—Vertical. ν_{ref}=377.103258084 THz. The color scale shows the normalized signal magnitude.

closely spaced neighboring peaks to be partially canceled. The simulation also shows that the FWM signal is zero for $F = 1$ to $F' = 1$ transition for ^{87}Rb. For the HHHH case all of the Feynman diagrams have the same sign and hence the peaks do not cancel each other. The results of the simulation, shown in Fig. 6.10(c) and (d), are in good agreement with the experimental results. The slight mismatch between theory and experiment is attributed to laser pulse propagation effects.

The two-dimensional plots in Fig. 6.9 show additional interesting behavior. The strengths of the off-diagonal peaks are not the geometric mean of its corresponding diagonal peaks strengths that is expected for a three-level system. Furthermore, some of the peaks are even weaker than their corresponding diagonal peaks. For instance the peak around (3,-6.5) GHz on Fig. 6.9 (b) is much weaker compared to peaks at (3,-3) GHz and (6.5,-6.5) GHz that correspond to the $F = 3$ to $F' = 3$ and $F = 2$ to $F' = 3$ transitions in ^{87}Rb, respectively. This difference can be explained with the fact that the $F = 2$ state has five magnetic sub-levels whereas $F = F' = 3$ has seven sub-levels. In the linear polarization basis (π) all the sub-levels of the $F = 2$ and $F = 3$ states contribute for the diagonal peaks, except $m_F = 0$ for the $F = 3$ to $F' = 3$ transitions whose Clebsh-Gordan coefficient are zero. However, only $m_F = -2, -1, 1, 2$ sub-levels of the $F = 2$ and $F = 3$ hyperfine states contribute for the off-diagonal peak. The theoretical simulation shows good agreement with the experimental results for this and other off-diagonal peaks as well.

We also would like to note that the two-dimensional spectra shown in Fig. 6.9 (a) and (b) were generated in under four minutes. Similar resolution either is not achievable or requires several hours or even days of acquisition time with currently available methods.

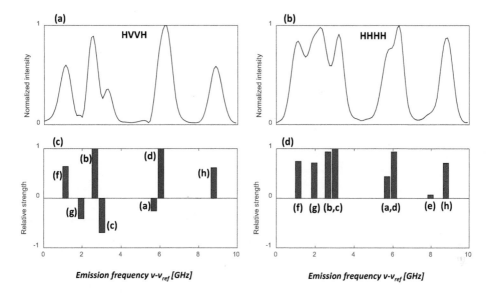

Fig. 6.10 Diagonal slices of figure 6.9 (a) and (b) respectively. (c) and (d) theoretical simulations showing the strength of the FWM signal at each hyperfine resonance (a-h) for HVVH and HHHH cases.

6.4 Frequency comb-based double-quantum 2D spectroscopy

As discussed in the previous section, frequency comb-based single-quantum two-dimensional coherent spectroscopy is a very powerful method. However, it cannot provide complete information about the sample. Single-quantum two-dimensional coherent spectroscopy does not provide a background-free signature of many-body interactions (dipole-dipole interactions) and collective effects in an atomic vapor, but rather these appear as small modifications to resonance parameters. These interactions are extremely weak and single-quantum two-dimensional spectroscopy is not sensitive enough to isolate the FWM signal that is due to interactions from the signal that is produced by a small fraction of the atoms. This can clearly be seen from the Fig. 6.9 (a) and (b) above. The FWM signal due to the couplings of two different atoms via dipole-dipole interactions is nonzero, however, its strength is very weak and it is not noticeable in the graph. To probe these processes the measurement of a double-quantum two-dimensional spectrum is required.

Frequency comb-based double-quantum two-dimensional coherent spectroscopy [234, 235] is an extension of single-quantum two-dimensional coherent spectroscopy. Experimentally its spectrum can be generated with the same experimental apparatus shown above (Fig. 6.7 (a)) but swapping the time order of the excitation pulses such that the AOM shifted pulse arrives last.

The generation of a double-quantum FWM signal in the time domain is shown in Fig. 6.11 (a). The first pulse excites the coherence between the ground and excited states and then converts it into the coherence between the ground and doubly excited states, otherwise known as a "double-quantum coherence." The double-quantum coher-

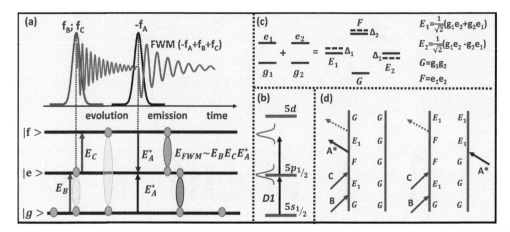

Fig. 6.11 The generation of a FWM signal in the double-quantum excitation scheme. $|g\rangle$ - ground state, $|e\rangle$ - excited state, $|f\rangle$ – doubly excited state. (b) Energy level diagram of Rb showing no energy level at 2 X D1 frequency. (c) Energy level diagram of two combined atoms without interaction (solid lines) and with interaction (dashed lines). (d) Double-sided Feynman diagrams of the double-quantum FWM signals.

ence evolves with the frequency that corresponds to the energy difference between the ground and doubly excited states. The second pulse then converts the double-quantum coherence either back to the single coherence between the ground and excited states or to the coherence between the excited and doubly excited states. This coherence emits the FWM signal that is detected as a function of the emission time and the evolution time between the excitation pulses. However, in the experiment, the laser spectrum was filtered such that it could not excite any doubly excited states in individual Rb atoms, such as the 5d states, as shown in Fig. 6.11 (b). In this case the generation of a double-quantum FWM signal can be explained by introducing a combined atom picture shown in Fig. 6.11 (c), where the doubly excited state indicates that both atoms are in the excited state. But it is critical to note that the combined picture itself, as shown in Fig. 6.11 (c), cannot produce any FWM signal as the contributions shown in Fig. 6.11 (d) (described using double-sided Feynman diagrams) have opposite signs and equal strengths. However, in the presence of interactions (even very weak interactions such as dipole-dipole interactions) singly and doubly excited states experience slight energy shifts (dashed lines shown in Fig. 6.11 (c)) or changes in their linewidth. These changes lead to partial cancelation in the contributions described above and hence the generation of a FWM signal.

Figure 6.12 (a) and (b) show double-quantum two-dimensional spectra measured by the cross-linearly (HVV-H) and co-linearly (HHH-H) polarized excitation pulses, respectively. The diagonal peaks (along the (0,0) and (10,20) GHz line) correspond to coupling between the same hyperfine energy levels (a–h lines shown in Fig. 6.8 (a)) of two atoms whereas the off-diagonal peaks show coupling between different hyperfine energy levels of two atoms of the same and different isotopes [234]. It is important to emphasize that the signal in this experiment is only due to interactions between

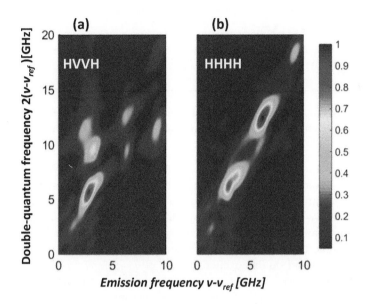

Fig. 6.12 (a) and (b) Double-quantum spectrum generated by cross-linearly and co-linearly polarized excitation pulses. ν_{ref}=377.103258084 THz. The color scale shows the normalized signal magnitude.

atoms and contains no information about the signal that is due to an individual atom. Hence frequency comb-based double-quantum MDCS excels in isolating and identifying many-body interactions with extremely high sensitivity.

In Fig. 6.12 (a) and (b) it is clear that the peaks are diagonally elongated, suggesting that the emission and double-quantum frequencies are correlated. The elongation had never been observed for Doppler broadened atomic systems (due to resolution limitations of other MDCS methods). The correlation indicates that the FWM signal is predominantly from those two atoms that have near zero relative velocity.

This point can be demonstrated by a simulation based on optical Bloch-equations for two coupled three-level V-type systems. The energy level diagram of two combined V-type systems (without interaction) is shown in Fig. 6.13, which can also be described as a superposition of the states created by coupled two-level systems shown in Fig. 6.13 (1,2,3,4). In our simulation we used infinitely short-pulses (delta-function pulses) $E(t) \approx E_0 e^{i\omega t} \delta(t)$ and all of the excitation pulses were co-polarized. At first we treated the systems to be homogeneously broadened. Under these conditions the third-order polarization (one of the pathways) for figure Fig. 6.13 (1) created by the sequence of MDCS pulses is:

$$P(t,\tau) = (-i/\hbar)^3 E_0^3 \mu_{ge}^4 exp[-i\omega_t t - i\omega_\tau \tau - \gamma_t t - \gamma_\tau \tau] \qquad (6.3)$$

where E_0 is the magnitude of the excitation pulses (assumed to be the same for all three pulses), μ_{ge} is the transition dipole strength, $\omega_\tau = 2\pi\nu_\tau$ is a double-quantum angular frequency, $\omega_\tau = \omega_t + \omega_A$ where ω_t is the emission angular frequency and ω_A is the resonant frequency of the phase conjugated pulse. γ_t and γ_τ are the dephasing

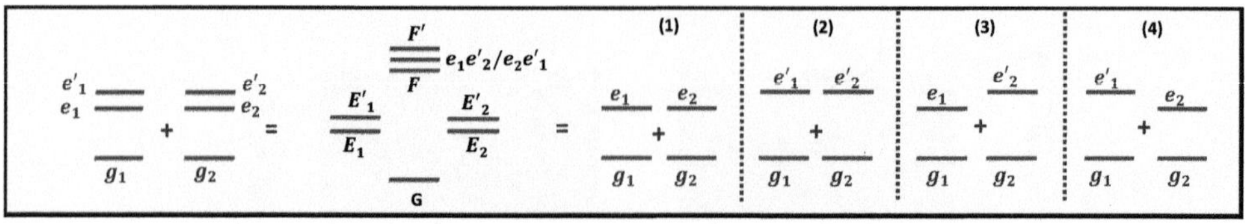

Fig. 6.13 Energy level diagram of two combined V-type systems which is also represented as a superposition of states created by two-level systems (1), (2), (3), and (4). Adapted from Ref. 237.

rates of the single and double-quantum coherences, respectively. We note that in the model γ_t and γ_τ describe overall dephasing rates and we do not distinguish dephasing rates caused by spontaneous decay, collision broadening, etc. In our model $\gamma_\tau = 2\gamma_t$ but one can model many-body interactions by including additional dephasing rates that distinguishes single and double-quantum coherences. In our simulation we modeled the interactions between the systems by including the energy shifts Δ_1 and Δ_2 (described below) for the single and double excited states.

To model a real system, inhomogeneous broadening was incorporated into the simulation by integrating the polarization over a generalized two-dimensional Gaussian function [71, 237]:

$$f(x,y) = \frac{1}{2\pi\sigma_x\sigma_y\sqrt{1-r_c^2}} e^{-\frac{(\frac{x-\nu_x}{\sigma_x})^2 - 2r_c(\frac{x-\nu_x}{\sigma_x})(\frac{y-\nu_y}{\sigma_y}) + (\frac{y-\nu_y}{\sigma_y})^2}{2(1-r_c^2)}} \tag{6.4}$$

here ν_x, ν_y, σ_x, σ_y correspond to the centers and widths of two interacting inhomogeneously broadened resonances and r_c is a correlation parameter. $r_c = 1$, $r_c = 0$, and $r_c = -1$ implies that the resonances are perfectly correlated, uncorrelated, and anti-correlated, respectively.

The integration yields

$$P(t,\tau) = (-i/\hbar)^3 E_0^3 \mu_{ge}^4 \, exp[-i\omega_t t - i\omega_\tau\tau - \gamma_t t - 2\gamma_t\tau - \frac{1}{2}(\tau^2(\sigma_A^2 + \sigma_t^2 + 2r_c\sigma_A\sigma_t)$$
$$+ 2t\tau(r_c\sigma_A\sigma_t + \sigma_t^2) + t^2\sigma_t^2)]. \tag{6.5}$$

If we assume that $\sigma_A = \sigma_t \equiv \sigma$ and include the energy shifts due to interactions $\omega_t = \omega_{ge} \pm \Delta_1$ and $\omega_\tau = 2\omega_{ge} + \Delta_2$ (where ω_{ge} is the transition frequency between the ground and the excited states), then all the polarization terms that are contributing in the generation of the FWM signal for the system shown in Fig. 6.13 ((1)) are

$$P_I(t,\tau) = (-i/\hbar)^3 E_0^3 \mu_{ge}^4 \, exp[-i(\omega_{ge} + \Delta_1)t - i(2\omega_{ge} + \Delta_2)\tau - \gamma_t t - 2\gamma_t\tau$$
$$-\frac{1}{2}(2\tau^2\sigma^2(1+r_c) + 2t\tau\sigma^2(r_c+1) + t^2\sigma^2)]$$

$$P_{II}(t,\tau) = -(-i/\hbar)^3 E_0^3 \mu_{ge}^4 \, exp[-i(\omega_{ge} + \Delta_2 - \Delta_1)t - i(2\omega_{ge} + \Delta_2)\tau - \gamma_t t - 2\gamma_t\tau$$
$$-\frac{1}{2}(2\tau^2\sigma^2(1+r_c) + 2t\tau\sigma^2(r_c+1) + t^2\sigma^2)]. \tag{6.6}$$

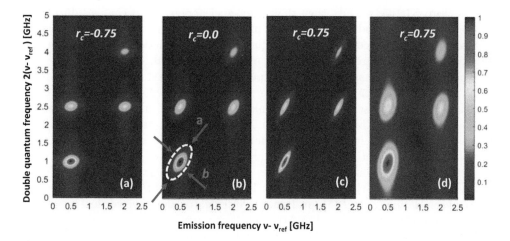

Fig. 6.14 Theoretical simulation of double-quantum spectra. (a) r_c =-0.75, (b) r_c =0.0, (c) r_c =0.75, (d) r_c =0.75 and increased decay rate. The color scale shows the normalized signal magnitude. ν_{ref}—Arbitrary reference. Adapted from Ref. 237.

A two-dimensional spectrum is then generated by summing all the polarization terms and taking a two-dimensional Fourier transform with respect to t and τ. In our calculation we used σ=600 MHz (corresponding to Doppler Broadened atomic samples) and γ_t=6 MHz.

The simulated double-quantum 2D spectra are shown in Fig. 6.14. $r_c = 0.0$ corresponds to uncorrelated systems, $r_c = -0.75$ and $r_c = 0.75$ correspond to partially anti-correlated and correlated systems, respectively. The peaks in each figure are diagonally elongated and the effects of correlation are obvious. To give quantitative information, we used the ellipticity as a metric to describe the peak elongation. The ellipticity is

$$E = \frac{a^2 - b^2}{a^2 + b^2}, \tag{6.7}$$

where a and b are the sizes of the ellipse along the major and minor axes, as shown in Fig. 6.14.

The measurements showed that for $r_c = 0.0$ the ellipticity is 0.5 and it approaches to 0 and 1 for $r_c = -1.0$ and $r_c = 1.0$, respectively. It is important to note that the correlation parameter gives insight into the many-body interactions. For example, for Doppler broadened atomic systems near-perfect correlation implies that the generated FWM is due to the coupling of resonances between two atoms that have near zero relative velocity. On the other hand $r_c = 0.0$ and $r_c = -1.0$ correspond to coupling of the resonances of the atoms that have any relative velocity and opposite velocity, respectively. The ellipticity of the experimental results in Fig. 6.12 (a) and (b) is about $E = 0.85$, which corresponds to $r_c = 0.75$. This indicates that the FWM signal is due to atoms that have near zero relative velocity. A plausible explanation of the high correlation could be the fact that the dipole-dipole interaction is proportional to $(1/r^3)$, where r is the inter-nuclear separation between the atoms. If two atoms have nonzero relative velocity then their inter-nuclear separation changes during the time between second and third excitation pulses (that is scanned over 1 ns). For high relative velocities this could causes the dipole-dipole interaction to degrade rapidly $(1/r^3)$ and hence to decrease the strength of the FWM signal.

We also note there has been MDCS experiments performed on Doppler broadened atomic samples that showed that the peaks were not elongated along the diagonal line (the elongation was obscured and the peaks were elongated more along the vertical line) [80, 129]. But in the experiments an argon (Ar) buffer gas was introduced into the gas cell to artificially broaden (collisional broadening) the resonances to match the spectrometer resolution. To model that case, we increased the decay rates (by a factor of 20 which is similar to the values of their experimental parameters) in our simulation. The results that are plotted in Fig. 6.14 (d) show that even with the high degree of correlation $(r_c = 0.75)$, peaks now are elongated along the vertical line, which is similar to the results observed in [80, 129].

Lastly, we want to add that this model can be extended to semiconductor materials as well. For quantum wells (and quantum dots) r_c is expected to be close to zero (or partially correlated). This is because in semiconductor materials a double-quantum FWM signal is due to the coupling of the excitons that are located in nearby quantum wells and the thickness of wells are most likely random. In this case the peaks are expected to be elongated along the diagonal (ellipticity $= 0.5$) but the experiments showed that the peaks are tilted toward the vertical axis [258, 286]. This lack of correlation can be explained with the fact that unlike atomic systems, the excitons experience additional dephasing due to exciton-exciton and exciton- free carrier scattering (which is a strong function of the temperature) reported in Ref. 399. This scattering causes the 2D peaks to be tilted similarly to the results that we showed for atomic systems in Fig. 6.14 (d).

In summary, frequency comb-based double-quantum multidimensional spectra identified the collective hyperfine resonances in atomic vapor (containing two different isotopes) that were induced by dipole-dipole interactions. In addition, the measured spectra gave insight of the effects of thermal motion on dipole-dipole interactions. This information is not accessible with other MDCS methods. The combination of single and double-quantum spectra makes frequency comb-based multidimensional coherent spectroscopy an extremely powerful tool for obtaining complete spectroscopic information about atomic and molecular systems.

6.5 Tri-comb spectroscopy

Frequency comb-based single and double-quantum spectroscopy (described above) enables the measurement of multidimensional coherent spectra rapidly and with high-resolution. However, the experimental setup contains a mechanical stage, which still limits both the resolution and acquisition speed. To fully leverage the advantages provided by frequency combs, a novel approach to multidimensional coherent spectroscopy that utilizes three frequency combs has been developed. This novel approach, which was named tri-comb spectroscopy (TCS) [236, 241], contains no mechanical moving elements and can measure comb-resolution multidimensional coherent spectra in under half a second.

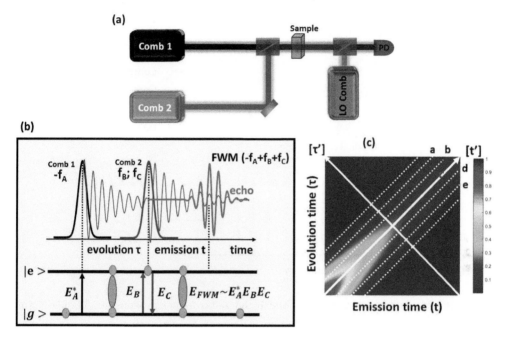

Fig. 6.15 (a) Schematic diagram of tri-comb spectroscopy. Comb 1 and Comb 2 are used for the generation of a photon echo FWM signal which is then sampled using a LO comb. (b) photon echo excitation scheme (please see the text for details). (c) Cartoon showing the magnitude of a FWM signal as a function of emission and evolution times. The color scale shows the normalized signal magnitude.

The experimental setup for tri-comb spectroscopy is pictorially shown in Fig. 6.15 (a). The setup uses three frequency combs (Comb 1, Comb 2, and LO Comb) with slightly different repetition rates and locks the phases of the repetition frequencies to a four-channel direct-digital synthesizer (DDS). Path length and offset frequency fluctuations for each comb are measured and corrected. Pulses from Comb 1 and Comb 2 are used for the generation of a photon echo, as shown in Fig. 6.15(b). The emitted FWM signal is then sampled and spectrally isolated in the RF domain after interfering with the LO Comb on a photodetector [241]. The output of the detector is digitized using a fast data acquisition board. Figure 6.15 (c) is a simulation showing the magnitude of a photon echo FWM signal as a function of the evolution and emission times. The signal is nonzero only near the diagonal line, hence to optimize the acquisition speed, the relative repetition frequencies between Comb 2 and LO comb are set to be exactly equal to the relative repetition frequency between Comb 1 and Comb 2. This arrangement enables the signal to be sampled along the echo (along the diagonal line) and not in the region where the signal is zero. In order to sample points off the diagonal line, the phase of the DDS is stepped, serving as the reference for the LO comb. This shifts the timing of LO pulses but without using a mechanical delay line. The FWM signal is measured along the lines parallel to the diagonal (dashed white lines (a,b,c,d,e) shown

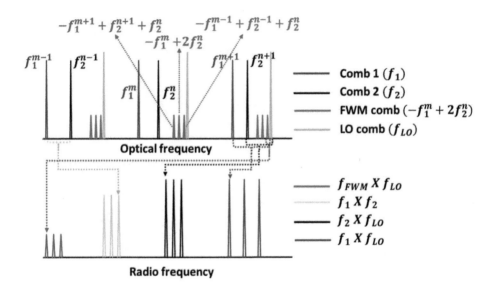

Fig. 6.16 Tri-comb spectroscopy in optical and radio frequency domain. Adapted from Ref. 241.

in Fig. 6.15 (c)). After digitizing the FWM signal, a multidimensional spectrum can be generated by calculating a two-dimensional Fourier transform with respect to t' and τ'.

Figure 6.16 shows a schematic diagram for tri-comb spectroscopy in the optical and radio frequency domains. In the top figure (optical frequency domain) the blue, black, and green lines correspond to comb lines for Comb 1, Comb 2, and the LO comb respectively while the red lines correspond to FWM comb lines. Each FWM line has contributions from multiple combinations of m and n. The space between adjacent lines in a given group of FWM comb lines (for example the middle group of three red lines) is the difference between the repetition frequencies of the excitation combs ($f_{rep_1} - f_{rep_2} = 274$ Hz) whereas the repetition frequency for FWM comb is equal to the repetition frequency of the LO comb ($-f_{rep_1} + 2 f_{rep_2} = -(f_{rep_2} + \Delta) + 2 f_{rep_2} = f_{rep_2} - \Delta = f_{rep_{LO}}$). The bottom figure shows the separation of linear and FWM signals in the radio frequency domain. The red, black, and blue lines are the result of heterodyning FWM, Comb 2, and Comb 1 teeth with the LO Comb respectively whereas the cyan lines correspond to beat signals between Comb 1 and Comb 2. The red comb structure in the bottom figure corresponds to a Fourier transform of the FWM time domain signal with respect to $[t']$ axis in Fig. 6.15 (c) for one DDS phase. A second axis for a two-dimensional spectrum is constructed by taking a Fourier transform of the red comb structure with respect to $[\tau']$ (DDS phase).

To demonstrate the resolution and acquisition speed improvement that can be achieved with TCS, the measurement shown in Fig. 6.9 (a) is repeated using TCS. The results are shown in Fig. 6.17 (a). The two-dimensional spectrum is tilted by 45 degrees to show the spectrum in the ν_t and ν_τ coordinate system. Comparing Fig. 6.9

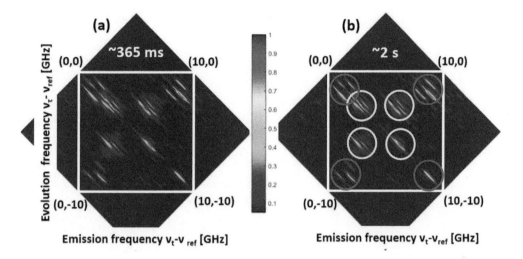

Fig. 6.17 Two-dimensional spectra generated using (a) 365 ms (b) 2 sec data records. The color scale shows the normalized signal magnitude. Yellow (^{85}Rb) and red (^{87}Rb) circles show which resonances are coupled to each other. Adapted from Ref. 236.

(a) to Fig. 6.17 (a) shows that the same results are reproduced with an improved the cross-diagonal resolution by a factor 4. The spectrum shown in Fig. 6.17 (a) is generated by a 365 ms data set, which is 600 times improvement compare to Fig. 6.9 (a). Figure 6.17 (b) shows the same spectrum with the acquisition time of two seconds which clearly shows an improvement in signal to noise, however, all the Rb resonances can be identified in the 365 ms data.

It is important to note that MDCS (unlike linear optical methods) can identify whether multiple resonances belong to the same or different species, which is critical for chemical sensing applications when the sample contains a mixture of different species (e.g., ^{85}Rb and ^{87}Rb). To demonstrate this point, in Fig. 6.18(a) we show a cartoon of a linear transmission spectrum (magnitude) for a sample that contains a mixture of different species with multiple resonances. Using this spectrum it is difficult to tell how many species are present in the sample. Based on the number of resonances shown in the spectrum it could be any number from 1 to 6. In Fig. 6.18 (b) a cartoon of a MDCS spectrum for the same sample is shown. Diagonal peaks (along the dashed line) correspond to absorption resonances, however, the off-diagonal peaks on the 2D spectrum show which resonances are coupled. For example, the 1st and 2nd peaks show no coupling peak which means that the resonances originate from the different species. By contrast, the 1st and 3rd peaks are coupled with each other, which is an indication that these resonances originate from the same species. Similar analysis can be performed for all the peaks on the 2D spectrum. With this information one can decompose the 2D spectrum into the separate spectra of individual species and plot them separately, as shown in Fig. 6.18 (c,d). The separated 2D spectra can be used to measure the locations (and also homogeneous/inhomogeneous linewidths) of each

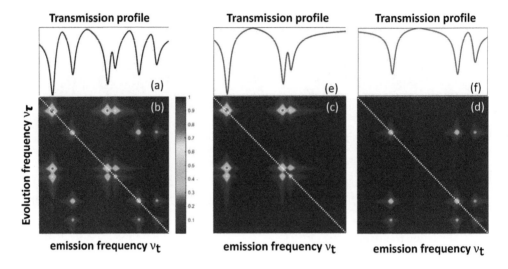

Fig. 6.18 (a) Transmission profile of a mixture. (b) Corresponding 2D spectrum. Color scale shows normalized signal magnitude. (c) and (d) 2D spectra of individual species. (e) and (f) reconstructed 1D spectra from (c) and (d). Adapted from Ref. 236.

resonance along the diagonal line and generate the one-dimensional spectra shown in Fig. 6.18 (e) and (f).

Following this procedure one can identify all possible couplings between the resonances for each isotope that appear at unique locations in Fig. 6.17 (b). In the figure the peaks that are coupled with each other are marked using the red and yellow circles. It is clear that there are no coupling peaks between the resonances marked with red and yellow circles indicating that the sample contains two different species (isotopes (^{85}Rb-yellow) and (^{87}Rb-red)). This information was not clear from the measured linear transmission spectrum (Fig. 6.8 (b)). This information is extremely valuable for chemical-sensing applications especially when probing a mixture without prior knowledge of its constituent species.

TCS contains no mechanical moving parts and enables the measurement of comb resolution multidimensional spectra in under half a second. With the development of micro-resonators, TCS can become a field deployable device for chemical sensing (described above) and other applications outside the laboratory. TCS also has excellent potential to be used for real-time medical imaging applications. In the future the method can be extended to quad-comb spectroscopy to measure the full Hamiltonian of the sample of interest [29, 240].

The remarkable improvement in spectral resolution, without the concomitant increase in acquisition time that would be required by standard techniques, achieved by using frequency combs to implement MDCS is illustrated in Fig. 6.19. Using traditional methods based on mechanical scanning delay stages usually results in the acquisition time increasing in proportion to the reduction in spectral resolution because the delay range determines the spectral range and a longer delay range takes longer given a constant velocity for the delay stage. Thus, the product of the spectral resolution

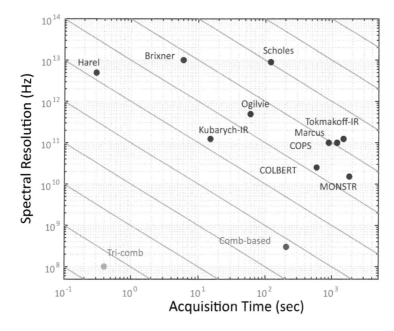

Fig. 6.19 Plot of spectral resolution as a function of the acquisition time for a selection of the approaches to implementing 2D coherent spectroscopy. The diagonal lines represent a constant product of acquisition time and spectral resolution. The points references corresponding to the labeled points are Harel [152], Brixner [101], Kubarych [369], Oglivie [124], Scholes [442], COPS [285], Marcus [391], Tokmakoff-IR [392], COLBERT [413], MONSTR [43], Comb-based [232], and Tri-comb [241].

and acquisition time is typically a constant and consequently represents a reasonable "figure-of-merit," where a lower figure-of-merit represents proportionally shorter acquisition time. The green lines in Fig. 6.19 represent a constant figure-of-merit. The traditional techniques follow this expectation with most having a figure-of-merit between 10^{13} and 10^{14} and none below 10^{12} (note that the data point labeled "Harel" [152] does not use a scanning stage, but rather time-of-flight within large diameter crossing beams). In comparison both the comb-based MDCS [232] and tri-comb spectroscopy [235] provide over an order of magnitude improvement in spectral resolution while not requiring longer acquisitions times. The improvement in figure-of merit is close to 100 for the comb-based MDCS and approximately 20000 for tri-comb spectroscopy. Of course, the spectral resolution should be matched to the sample being studied, there is no benefit in employing high-resolution techniques unless they are needed.

7

Two-dimensional spectroscopy of semiconductor quantum wells

Gallium arsenide (GaAs) quantum wells were the first semiconductor materials to be studied using MDCS, and they continue to be among the technique's most heavily scrutinized systems. A quantum well is a thin sheet of a semiconductor, typically on the order of 10 nm, sandwiched between slabs of a wider bandgap semiconductor to make the potential resemble a finite square well potential in the growth direction. Quantum wells are useful because the well thickness can be readily engineered in the growth process to tune the resonant frequencies of excitons within the well [118]. Beyond this, the quantum confinement in the growth direction has additional advantages for the study and manipulation of excitons in semiconductors because (1) it increases the exciton binding energy, spectrally separating excitons from continuum electrons and holes and increasing the maximum temperature at which excitons remain bound, (2) it can split degenerate excitons into distinct energy bands in a controllable fashion, (3) it increases the exciton dipole moment, thereby increasing the overall strength of light-matter interactions, and (4) the broken translational symmetry along the quantum well growth direction also removes (or in the case of multiple-quantum wells, modifies) the exciton-polariton dispersion in this direction [90, 311], leading to interesting tunable polaritonic effects.

Perhaps the largest contribution that MDCS has achieved toward understanding the mechanisms driving exciton formation and dynamics in GaAs quantum wells has been its ability to probe many-body effects. Although such effects can in part be probed using one-dimensional techniques, the techniques leave an incomplete picture of microscopic mechanisms. For example, the classic signature of exciton-exciton correlations in a time-integrated two-pulse four-wave-mixing experiment is the appearance of a signal at negative time delays [211], but it is unclear from such a measurement whether the signal originates from local fields [211], biexcitons [38], excitation-induced dephasing [158, 426], or excitation-induced shift [343]. Both because of its phase-resolved nature and because spectral features are spread out across multiple dimensions instead of just one, MDCS has proven capable of providing a much more stringent constraint on many-body theories [153, 192].

This chapter reviews the ways that MDCS has impacted semiconductor physics. The focus shall be on multidimensional measurements and therefore neglects some of the time-resolved four-wave-mixing literature upon which MDCS is based. The interested reader is directed to Refs. 73 and 52, where time-resolved four-wave mixing literature is more thoroughly discussed.

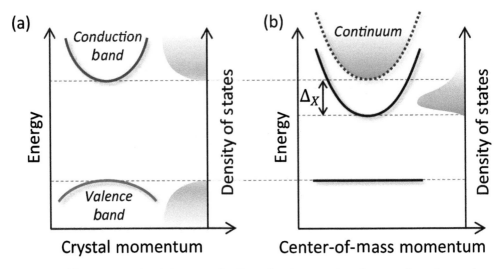

Fig. 7.1 (a) Electronic band diagram of a direct bandgap semiconductor, where the conduction band minimum and valence band maximum are aligned in crystal momentum space. (b) In the excitation picture, optically excited electron-hole pairs form bound hydrogenic states called excitons with zero center-of-mass momentum and a binding energy ΔX relative to the continuum of unbound, free electron-hole pairs. The density of states is concentrated near the exciton resonance.

7.1   Introduction to semiconductor optics

Semiconductors are generally classified as having either a direct or indirect bandgap in their energy-momentum dispersion. Direct gap semiconductors absorb and emit light efficiently, since the minimum of the conduction band and the maximum of the valence band reside at the same point in the crystal momentum k-space, as shown in Fig. 7.1(a). The canonical direct gap semiconductor for coherent spectroscopy experiments is GaAs and its compounds, which comprise materials from groups III and V of the periodic table. The fundamental bandgap of GaAs appears at the Γ-point in k-space and consists of a single conduction band with effective mass $m_c = 0.0662 m_e$, a heavy-hole valence band with $m_{hh} = 0.34 m_e$, and a light-hole valence band with $m_{lh} = 0.094 m_e$ (only one valence band is shown in Fig. 7.1 for simplicity) [248]. The most widely studied semiconductor nanostructures are based on III-V compound heterostructures, including ternary alloys such as InGaAs and AlGaAs, owing to their amenable optical properties for efficient light-matter interaction, ability to be grown in stacked layers with atomic precision using molecular beam epitaxy, and compatibility with nanofabrication processing techniques.

Because optically excited semiconductors are inherently a many-body system, Coulomb interactions between electrons and holes give rise to pronounced excitonic effects that dominate the band-edge optical response. At cryogenic temperatures, heavy-hole and light-hole excitons appear as sharp peaks in the linear optical spectrum near 1.515 eV, which is below the direct bandgap of GaAs at 1.519 eV by the exciton binding energy $\Delta X = 4.1$ meV, as shown in Fig. 7.1(b). In bulk GaAs, the valence bands are degenerate at $k = 0$; however, quantum confinement and mechanical strain lift the degeneracy, typically leading to 5-10 meV splitting between heavy-hole and light-hole exciton states. A prototypical example of a semiconductor nanostructure is a quantum well (QW) formed by sandwiching a thin GaAs layer between two layers of $Al_x Ga_{1-x} As$, as shown in Fig. 7.2(a). A typical alloy ratio of $x = 0.3$ results in a 245 meV deep QW in the conduction band and 130 meV for the valence band. For a sufficiently thin GaAs layer (e.g., 10 nm), quasi-2D quantum confinement of electrons and holes increases the exciton binding to $\Delta X \approx 10$ meV. The quantum confinement also lifts the degeneracy between the heavy-hole and light-hole bands. Each band consists of two spin states. In the electron picture, as shown in Fig. 7.2(c), the conduction band (CB) and the light-hole (lh) valence band (VB) each have two spin states with spin $\pm 1/2$, while the heavy-hole (hh) valence band has two spin states with spin $\pm 3/2$. The spin states can be selectively excited with circularly polarized light according to the selection rule. Both heavy-hole excitons (X_{hh}) and light-hole excitons (X_{lh}) can be optically excited. A typical linear absorption spectrum of a GaAs QW at a temperature of 6 K is shown in Fig. 7.2(d), which shows that the resonance energy of X_{lh} is about 10 meV higher than that of X_{hh}.

An important component of semiconductor QWs is the interface between the well and the barrier materials. Due to the stochastic nature of the epitaxial growth process, monolayer well-width fluctuations introduce interfacial surface roughness that acts as shallow in-plane disorder potentials for excitons. Well-width fluctuations lead to an inhomogeneous distribution of exciton frequencies that broadens the optical linewidth.

The basic concepts of excitons in GaAs heterostructures also applies to other semiconductors if the exciton Bohr radius extends over many lattice sites (Wannier-Mott-type), which is true for the systems discussed in this chapter. Although historically III-V semiconductor heterostructures have been the material-of-choice for coherent spectroscopy experiments, recent advances in material growth, processing, and semiconductor fabrication technologies have afforded new opportunities for exploring coherent-light matter interaction in more complex heterostructures. A natural extension of a single QW is the asymmetric double QW (DQW) in which electron and hole wavefunctions become delocalized across both wells separated by a narrow barrier. Experiments on InGaAs/GaAs DQWs reveal unexpected interactions associated with many-body effects even in quantum-mechanically isolated QWs. Semiconductor quantum wells can also be embedded in a microcavity, providing an excellent test of many-body physics in the regime of strong light-matter interaction.

When confined to a truly 2D system, interesting exciton physics emerge. 2D excitons have only recently been realized in monolayer transition metal dichalcogenides (TMDs), which are atomically thin semiconductors with a direct gap at the K point in k-space. The heavier electron and hole effective masses in TMDs (both on the order of $0.5m_e$) compared to III-V materials, in combination with reduced dielectric screening of the Coulomb interaction in two dimensions, leads to an enhancement of the exciton binding energy to hundreds of meV, making them stable even at room temperature. 2D materials and their heterostructures offer a new avenue for controlling the optical and electronic response, which has been leveraged to tune the exciton energy and charge

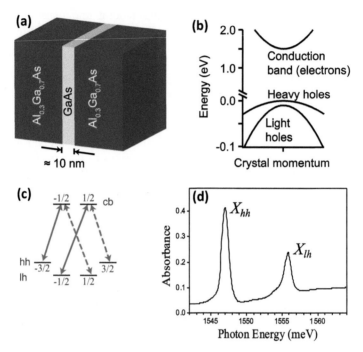

Fig. 7.2 (a) Typical structure of a semiconductor quantum well. A narrower bandgap material (in the illustrated case, GaAs) is sandwiched on both sides by a wider bandgap material ($Al_{0.3}Ga_{0.7}As$) to produce one-dimensional quantum confinement. (b) Changing the quantum well thickness tunes the resonant frequencies of the excitons within the quantum well. For GaAs, the relevant transitions are between the electronic conduction band and either the heavy-hole (hh) or light-hole (lh) valence band. (c) Quantum well level structure, showing the angular momentum states of the conduction band (CB) and the hh and lh valence bands (VB). Allowed transitions for circular optical polarization are shown, with solid and dashed lines denoting opposite handedness. (d) Typical linear absorption spectrum of the hh and lh excitons at 6 K.

transport properties in single-walled carbon nanotubes. The application of MDCS to TMDs is presented in Chapter 10.

7.2   Many-body signatures in one-quantum 2D spectra

Although one-dimensional techniques such as transient four-wave-mixing spectroscopy can detect features of many-body effects, it is challenging to identify the underlying microscopic mechanisms. The interpretation of one-dimensional spectra relies on comparing a trace to the simulation, where signatures of many-body effects are often reproduced with a phenomenological model based on the modified optical Bloch equations. In contrast, 2D spectra can provide much more detailed information of many-body effects, such as spectral pattern, phase information, and polarization dependence. Some 2D spectral features can be attributed to specific microscopic processes involving

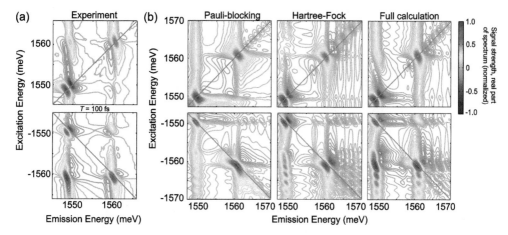

Fig. 7.3 Demonstration of many-body effects in gallium arsenide quantum wells. **(a)** Experimental MDCS measurements of the heavy-hole and light-hole exciton for a 10-period GaAs/Al$_{0.3}$Ga$_{0.7}$As multiple-quantum well with 10-nm wells and 10-nm barriers. Data correspond to the real part of a rephasing spectrum (lower panel) and nonrephasing spectrum (upper panel), with co-circular polarization, at an intermediate time delay $T = 100$ fs. **(b)** Simulations of the spectra from (a), with increasingly sophisticated theoretical treatments running from left to right. Adapted from Ref. 462.

many-body interactions/correlations and successfully reproduced only by microscopic theories.

The first studies using MDCS to study many-body physics in semiconductor quantum wells were conducted in the mid-2000s in a series of experiments [34, 220, 462] examining a 10-period GaAs quantum-well sample consisting of 10-nm-thick GaAs layers separated by 10-nm-thick Al$_{0.3}$Ga$_{0.7}$As barriers. The band structure cartoon in Fig. 7.2(b) illustrates the relevant optical transitions, which are a "heavy-hole" exciton transition at about 1550 meV, and a "light-hole" exciton transition about 10 meV above this.

As shown in Fig. 7.3(a), MDCS reveals a rich array of spectral features for these two transitions [463]. The two spectra correspond to the real part of the spectral response for rephasing (bottom panel) and nonrephasing (top panel) pulse sequences. Both the rephasing and non-rephasing spectra in Fig. 7.3(a) exhibit prominent diagonal peaks at about 1549 meV and 1560 meV, which are signatures of absorption at the heavy-hole and light-hole frequencies. The spectra also exhibit cross peaks, in the rephasing spectrum at (1549, -1560) meV and (1560, -1549) meV, and in the nonrephasing spectrum at (1549, 1560) meV and (1560, 1549) meV, which are an indication of coherent coupling between the two resonances.

Beyond this, the spectra exhibit a number of other features that indicate a clear influence on the system from many-body effects. For example, the heavy-hole diagonal peak in both the rephasing spectrum and in the nonrephasing spectrum exhibits a phase shift relative to the phase that would be expected for the peak resulting from Pauli blocking in the absence of many-body effects for a simple two-level system (see Fig. 3.1) and as observed in atomic vapor (see Fig. 5.4). The MDCS spectrum from a simple three-level system where the nonlinear response is driven by Pauli blocking should be purely real and purely positive at the point of maximum amplitude. In addition, the cross peaks that are clearly evident in both the upper and lower panels of Fig. 7.3(a) should not normally be expected to have a higher peak amplitude than the peak amplitudes of the diagonal peaks they connect. In fact, as a result of angular momentum selection rules [Fig. 7.2(c)], such cross peaks should be absent for co-circularly polarized excitation beams without taking the Coulombic interactions between excitons into account because the heavy-hole and light-hole excitons are generated from entirely independent single-particle electronic states. Finally, the vertical stripes extending below the excitonic resonances are due to the continuum of free-electron hole pairs. Free-electron hole pairs correspond to transitions from the valence band to the conduction band, which are nearly vertical because the photon momentum is negligible compared to that of the electrons and holes. Since there is continuum of these transitions with varying energy, as determined by the dispersions of the valence and conduction bands, this process has often been treated as an inhomogeneously broadened set of two-level transitions, which would lead to a photon echo [25]. However, a photon echo corresponds to a diagonal feature in a 2D coherent spectrum, thus there is clearly a discrepancy.

These observed effects of many-body interactions on the 2D spectrum of the excitonic resonances in a quantum well can be explained phenomenologically. The lineshape of the diagonal peaks occurs because the response is dominated by an excitation-induced shift (EIS) [343], although there is some contribution from excitation-induced dephasing (EID) [426] as well [220]. The presence and anomalous strength of the cross-diagonal peaks can also be explained by EIS and EID, but in this case the EIS and EID are between two difference "species," i.e., the heavy-hole and light-hole excitons. Specifically, if the excitation of a population of light-hole (heavy-hole) excitons causes the resonance energy or dephasing rate of heavy-hole (light-hole) excitons to change, a peak at the absorption energy of the light-hole (heavy-hole) exciton and emission energy of the heavy-hole (light-hole) exciton will appear. If this process dominates, it also explains the asymmetry between the cross peaks, as the relevant coefficients for the EIS and EID processes need not be symmetric. The vertical features due to the continuum can also be explained in the same phenomenological framework. Namely, if the excitation of free carriers causes the resonance energy of an exciton (either heavy-hole or light-hole) to shift or its dephasing rate to change, a feature at the absorption energy of the free carriers and emission energy of the exciton will appear, which will be a "stripe" rather than a peak because there is a continuum of free-carrier energies [34]. Note that there is, in principle, still a diagonal contribution, corresponding to a photon echo, from the free-carrier transitions, however, it can be much weaker, depending on the strength of the EID and EIS, and thus not observable. While this phenomenological description gives intuitive insight into the origin of the signals, it is unsatisfying in that the coefficients describing EID and EIS can be independently adjusted to match the spectra, whereas they should be linked as they are the real and imaginary changes in the excitonic self-energy due to many-body interactions, and moreover should be set by the Coulomb interaction, rather than being an adjustable parameter.

A microscopic calculation of these effects can be performed starting from the semi-conductor Bloch equations [153] to provide insight into their specific nature. A comparison between the experimental data from Fig. 7.3(a), and the left, center, and right columns of Fig. 7.3(b), which correspond to increasingly sophisticated theoretical treatments of the system based on a microscopic many-body theory using a 1D tight-binding model, provide insight into the nature of the interactions. As illustrated by the left column of Fig. 7.3(b), keeping only the Coulomb terms responsible for exciton formation and ignoring higher-order terms that result in exciton-exciton interactions such that Pauli blocking alone gives rise to the nonlinear response cannot adequately replicate the experimental features from Fig. 7.3(a). The center column, which incorporates Hartree-Fock terms, i.e., the hierarchy of equations descrbing the many-body interactions is truncated based on being first order in the Coulomb interaction, in addition to the effect of Pauli blocking, does a somewhat better job. A much more satisfactory agreement—even taking into account the inherent limitations of a 1D model—can be obtained by employing a dynamics-controlled truncation scheme incorporating Coulombic correlations beyond a mean field and truncating the hierarchy based on keeping terms that are third-order in the laser field [16, 17, 223]. Results of this "full" calculation are displayed in the column at right.

The bulk of the data displayed in references 220, 461, 462 were acquired using either co-linear or co-circular polarization, which emphasizes the influence of excitation-induced dephasing and excitation-induced shift many-body effects. Probing a GaAs quantum well sample with cross-polarized MDCS pulses (in either cross-linear or cross-circular orientation) opens the possibility of observing and characterizing biexcitons, which are four-particle bound states consisting of two electrons and two holes. Figure 7.4 shows two measurements on GaAs using co-circular ($\sigma^+\sigma^+\sigma^+\sigma^+$) and cross-linear (XYYX) geometries [42]. Whereas co-circular polarization suppresses the biexciton resonance due to spin dependent selection rules (the biexciton corresponds to an antisymmetric spin configuration), biexcitons are unmistakably visible in the cross-linear polarization spectrum as a shifted peak to the left of the heavy-hole direct peak, which is itself suppressed in a cross-polarized geometry. Measurements of the real part of the spectrum show that the sign of the biexciton resonance is negative relative to that of the diagonal resonance, which is an expected consequence of the fact that the biexciton shows up in the spectrum as an excited-state absorption feature.

Although the model behind Fig. 7.3(b) was constructed using Fermionic creation and annihilation operators, it is worth noting that the strongly dispersive character of the diagonal peaks in Fig. 7.3(a) can be perhaps more compactly understood through a treatment of excitons as bosons, where the dispersive character of the lineshape arises by a near, but not quite total, cancellation between the interaction pathway for creating an exciton out of the vacuum and the interaction pathway for boosting an exciton from a singly occupied mode to a doubly occupied mode. Many-body effects break the symmetry of these two processes such that the latter process occurs at a slightly higher frequency than the former process, resulting in an asymmetric phase. Quantitative fits to co-circularly and cross-circularly polarized MDCS measurements in

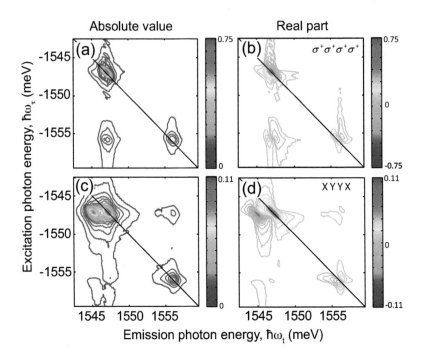

Fig. 7.4 Biexcitons in GaAs. (a) Absolute value and (b) real part of a rephasing MDCS measurement of a GaAs multiple-quantum well with co-circular polarization, in which the biexciton resonance is suppressed. (c) Absolute value and (d) real part of a rephasing MDCS measurement using cross-linear polarization (XYYX), in which the biexciton is accentuated. Adapted from Ref. 42.

GaAs quantum well have been performed, demonstrating that the phase and linewidth of MDCS measurements of quantum wells under both polarization schemes can be understood within a single theoretical framework [360].

Another area in which MDCS has impacted the field of quantum wells has been in its ability to examine not just the static, but also the dynamical evolution of a bath of generated excitons. Dynamics can be observed and characterized by plotting single-quantum spectra (excitation vs. emission frequency) for a variety of different delays of the mixing time T. Since the spectral features contributed by different many-body processes are separated in 2D spectra, the dynamics associated with different processes can be studied by measuring the time evolution of different spectral features.

This experimental scheme was used to characterize the evolution of many-body effects in GaAs/AlGaAs quantum wells [414]. The 2D correlation spectra were measured on a 10-nm GaAs quantum well sample at 10 K with co-circularly polarized excitation pulses. The real part of the spectra are shown in Fig. 7.5 for different waiting times. The spectrum at a short waiting time (0.20 ps) exhibits all of the many-body spectral features discussed above. The H diagonal peak highlighted in the red square has a dispersive lineshape due to many-body interactions. The cross peaks, labeled X^u and X^l, between the H and L features are due to Coulomb correlations between the hh and

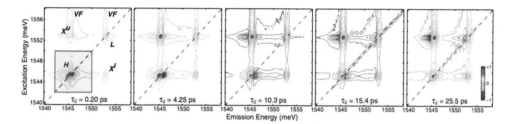

Fig. 7.5 The real part of the 2D correlation spectra measured with co-circular polarizations at different waiting times (τ_2). The spectra were obtained on a 10-nm GaAs quantum well at a sample temperature of 10 K. Adapted from Ref. 414.

lh excitonic states. The vertical features (VF) are a result of the scattering between excitons and free carriers.

As the waiting time increases, the lineshape of the H diagonal peak changes. The positive part becomes stronger than the negative part and shifts toward the diagonal line. The H peak becomes absorptive in lineshape and is centered on the diagonal line at very long waiting times. This transition from dispersive to absorptive lineshape occurs within about 15 ps. In contrast, the cross peaks and vertical features persist at much long waiting time, up to 40 ps in this measurement. The time evolution of these spectral features reveal key insights into the dynamics of many-body interaction in GaAs quantum wells. The fast transition from dispersive to absorptive indicates the fast dynamics of the underlying process. The many-body terms contributing the cross peaks and vertical features do not vanish at long waiting times. The observed dynamics suggest that the dispersive lineshape is due to polarization-polarization scattering and the polarization decay on the ps time scale. The persistent cross peaks and vertical features at long waiting times are the result of many-body interactions involving exciton populations that have a longer lifetime.

To model many-body interactions in semiconductor quantum wells, many microscopic theoretical approaches assumed the coherent limit in which the exciton polarization dephasing rate is linked to the exciton population decay rate. In this limit, effects due to exciton coherences and exciton populations cannot be separated. These two kinds of interactions can be isolated in 2D spectra at waiting times long enough that the exciton polarization has decayed. The observed dynamics in this experiment suggests that theoretical models should include dynamics of many-body interactions without assuming the coherent limit.

In another experiment, two-dimensional rephasing spectra have also been measured as a function of the waiting time T to characterize spectral diffusion effects in GaAs/AlGaAs quantum wells [358, 359]. Spectral diffusion is a process by which memory of resonance characteristics gets lost over time following excitation of a sample by the pump pulse, due (for example) to phonon-assisted exciton spatial migration. The result in MDCS is that inhomogeneously broadened spectral peaks are initially diagonally elongated at small values of T, but become increasingly round at larger values of T.

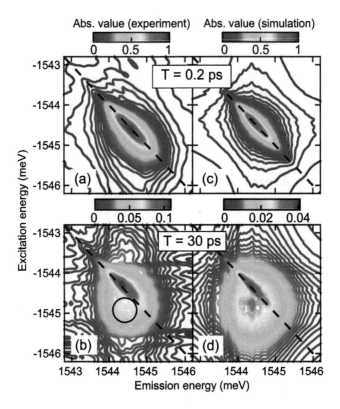

Fig. 7.6 Spectral diffusion for the heavy-hole exciton in a $GaAs/Al_{0.3}Ga_{0.7}As$ multiple-quantum well with 10-nm wells and 10-nm barriers. The sample was measured at 5 K. (a)–(b) Experimentally measured spectra. (c)–(d) Simulations. Adapted from Ref. 359.

Spectral diffusion processes are most commonly treated within the Gauss-Markov approximation, which assumes exponential decay dynamics in going from diagonally elongated features to rounder features. By characterizing the spectral features of GaAs quantum wells on T at various temperatures, it was discovered that the Gauss-Markov approximation breaks down for temperatures lower than 70 K [358]. For the lowest temperatures, the shape of the spectral peaks at large T delays become asymmetric to the point where the concept of a cross-diagonal linewidth becomes itself ill-defined outside the context of complicated lineshape features [359]. An example of this, measured on a GaAs quantum-multiple quantum well maintained at 5 K, is depicted in Fig. 7.6. The lineshape of the heavy-hole exciton absolute value spectrum for $T = 0.2$ ps is nearly Gaussian along the diagonal direction, and nearly proportional to the square root of a Lorentzian along the cross-diagonal direction [Fig. 7.6(a)]. By contrast, the lineshape at $T = 30$ ps [Fig. 7.6(b)] is noticeably skewed toward the lower-left portion of the spectrum (see the solid black circle), and has become almost triangular-shaped in its two-dimensional profile [Fig. 7.6(b)]. The authors used the measurements to demonstrate that the strong-redistribution approximation for spectral diffusion (in which it is assumed that diffusive processes to higher energies are as likely to occur as those to lower energies) breaks down in addition to the Gauss-Markov approximation at the lowest temperatures. A theoretical simulation could nevertheless still be achieved by solving the two-dimensional Schrödinger equation under random realizations of disordered confinement potentials, and analyzing the average of resulting dynamic localization [Figs.7.6(c) and 7.6(d)].

One of the challenges of using MDCS to study transport, such as the migration of excitons that gives rise to spectral diffusion, is the need for theoretical treatments to address both the complex optical spectroscopic method and simultaneously the transport phenomena of interest. One approach to overcoming this challenge is to apply algorithms developed for "deblurring" images to remove the effects of the spectroscopic measurements and reveal the underlying transport [320].

7.3   Many-body signatures in double- and multi-quantum 2D spectra

Single-quantum spectra have considerable power to clarify and elucidate many-body interactions in semiconductors, as already discussed previously. For certain types of information, however, even more information can be gleaned by arranging pulses into a double-quantum configuration, as discussed in Chapters 2 and 3 and depicted in Fig. 2.4. In the case of GaAs quantum wells, there are no single-exciton states at the double-quantum frequency, so double-quantum measurements give an exclusive and background-free measurement of excitonic interactions. Double-quantum MDCS measurements on GaAs quantum wells were preceded by transient four-wave mixing studies [6, 52, 115, 194, 204, 205, 211], but the extension of the technique into the multidimensional realm has uncovered a number of new and interesting phenomena.

In 2009, Stone *et al.* used double-quantum MDCS to directly generate and probe the lifetimes of coherent superposition states between the ground state of GaAs and the biexciton state [374]. The initial results were puzzling because measurements of the biexciton binding energy made by comparing the vertical shift of the double-quantum peak to twice the projection of the peak onto the single-quantum excitation axis gave a different answer from both complementary biexciton binding energy measurements [31, 297, 428] and from the binding energy as measured by taking the difference between the biexciton-to-exciton emission peak and the exciton-to-ground-state emission peak. Shortly thereafter, it was realized that the double-quantum signal was richer in content than had initially been assumed, with a three-dimensional double-quantum measurement disentangling biexciton coherences that had previously been overlapping [412], and with the real part of double-quantum measurements [76, 176, 375] revealing that unbound two-exciton states play a significant role in the double-quantum signal in addition to bound biexcitons.

In GaAs quantum wells, two excitons can form a bound state, known as the biexciton, when the two excitons have opposite electron spins, or an unbound two-exciton state when the spins are the same. The energy level diagram in the exciton picture, as shown in Fig. 7.7, displays one- and two-exciton states of the hh and lh excitons. The two-exciton states have an energy that is the sum of the individual exciton energies plus a shift. The biexciton states are shifted lower by the biexciton binding energy.

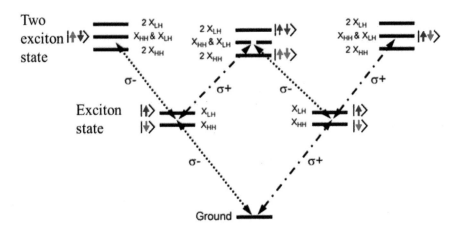

Fig. 7.7 Energy level diagram showing one-exciton and two-exciton states in GaAs quantum wells. Electron spins are labeled for one-exciton states and bound biexciton states.

The unbound two-exciton states are shifted higher by the scattering energy. Double-quantum 2D spectra of these two-exciton states have been observed and shown that the dominant signal is due to mean-field many-body interactions, rather than higher order correlations terms, including biexcitons [176]. This conclusion differs from the single-quantum spectra, and shows that the different types of spectra are sensitive to different many-body effects.

Experimental double-quantum 2D spectra of GaAs quantum wells are shown in Fig. 7.8. The 2D spectra in Figs. 7.8(a) and 7.8(b) are the results when the laser spectrum is centered at the hh exciton resonance. For the cross-circular excitation $(\sigma^-\sigma^+\sigma^+\sigma^-)$, the spectrum has two peaks: an absorptive peak (labeled "A") corresponding to a biexciton state, and a dispersive peak (labeled "B") due to an unbound two-exciton state. The spectrum with the co-circular polarizations $(\sigma^+\sigma^+\sigma^+\sigma^+)$ has only a dispersive peak associated with an unbound two-exciton state. These spectral peaks were predicted by previous theoretical calculations [452, 453], and the bound biexciton contribution was expected to dominate the cross-circular spectra. However, the dominant feature in experimental spectra is peak B due to the unbound two-exciton state. Peak B in both spectra has a dispersive lineshape, indicating its origin of many-body interactions. These results show that the double-quantum signal is primarily contributed by mean-field many-body interactions, rather than bound biexcitons.

Both hh and lh exciton resonances can be excited if the laser spectrum is centered between the hh and lh excitation energies. The resulting double-quantum 2D spectra are shown in Figs. 7.8(c) and 7.8(d) for the cross-circular $(\sigma^-\sigma^+\sigma^+\sigma^-)$ and co-circular $(\sigma^+\sigma^+\sigma^+\sigma^+)$ polarizations, respectively. Besides peaks A and B from hh excitons, there are also contributions from lh excitons and mixed states of hh and lh excitons. Peak C has contributions from two unbound lh excitons, while no biexcitons are observed for the lh exciton. In addition, peaks D and E are a result of the unbound two-exciton states consisting of a hh exciton and a lh exciton.

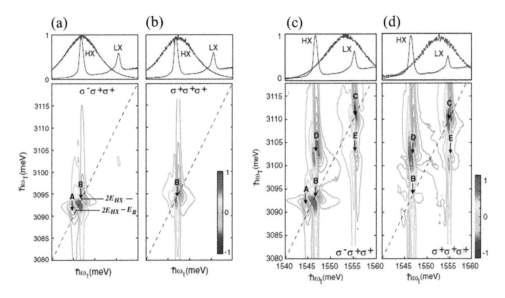

Fig. 7.8 Experimental double-quantum real 2D spectra for (a) cross-circular and (b) co–circular excitations with the laser spectrum centered at the hh exciton, and spectra for (c) cross-circular and (d) co-circular excitations with the laser spectrum centered between the hh and lh excitons. The corresponding absorption spectra and laser spectra are also shown in the top panels. Adapted from Ref. [176].

The dominant role of mean-field many-body interactions can be confirmed the simulation based on a three-band 1D tight-binding model. In the simulation, most parameters are determined from the known effective masses and the linear absorption spectrum. The biexciton binding energy and dephasing rates are adjusted to fit the experimental 2D spectra. Double-quantum 2D spectra are then calculated from the third-order nonlinear signal by solving the equations of motion. Two simulated 2D spectra are shown in Fig. 7.9 for the cross-circular and co-circular excitation with the laser spectrum centered at the hh exciton energy. Both spectra are in good agreement with the corresponding experimental spectra. The model successfully reproduces the dominant contribution from unbound two-exciton states, the polarization dependence of peak A, and the lineshape of all peaks. To understand the origin of the observed spectral features, the simulation is repeated neglecting higher-order correlations beyond mean field. The results show that the bound biexciton peak disappears, while the peaks corresponding to the unbound two-exciton states remain, with little change in lineshape and strength. This result suggests that the mean-field terms dominate the double-quantum signals in these spectra, however, higher-order correlations are required to match the bound biexciton peak.

An important aspect of these double-quantum spectra is that the mean-field contributions are revealed because the stronger signal from correlations terms beyond mean field are energetically shifted. In the single-quantum spectra, these terms are energetically degenerate and thus the mean-field terms are masked by the stronger

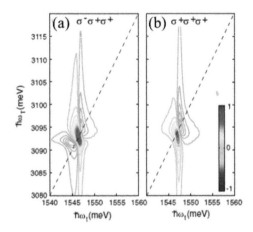

Fig. 7.9 Simulated two-quantum real 2D spectra for (a) cross-circular and (b) co-circular excitations with the laser spectrum centered at the hh exciton. Adapted from Ref. [176].

correlation terms. Thus simultaneously reproducing the single- and double-quantum spectra provides a more stringent test of the accuracy of a microscopic many-body theory.

Double-quantum spectra are just the first of many possible multiple-quantum states that can be probed using MDCS. In follow-up experiments to the results summarized above, Turner and Nelson conducted a series of experiments in which they examined polarizations up to seventh-order in GaAs, revealing tri-exciton coherences in which bound states form among three excitons [411]. Examples of these higher-order electronic correlations are displayed in Fig. 7.10, where three-quantum coherences are displayed in Figs. 7.10(a)–7.10(d), and a four-quantum coherence is displayed in Fig. 7.10(e). These 2D spectra are the fifth-order and seventh-order nonlinear responses measured in two different phase-matched directions, as shown in Fig. 7.10(f). The spectra in Figs. 7.10(a)–7.10(c) have spectral features attributed to bound triexcitons. The three-quantum 2D spectrum in Fig. 7.10(d) measured with co-circular pulses has only signal from the exciton-free-carrier scattering. The four-quantum 2D spectrum in Fig. 7.10(e) demonstrates a lack of bound-state correlations beyond those at the three-quantum level. Multiple-quantum spectra have also revealed how many-body interactions play a role in coupling in doped QWs, where neutral and charged excitons couple through their mutual interaction with the electron gas [271]. In modulation doped QWs, quantum coherence between Mahan excitons arising from interactions of charged holes with electrons at the Fermi edge in the conduction band has been observed. Collective excitations of the many-electron system lead to distinct signatures in multiple-quantum spectra, where coherence was revealed to persist due to reduced screening in the system [304].

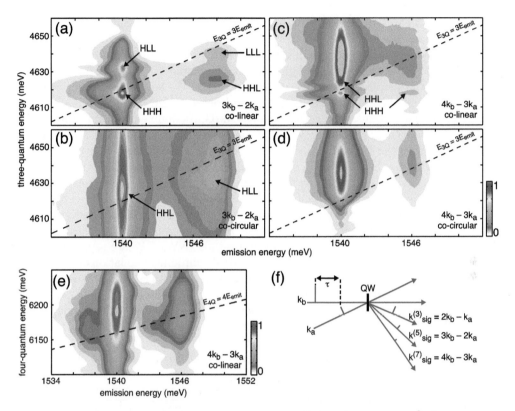

Fig. 7.10 Higher-order multiple-quantum spectra in GaAs. Three-quantum coherences, resulting from fifth-order polarization with (a) co-linear and (b) co-circular excitation, and seventh-order polarization with (c) co-linear and (d) co-circular excitation. (e) Four-quantum coherence, from a seventh-order polarization with co-linear excitation. (f) Phase-matching geometry showing the phase-matched directions for the fifth- and seventh-order nonlinear signal. Figure credit: Daniel Turner.

7.4 Two-dimensional spectroscopy of coupled quantum wells

Problems involving double-well potentials are one of the most fundamental topics in quantum mechanics. The double-well potential is a good model for understanding many physical processes such as defect diffusion in solids, structure relaxation in glasses, inversion of chiral molecules, proton tunneling between DNA base pairs, and molecular energy transfer. Semiconductor double-quantum wells provide an ideal physical model for double-well potentials since quantum well and barrier sizes can be precisely engineered and systematically varied to compare with theory. Double-quantum wells can also serve as a model system for other problems such as an engineered diatomic molecule without the complexity of nuclear motion, and an energy transfer process with only electronic couplings but no vibrational couplings. Although a typical 10-nm-thick GaAs QW is wider than the exciton Bohr radius ($6--7$ nm), the electronic transitions are well-described by quantum mechanics in two dimensions.

QW "layers" separated by a narrow barrier represent an interesting model system for studying long-range interactions between layers in a well-controlled environment. The thickness and separation of each QW can be tailored to mimic processes in other systems, such as energy transfer in light-harvesting complexes.

As one of the simplest examples, semiconductor DQWs have attracted significant experimental and theoretical interest for several decades. DQWs can be realized in epitaxially grown systems. One system is InGaAs/GaAs double-quantum wells, with $In_{0.05}Ga_{0.95}As$ serving as quantum well material, surrounded by layers of wider-band-gap GaAs forming the barrier. InGaAs/GaAs serves as a particularly nice material in these types of studies because strain effects shift the light-hole exciton out of the quantum well, simplifying the spectra corresponding to the remaining heavy-hole excitons in the wells. An understanding of coupling between the wells can be facilitated by making the two wells asymmetric in their thickness, such that excitons corresponding to the wide well have lower frequencies than excitons primarily situated in the narrow well. For example, a 10-nm-wide (WW) and an 8-nm-wide (NW) QW are separated by a barrier, as depicted in Fig. 7.11(a). One-dimensional Schrödinger equation calculations of the single-particle electron and hole wavefunctions show that for a narrow barrier (≤ 10 nm), the hole wavefunctions are localized to their respective QWs, whereas the electron wavefunctions are partially delocalized across the two wells, as shown in Fig. 7.11(b). This type of coupled DQW system has been studied to understand the roles of phonon-assisted tunneling [293], carrier percolation through the barrier [188], dipole-dipole interactions [20], and resonant energy transfer in the inter-well coupling dynamics [89].

The role played by many-body effects, which govern exciton physics in a single QW, has been more difficult to identify in DQWs until recently. Using optical 2DCS, researchers have shown that inter-well coupling in an InGaAs/GaAs asymmetric DQW nanostructure originates from many-body interactions [286]. Analysis of zero-, single-, and double-quantum 2D spectra provides a comprehensive picture of the exciton dynamics and coupling mechanisms. In the single-quantum rephasing spectrum shown in Fig. 7.11(c), coupling between excitons primarily localized to their respective QW is apparent from the off-diagonal peak CP. The real part of the spectrum (bottom panel) reveals the complex lineshape of the peak, which is particularly sensitive to many-body interactions. The dispersive lineshape is indicative that EIS plays a key role in the inter-well coupling.

Further evidence of many-body effects as the coupling mechanism is provided by the non-zero signal in the double-quantum spectrum shown in Fig. 7.11(d). The two peaks on the diagonal line, $\hbar\omega_2 = 2\hbar\omega_3$, arise from coherent interactions between two excitons in the same QW. The appearance of the off-diagonal peak 2CP at the double-quantum energy $E_{WW} + E_{NW}$ necessarily requires coherent coupling between excitons in the two QWs. Through density matrix calculations, it is shown that hybridization of the electronic states due to wavefunction tunneling cannot account for the interaction; instead, many-body effects are essential for understanding the coherent inter-well coupling. An analysis of the peak shapes at lower excitation densities ($< 10^8$ excitons/cm^2) than the experiments discussed above ($> 10^9$ excitons/cm^2) reveal that carrier interactions are important at lower densities than previously thought [400].

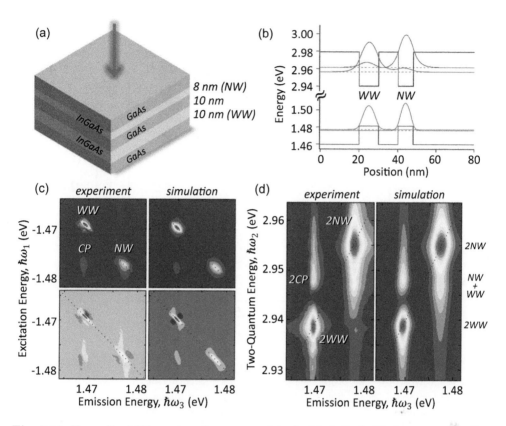

Fig. 7.11 Normalized 2D spectra of an asymmetric double InGaAs/GaAs quantum well at 15 K. (a) Cartoon of the double-quantum well structure. (b) Single-particle energy levels and wavefunctions calculated from the one-dimensional Schrödinger equation. (c) Magnitude (top) and real part (bottom) of the experimental (left) and simulated (right) rephasing one-quantum spectrum. The spectra feature two peaks on the diagonal associated with excitons in the narrow well (NW) and wide well (WW) and an off-diagonal coupling peak (CP). (d) Experimental (left) and simulated (right) double-quantum spectrum indicating coherent interactions between excitons confined to the two wells. Simulations are based on the optical Bloch equations with many-body interactions included phenomenologically. Adapted from Ref. 286.

Concepts developed to understand inter-well coupling have also been helpful in characterizing interactions in other material systems including modulation-doped QWs [271].

In another experiment, Tollerud *et al.* performed measurements on uncoupled In-GaAs quantum wells, and were able to identify cross peaks between bright excitons and optically dark indirect barrier excitons within the same well that would not otherwise have been visible [401]. This extra degree of visibility, which is illustrated in Fig. (7.12), originates from the fact that the brightness of the optically dark excitons is proportional to the fourth power of the dark-state dipole moment, whereas the cross-

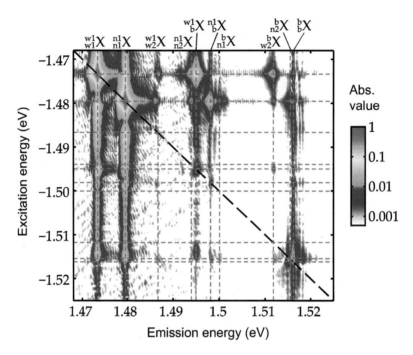

Fig. 7.12 Observation of optically dark indirect barrier excitons through multidimensional coherent spectroscopy. The brightest peaks at upper left corner of the spectrum correspond to optically accessible heavy-hole exciton resonances for an uncoupled asymmetric InGaAs/-GaAs double-quantum well. The cross peaks accentuate interactions with optically dark excitons. Adapted from Ref. 401.

peak between the dark state and a bright state is proportional to the square of the dark-state dipole moment multiplied by the square of the much more optically active bright state dipole moment.

The ability of to isolate otherwise murky spectral features though cross peaks is not restricted to single-quantum spectroscopy techniques. Tollerud and Davis demonstrated a optical MDCS analog of double-quantum heteronuclear NMR [402], in which they examined coupling interactions in coupled asymmetric InGaAs/GaAs double-quantum wells (8-nm and 10-nm well thickness, 10-nm barrier) using double-quantum spectroscopy, but in which they tuned the bandwidth of their excitation pulses to select out specific excitation pathways. The authors found that the technique allowed them to identify a double-quantum signal associated with GaAs defects that had previously been confused with other interaction-induced effects.

7.5   Quantum well exciton-polaritons in microcavities

The resonant interaction between semiconductor excitons and an optical field inside a microcavity can strongly modify the light-matter interaction. In the strong coupling regime, the exciton and cavity modes form new admixed eigenstates known as upper (UP) and lower (LP) exciton-polaritons that exhibit an anti-crossing in their dispersion

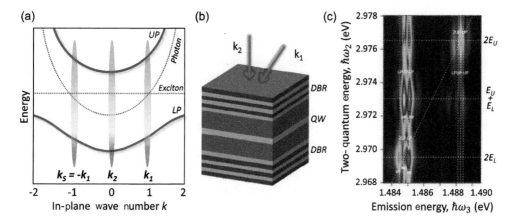

Fig. 7.13 Two-quantum spectrum (magnitude) of exciton-polaritons in a single $In_{0.04}Ga_{0.96}As$ quantum well in a GaAs/AlGaAs microcavity at 4 K. (a) Diagram depicting the lower-polariton (LP) and upper-polariton (UP) dispersion curves for a negative cavity-exciton detuning. The dashed black curves correspond to the bare exciton and cavity energy-momentum dispersion. The pump, probe, and signal wavevectors are indicated. (b) Semiconductor microcavity structure and excitation pulse geometry in the pump-probe configuration, with $\mathbf{k}_2 = 0$ μm^{-1} and $\mathbf{k}_2 = 0.96$ μm^{-1}. (c) Two-quantum spectrum revealing coherent cross-interactions between upper- and lower-polaritons (peaks at $\hbar\omega_2 = E_U + E_L$) and coherent self-interactions between lower polaritons (at $\hbar\omega_2 = 2E_L$) and upper polaritons (at $\hbar\omega_2 = 2E_U$). Adapted from Ref. 387.

curves, as shown in Fig. 7.13(a). Exciton-polariton states can be created by embedding QDs or a QW within a semiconductor microcavity [438]. A typical group III-V microcavity system consists of a single or few layers of $In_xGa_{1-x}As$ QWs sandwiched between GaAs/AlAs distributed Bragg reflectors (DBRs) acting as the cavity end mirrors. Due to their mixed light-matter character, exciton-polaritons have extremely low effective mass that facilitates Bose-Einstein condensation [179] and superfluidity [415] in the solid-state. Because they retain an excitonic component, exciton-polaritons are strongly influenced by the many-body interactions inherent to semiconductor nanostructures. Transient FWM experiments reveal that two-exciton Coulomb correlations clearly play a role in exciton-polariton scattering and decoherence [37, 195, 197, 245].

Substantial progress in quantitatively understanding the details of higher-order correlations and coherent exciton-polariton scattering has been attained in the last few years with non-collinear optical 2DCS techniques. A typical experimental configuration is shown in Fig. 7.13(b), where two excitation pulses $\mathcal{E}_1$ and $\mathcal{E}_2$ with wavevectors $\mathbf{k}_1$ and $\mathbf{k}_2$ interact with the nanostructure to generate a nonlinear signal in the phase-matched direction $2\mathbf{k}_2 - \mathbf{k}_1$ (FWM), $3\mathbf{k}_2 - 2\mathbf{k}_1$ (six-wave mixing), or $4\mathbf{k}_2 - 3\mathbf{k}_1$ (eight-wave mixing). A representative double-quantum 2D spectrum obtained for the "negative" time-ordering is shown in Fig. 7.13(c). The experiment is performed on a microcavity containing a single 8-nm wide $In_{0.04}Ga_{0.96}As$ QW with a cavity-exciton detuning $\delta = -0.38$ meV and a sample temperature of 4 K [387]. The wave numbers of the excitation beams and the FWM signal are shown schematically in Fig. 7.13(a). The fact that the two-quantum signal is non-zero indicates that two-polariton inter-actions are coherent. The grouped peaks on the diagonal near 2EL and 2EU arise from lower and upper polariton-polariton self-interactions, respectively that lead to a slight blue-shift of the two-polariton state compared to the sum of the individual po-lariton energies. Similarly, the off-diagonal peaks arise from polariton-polariton cross-interactions. Non-perturbative numerical simulations of the FWM signal attribute the fine structure within each peak group to non-resonant polariton-polariton scattering within the energy-momentum dispersion curves.

Interestingly, higher-order multi-quantum 2D spectra from six- and eight-wave mix-ing experiments reveal that up to at least four exciton-polaritons can be correlated—a higher order than what has been observed in bare QWs, demonstrating that higher-order correlations are mediated via the cavity [439]. A map of the cavity-exciton detun-ing on the relative peak amplitudes, homogeneous dephasing rate, and inhomogeneous broadening from the 2D spectra provides additional insight into how Coulomb correla-tion strengths are modified as the lower- and upper-polariton branches intersect [441]. Using an InGaAs QW microcavity structure embedded in a p-i-n diode to perform multi-quantum photocurrent spectroscopy, Autry *et al.* accessed the exciton-polariton ladder of states, which revealed doubly and triply avoided crossings for pairs and triplets of exciton-polaritons [14]. These higher-order avoided crossings demonstrated a characteristic semiclassical spectrum of two coupled quantum anharmonic oscilla-tors with the anharmonicity arising from polariton-polariton many-body interactions, which would otherwise be masked in linear spectroscopy. Applying similar methods to study coherent many-body phenomena in other material systems that exhibit strong excitonic interactions and light-matter coupling, such as transition metal dichalco-genides, might facilitate the realization of polariton condensation and superfluidity at room temperature.

8

Three-dimensional coherent spectroscopy

Although the title of this book has the word "Multidimensional," virtually all of the spectra presented so far have been only 2D. In principle it is a straightforward extension to produce higher-order spectra, although their acquisition can be extremely time-consuming and representing them becomes a challenge. However, several approaches to three-dimensional (3D) spectroscopy have been implemented and applied to different systems. To the best of our knowledge higher order than 3D spectroscopy has not been implemented. This chapter will provide a brief introduction to 3D coherent spectroscopy and then look at its use in atomic vapors and semiconductor quantum wells in more detail.

The first demonstration of 3D spectroscopy used an IR fifth-order response [96] in a five-beam apparatus to reveal heterogeneous structural relaxation dynamics in liquid water [130]. 3D electronic spectroscopy has also been demonstrated using the fifth-order response [117]. However, 3D electronic spectroscopy has tended to focus more on the third-order response [412], where it has been used to extract more information about the excitonic basis of light-harvesting materials [154] and fully determine the Hamiltonian of an atomic vapor [214]. Adding a third dimension is also very useful for multidimensional spectroscopy based on phase retrieval [83].

The use of 3D coherent spectroscopy is motivated by a number of advantages it provides compared to 2D coherent spectroscopy. Some of these are improvements of capabilities that are already present in 2D coherent spectroscopy, such as disentangling congested spectra, and some are completely new capabilities, such as measuring populations of multiply excited states, that cannot be realized in 2D coherent spectroscopy.

It is worthwhile to briefly address what is meant by "3D coherent spectroscopy." In principle a "3D spectrum" is simply one that is a function of three variables. The same is true in 2D, however, a 2D coherent spectrum is clearly different, and contains more information, than a 2D incoherent spectrum that is simply a series of power spectra as a function of excitation wavelength. Furthermore, simply taking a series of one-dimensional power spectra and measuring them as a function of an additional time variable, and possibly taking a Fourier transform, is different than a 2D coherent spectrum. The key difference is that a 2D coherent spectrum is produced by measuring the 1D *complex* spectra, which means that they contain phase information. The same is true for constructing 3D spectra; they are produced by taking a series of complex 2D spectra as function of a third time delay, and then taking a Fourier transform.

Clearly the distinction is whether the lower-dimensional spectra are power spectra or complex spectra, i.e., including phase information. By using complex spectra, it is possible to track the direction of the phase evolution of any terms that oscillate during the third time delay, and thus determine their frequency(s) including the signs. If power spectra are used, all that can be determined are the absolute values of any frequency differences, which provide much less separation between peaks. In addition, the use of complex spectra means that the phases of the signal are obtained, which can contain very useful information. Furthermore, the use of a "prepulse" that incoherently prepares the system, followed by the measurement of a 2D coherent spectrum, does not fall within the definition of a 3D coherent spectrum.

8.1   Fifth-order 3D infrared spectroscopy

Extending multidimensional spectroscopy into a third dimension was discussed theoretically for infrared and Raman excitation of molecular vibrations by Park and Cho [299]. Their results showed that three-mode coupling constants could be measured using 3D vibrational spectroscopy, while they are not accessible using 2D spectroscopy. This work was later generalized to properly describe the effect of bath degrees of freedom [61]. Further theoretical investigations by Hamm considered 3D spectra that are generated from a fifth-order nonlinear response [145]. The motivation of this study was to provide information about solvation effects beyond a linear response theory. Such non-Gaussian effects require the measurement of the three-point frequency fluctuation correlation function, which is inaccessible in third-order measurements. In addition, fifth-order spectroscopy can access $\nu = 2$ to $\nu = 3$ vibrational transitions, which are also not accessible in third-order experiments.

Some of the first 3D infrared measurements were published by Ding and Zanni [96]. This paper reported 3DIR of spectra carbonyl vibrations in three different model systems, $W(CO)_6$ in hexane, azide in an ionic glass, and an iridium dicarbonyl in a mixture of hexane and cholorform. These systems were chosen because the properties and dynamics of their carbonyl vibrations were well known. This work used the fifth-order response in a double-quantum three-pulse sequence. The fifth-order response was isolated by phase matching. These results extended earlier work using fifth-order 2DIR spectroscopy to study the same system [125] and helped set the foundations for interpreting 3DIR spectra.

One drawback of using only three pulses to generate a fifth-order signal is that a purely absorptive spectrum, as was demonstrated in 2DIR spectroscopy [184], cannot be generated. This difficulty was overcome using a 5-beam apparatus [130, 131]. Measurements of CO_2 in water agreed well with simulations. This work established a basis for studying isotope substituted liquid water using 3D spectroscopy [132]. These measurements revealed heterogeneous structural relaxation dynamics. The heterogeneity was attributed to subensembles of water molecules that did not interconvert on the half-picosecond timescale of the measurements.

Another advantage of a fifth-order signal is that it can produce population in the doubly excited states, whereas a third-order signal can only create a coherence between the singly and doubly excited states. This ability was used to study the vibrational dynamics of the OD stretch mode in ice [305]. Fifth-order 3DIR spectra were acquired using a five-pulse excitation sequence as function of the waiting time between the fourth and fifth pulses. By plotting the decay of peak volumes, the decay of the populations of the vibrationally excited states could be measured and compared with simulations. The results suggest that the second excited state lives for such a short time that the concept of a second excited state of the OD vibration becomes essentially meaningless as the mode is strongly mixed with lattice degrees of freedom.

Another approach to 3DIR spectroscopy uses a dual pulse shaper that generates a pair of pump beams and each pump beam is a pair of pulses [283]. The effect of the pump pulses on a probe beam is measured and used to construct a 3D spectrum, as was done earlier for 2D spectroscopy [141].

8.2   Fifth-order 3D electronic spectroscopy

Visible or near-IR light must be used to access electronic transitions, rather than the vibrational transitions excited by infrared light. Performing multidimensional coherent spectroscopy in the near-IR or visible regime is more challenging than in the infrared because of the difficulty in maintaining the phase stability between the excitation pulses. This difficulty is greatly increased when performing 3D spectroscopy. Not only is the acquisition time longer, thus phase stability must be maintained that much longer, but in addition phase stability is typically required for all time delays, whereas for 2D spectroscopy it is typically only needed between two pairs of pulses.

One demonstration of 3D spectroscopy in the visible regime used a fifth-order response from a three-pulse sequence [117]. The geometry and pulse sequence was similar to that used in the IR by Ding and Zanni [96], however, it used a single-shot method to acquire underlying one-dimensional spectra [152]. Using this method, a 3D spectrum, as shown in Fig. 8.1, could be acquired in eight minutes, which greatly reduced the issues of long term phase drift. The initial measurements were made on IR144 dye dissolved in methanol. The 3D spectrum revealed a vibronic cross peak that was fully resolved. This peak cannot be observed in 2D spectra because it is masked by the diagonal peak. Simulations based on a cumulant expansion truncated at second order reproduced the basic features in the 3D spectrum. Since fifth-order spectroscopies are sensitive to three-point frequency correlations, truncating the cumulant expansion at second order is typically insufficient, however, it is exact if the system displays Gaussian statistics. While the simulations displayed similar overall structure to the experiment, the spectral locations did not match. The disagreement between experiment and simulation were attributed to not having included the proper vibrational modes or incorrect coupling strengths.

Purely absorptive fifth-order electronic 3D spectroscopy was demonstrated using a five-pulse sequence [464]. This approach uses a pump-probe geometry as was first demonstrated for 2D spectroscopy [141]. In 2D spectroscopy, the pump is a two-pulse sequence and the appropriate terms are separated by the Fourier transform. To perform the fifth-order experiment the pump consisted of a four-pulse sequence, again generated by a pulse shaper. The method was demonstrated on chlorophyll *a* dissolved in methanol.

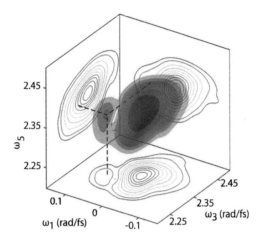

Fig. 8.1 Experimental fifth-order electronic 3D spectrum of IR144 produced using three excitation pulses. The dashed lines indicate a vibronic cross peak that is hiden in a 2D spectrum. The ω_1 axis is detected in a rotating frame. Adapted from Ref. 117.

8.3 Third-order 3D electronic spectroscopy

The previous two sections have described 3D spectroscopy based on the fifth-order nonlinear response, whereas 2D spectroscopy is usually based on the third-order response. It is natural to increase the order of the nonlinearity when increasing the dimensionality of the spectroscopy. In the most common pulse sequence used to generate a 2D spectrum, known as a rephasing spectrum (often designated as S_I), the two frequency dimensions correspond to times in which the system is in a coherent superposition state between two levels connected by a dipole-allowed transition. During the third time, known as the "waiting" time or "mixing" time, the system is usually described as being in a population state, which does not oscillate. Thus to have a third frequency dimension that corresponds to the system being in a dipole-allowed superposition state, the fifth-order nonlinear response must be used.

However, the description of the system being in a population state during the waiting time, T, is strictly only correct for a two-level system. If the system has two excited states, then it can be in a coherent superposition of those two states, which will oscillate, during T. These coherences can be observed by scanning T and displaying an alternate 2D spectrum [454]. But they will also result in a 3D spectrum that has peaks at finite ω_T. In addition, by using a different pulse ordering known as S_{III} because the conjugated pulse arrives third, it is possible to put the system in a "double-quantum coherence" during the time T [79, 176, 374], which will also generate peaks in third-order 3D spectra.

8.3.1 Three-dimensional spectra of atomic vapors

As an illustrative example, the D-lines of K atoms provide a three-level system that is appropriately simple and yet sufficiently complex to demonstrate the unique capa-

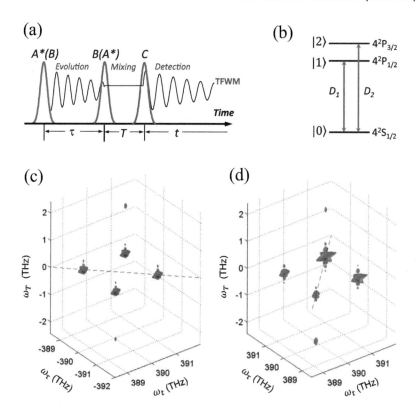

Fig. 8.2 (a) Excitation pulse sequence for 3D coherent spectrscopy. (b) Energy level diagram of a three-level system consisting of two D-lines for K atom. (c) Rephasing and (d) non-rephasing amplitude 3D spectra. Adapted from Ref. 214.

bilities of 3D coherent spectroscopy [214]. The experiment was performed in the box geometry with the excitation pulse sequence shown in Fig. 8.2(a). The TFWM signal is recorded in the frequency domain by using spectral interferometry while both τ and T are scanned. Fourier transforming the signal generates a 3D spectrum $S(\omega_\tau, \omega_T, \omega_t)$ in three frequency dimensions. Typical amplitude 3D spectra of K vapor are shown in Figs. 8.2(c) and (d) for the rephasing and non-rephasing excitation, respectively. The spectra are visualized with isosurfaces, a 3D analog of 2D contour lines. The solid red isosurface represents a higher amplitude than the semi-transparent red isosurface. The horizontal axes are the absorption (ω_τ) and emission (ω_t) frequencies. The vertical axis is the mixing frequency (ω_T) associated with the zero-quantum coherence between states $|1\rangle$ and $|2\rangle$. The dashed lines are diagonal lines ($\omega_t = |\omega_\tau|$, $\omega_T = 0$). Compared with a 2D spectrum, the 3D spectrum further unfolds the nonlinear optical response along a third dimension. The contributions from the zero-quantum coherences are isolated from the off-diagonal peaks (rephasing) or the diagonal peaks (non-rephasing) on the $\omega_T = 0$ plane.

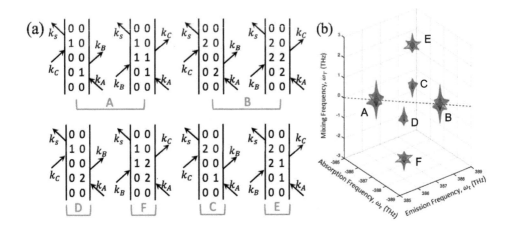

Fig. 8.3 (a) Double-sided Feynman diagrams showing eight possible excitation pathways in the experiment. (b) Simulated amplitude 3D spectrum including contributions from eight excitation pathways. Adapted from Ref. 397.

A critical challenge in 3D coherent spectroscopy is the interpretation of 3D spectral data. To understand how the information of a system's optical nonlinear response is manifested in 3D spectra, here we develop a theoretical model of 3D coherent spectroscopy in K vapor based on the optical Bloch equation (OBE). Based on the density matrix formalism, the light-matter interaction can be described by the equation of motion [341]

$$\dot{\rho} = -\frac{i}{\hbar}[H, \rho] - \frac{1}{2}\{\Gamma, \rho\}, \tag{8.1}$$

where $H = H_0 + H_I$, $[H, \rho] = H\rho - \rho H$ and $\{\Gamma, \rho\} = \Gamma\rho + \rho\Gamma$. The matrix elements of H_0 and H_I are $H_{0,ij} = \hbar\omega_i\delta_{ij}$ and $H_{I,ij} = -\mu_{ij}E(t)$ ($H_{I,ij} = 0$ for $i = j$), where $\hbar\omega_i$ is the energy of state $|i\rangle$, $E(t)$ is the electric field of the incident light, $\mu_{ij}(i \neq j)$ is the dipole moment of the transition between states $|i\rangle$ and $|j\rangle$, and δ_{ij} is the Kronecker delta function. The relaxation operator Γ has matrix elements $\Gamma_{ij} = \frac{1}{2}(\gamma_i + \gamma_j)$, where γ_i and γ_j are the population decay rates. In general, coherences can decay due to pure dephasing processes in addition to population decay. To include pure dephasing, the equation of motion can be modified as

$$\dot{\rho}_{ij} = -\frac{i}{\hbar}\sum_k (H_{ik}\rho_{kj} - \rho_{ik}H_{kj}) - \Gamma_{ij}\rho_{ij}. \tag{8.2}$$

The relaxation operator Γ is redefined to have matrix elements $\Gamma_{ij} = \frac{1}{2}(\gamma_i + \gamma_j) + \gamma_{ij}^{ph}$, where γ_{ij}^{ph} is the pure coherence dephasing rate ($\gamma_{ij}^{ph} = 0$ for $i = j$).

To calculate the third-order nonlinear response of the medium, the equation of motion can be solved perturbatively for $\rho_{ij}^{(3)}$. The polarization is then $P_{ij}^{(3)} = N\mu_{ij}\rho_{ij}e^{i\omega t}$, where N is the number density. In general, the third-order nonlinear response, $\rho_{ij}^{(3)}$,

Table 8.1 The frequency coordinates of each spectral peak in the 3D spectrum.

Group	Frequency Coordinates		
	ω_τ	ω_t	ω_T
A	$-\omega_{10}$	ω_{10}	0
B	$-\omega_{20}$	ω_{20}	0
C	$-\omega_{10}$	ω_{20}	0
D	$-\omega_{20}$	ω_{10}	0
E	$-\omega_{10}$	ω_{20}	ω_{21}
F	$-\omega_{20}$	ω_{10}	ω_{12}

consists of multiple terms contributed by different quantum pathways. The calculation can be aided with double-sided Feynman diagrams so that the contribution from each single pathway is calculated separately. For the three-level system shown in Fig. 8.2(b), there are eight possible excitation pathways contributing to the rephasing signal in the signal direction $-k_A + k_B + k_C$. These eight pathways are represented by the double-sided Feynman diagrams in Fig. 8.3(a). The eight pathways can be divided into six different groups labeled A through F. The two pathways in group A have an absorption frequency of $\omega_\tau = -\omega_{10}$, an emission frequency of $\omega_t = \omega_{10}$, and a mixing frequency of $\omega_T = 0$, resulting in a spectral peak at $(\omega_\tau = -\omega_{10}, \omega_t = \omega_{10}, \omega_T = 0)$ in the 3D spectrum. Similarly, the pathways in other groups lead to spectral peaks at different locations in the 3D spectrum. The specific frequency coordinates for the spectral peak associated with each group of pathways are listed in Table 8.1.

The third-order nonlinear response is calculated separately for each pathway represented by a double-side Feynman diagram. For the first diagram in group A,

$$\rho_{10,A1}^{(3)} = \frac{i\mu_{10}}{2\hbar} e^{i\mathbf{k}_C \cdot \mathbf{r}} \int_{-\infty}^{t} dt''' e^{-i(\omega_{10} - i\Gamma_{10})(t-t''')} \hat{E}_C(t''') e^{-i\omega t'''}$$

$$\frac{-i\mu_{10}}{2\hbar} e^{i\mathbf{k}_B \cdot \mathbf{r}} \int_{-\infty}^{t'''} dt'' e^{-\Gamma_{00}(t'''-t'')} \hat{E}_B(t'') e^{-i\omega t''}$$

$$\frac{-i\mu_{10}}{2\hbar} e^{-i\mathbf{k}_A \cdot \mathbf{r}} \int_{-\infty}^{t''} dt' e^{-i(\omega_{01} - i\Gamma_{10})(t''-t')} \hat{E}_A^*(t') e^{i\omega t'} \rho_{00}^{(0)}, \qquad (8.3)$$

where ω is the carrier frequency of all pulses, and t_1, t_2, and t_3 are the arrival times for pulses A, B, and C, respectively. The system is initially in the ground state, that is $\rho_{00}^{(0)} = 1$. For the cases where the pulse duration is much shorter than the characteristic time scale of the dynamics, the excitation pulse can be approximated with a Dirac delta function such that $\hat{E}_{A,B,C}(t) = \mathcal{E}_{A,B,C} \delta(t - t_{1,2,3})$, where $\mathcal{E}_{A,B,C}$ represents the amplitude of the pulse. Then equation (8.3) can be rewritten as

$$
\rho_{10,A1}^{(3)} = \frac{-i\mu_{10}^3}{8\hbar^3} e^{i(-\mathbf{k}_A + \mathbf{k}_B + \mathbf{k}_C)\cdot\mathbf{r}} \mathcal{E}_A^* \mathcal{E}_B \mathcal{E}_C \int_{-\infty}^{t} dt''' e^{-i(\omega_{10} - i\Gamma_{10})(t - t''')} \delta(t''' - t_3) e^{-i\omega t'''}
$$

$$
\int_{-\infty}^{t'''} dt'' e^{-\Gamma_{00}(t''' - t'')} \delta(t'' - t_2) e^{-i\omega t''} \int_{-\infty}^{t''} dt' e^{-i(\omega_{01} - i\Gamma_{10})(t'' - t')} \delta(t' - t_1) e^{i\omega t'}
$$

$$
= \frac{-i\mu_{10}^3}{8\hbar^3} e^{i\mathbf{k}_S\cdot\mathbf{r}} \mathcal{E}_A^* \mathcal{E}_B \mathcal{E}_C e^{-\Gamma_{10}(t - t_3 + t_2 - t_1)} e^{-\Gamma_{00}(t_3 - t_2)} e^{-i\omega_{10}t} e^{-i(\omega - \omega_{10})(t_3 + t_2 - t_1)}
$$

$$
\Theta(t_2 - t_1)\Theta(t_3 - t_2)\Theta(t - t_3), \tag{8.4}
$$

where $\Theta(x)$'s are Heaviside step functions. We define the time intervals as $\tau = t_2 - t_1$, $T = t_3 - t_2$, and $t = t - t_3$, and assume the excitation is on resonance ($\omega - \omega_{10} = 0$). The third-order polarization is

$$
P_{10,A1}^{(3)}(\tau, t, T) = N\mu_{10}\rho_{10,A1}^{(3)} e^{i\omega t}
$$

$$
= \frac{-iN\mu_{10}^4}{8\hbar^3} e^{i\mathbf{k}_S\cdot\mathbf{r}} \mathcal{E}_A^* \mathcal{E}_B \mathcal{E}_C \Theta(\tau)\Theta(T)\Theta(t) e^{-\Gamma_{10}(t+\tau) - \Gamma_{00}T}. \tag{8.5}
$$

This is the time-domain solution in the rotating frame. In general, the spectrum in the time domain can be constructed by stepping the time delays τ, T, and t. The time-domain spectrum can then be numerically Fourier transformed to obtain the frequency-domain spectrum. In the current case, the Fourier transform of Eq. (8.5) has an analytical form

$$
P_{10,A1}^{(3)}(\omega_\tau, \omega_t, \omega_T)
$$

$$
= \frac{-iN\mu_{10}^4}{16\sqrt{2}\pi^{\frac{3}{2}}\hbar^3} e^{i\mathbf{k}_S\cdot\mathbf{r}} \mathcal{E}_A^* \mathcal{E}_B \mathcal{E}_C \frac{1}{\Gamma_{10} - i(\omega_\tau + \omega_{10})} \cdot \frac{1}{\Gamma_{10} - i(\omega_t - \omega_{10})} \cdot \frac{1}{\Gamma_{00} - i\omega_T}. \tag{8.6}
$$

Similarly, the third-order polarization due to the second diagram in group A is

$$
P_{10,A2}^{(3)}(\omega_\tau, \omega_t, \omega_T)
$$

$$
= \frac{-iN\mu_{10}^4}{16\sqrt{2}\pi^{\frac{3}{2}}\hbar^3} e^{i\mathbf{k}_S\cdot\mathbf{r}} \mathcal{E}_A^* \mathcal{E}_B \mathcal{E}_C \frac{1}{\Gamma_{10} - i(\omega_\tau + \omega_{10})} \cdot \frac{1}{\Gamma_{10} - i(\omega_t - \omega_{10})} \cdot \frac{1}{\Gamma_{11} - i\omega_T}. \tag{8.7}
$$

The sum $P_{10,A}^{(3)} = P_{10,A1}^{(3)} + P_{10,A2}^{(3)}$ contributes to the 3D spectral peak at ($\omega_\tau = \omega_{10}, \omega_t = \omega_{10}, \omega_T = 0$). This result shows that the peak amplitude is proportional to μ_{10}^4, the peak position reveals the transition frequency, and the peak lineshape profiles in three frequency dimensions are determined by the relaxation rates. Therefore, all parameters, namely the resonance energies, dipole moments and dephasing rates, in the Hamiltonian matrix elements associated with these two pathways can be extracted from the spectral peak.

The same procedures can be applied to other pathways to calculate the corresponding third-order polarizations. For group B,

$$P_{20,B}^{(3)}(\omega_\tau, \omega_t, \omega_T) = \frac{-iN\mu_{20}^4}{16\sqrt{2}\pi^{\frac{3}{2}}\hbar^3} e^{i\mathbf{k}_S \cdot \mathbf{r}} \mathcal{E}_A^* \mathcal{E}_B \mathcal{E}_C$$

$$\frac{1}{\Gamma_{20} - i(\omega_\tau + \omega_{20})} \cdot \frac{1}{\Gamma_{20} - i(\omega_t - \omega_{20})} \cdot \left(\frac{1}{\Gamma_{00} - i\omega_T} + \frac{1}{\Gamma_{22} - i\omega_T}\right). \tag{8.8}$$

For group C,

$$P_{20,C}^{(3)}(\omega_\tau, \omega_t, \omega_T) = \frac{-iN\mu_{10}^2\mu_{20}^2}{16\sqrt{2}\pi^{\frac{3}{2}}\hbar^3} e^{i\mathbf{k}_S \cdot \mathbf{r}} \mathcal{E}_A^* \mathcal{E}_B \mathcal{E}_C$$

$$\frac{1}{\Gamma_{10} - i(\omega_\tau + \omega_{10})} \cdot \frac{1}{\Gamma_{20} - i(\omega_t - \omega_{20})} \cdot \frac{1}{\Gamma_{00} - i\omega_T}. \tag{8.9}$$

For group D,

$$P_{10,D}^{(3)}(\omega_\tau, \omega_t, \omega_T) = \frac{-iN\mu_{10}^2\mu_{20}^2}{16\sqrt{2}\pi^{\frac{3}{2}}\hbar^3} e^{i\mathbf{k}_S \cdot \mathbf{r}} \mathcal{E}_A^* \mathcal{E}_B \mathcal{E}_C$$

$$\frac{1}{\Gamma_{20} - i(\omega_\tau + \omega_{20})} \cdot \frac{1}{\Gamma_{10} - i(\omega_t - \omega_{10})} \cdot \frac{1}{\Gamma_{00} - i\omega_T}. \tag{8.10}$$

For group E,

$$P_{20,E}^{(3)}(\omega_\tau, \omega_t, \omega_T) = \frac{-iN\mu_{10}^2\mu_{20}^2}{16\sqrt{2}\pi^{\frac{3}{2}}\hbar^3} e^{i\mathbf{k}_S \cdot \mathbf{r}} \mathcal{E}_A^* \mathcal{E}_B \mathcal{E}_C$$

$$\frac{1}{\Gamma_{10} - i(\omega_\tau + \omega_{10})} \cdot \frac{1}{\Gamma_{20} - i(\omega_t - \omega_{20})} \cdot \frac{1}{\Gamma_{21} - i(\omega_T - \omega_{21})}. \tag{8.11}$$

For group F,

$$P_{10,F}^{(3)}(\omega_\tau, \omega_t, \omega_T) = \frac{-iN\mu_{10}^2\mu_{20}^2}{16\sqrt{2}\pi^{\frac{3}{2}}\hbar^3} e^{i\mathbf{k}_S \cdot \mathbf{r}} \mathcal{E}_A^* \mathcal{E}_B \mathcal{E}_C$$

$$\frac{1}{\Gamma_{20} - i(\omega_\tau + \omega_{20})} \cdot \frac{1}{\Gamma_{10} - i(\omega_t - \omega_{10})} \cdot \frac{1}{\Gamma_{21} - i(\omega_T - \omega_{12})}. \tag{8.12}$$

A complete 3D spectrum includes all contributions from Eqs. (8.6)–(8.12). The simulated 3D spectrum is shown in Fig. 8.3(b), where the magnitude of the spectrum is plotted as isosurfaces. The solid and semi-transparent red surfaces represent the isosurfaces with the values of 0.3 and 0.1, respectively, with the maximum normalized to 1. The following parameters are used in the simulation: $\omega_{10} = 386$ THz, $\omega_{20} = 388$ THz, $\Gamma_{00} = \Gamma_{11} = \Gamma_{22} = 0.1$ THz, $\Gamma_{10} = \Gamma_{20} = \Gamma_{21} = 0.05$ THz, and $\mu_{10} = \mu_{20}$.

The 3D spectrum has six isolated peaks corresponding to different excitation pathways in Fig. 8.3(a). Compared to 2D spectra, a 3D spectrum further unfolds the nonlinear optical response in the third dimension ω_T. The off-diagonal peaks in a rephasing 2D spectrum have contributions from two pathways [78]. One of them generates a population during T and does not oscillate, while the other generates a coherence that

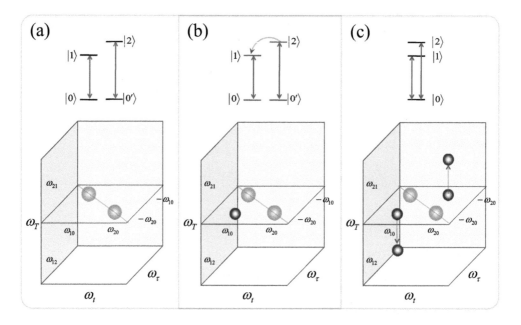

Fig. 8.4 Schematics of 3D spectral patterns of different energy level schemes: (a) two independent two-level systems, (b) two two-level systems coupled via an incoherent relaxation process, and (c) a three-level V system. Adapted from Ref. 214.

oscillates during T. In a 3D spectrum, these two contributions have difference frequencies in the ω_T directions and thus are separated. As a result, the contributions from different excitation quantum pathways are well separated in a 3D spectrum. Peaks D–F each corresponds to only a single-quantum pathway. Peak A and B each have contributions from two pathways that describe the same process for a closed system. This unique ability to unravel contributions from different excitation quantum pathways provides a global picture of the nonlinear response of a complex system, offering several advantages.

The first advantage is that the 3D spectrum allows the determination of a complete energy level scheme rather than just possible transitions that can be inferred from a 1D spectrum. For example, an absorption spectrum with two absorption peaks can be a result of a three-level V system with two excited states, two independent species of two-level systems, or two two-level systems coupled via an incoherent relaxation process. This ambiguity is absent in a 3D spectrum since the 3D spectral patterns are different for the three schemes. As shown in Fig. 8.4, 3D spectra of these energy schemes have different spectral features. In scheme (a), two transitions are independent and there is no coupling between them, so the corresponding 3DFT spectrum has only two diagonal peaks. In scheme (b), the excited states are coupled via incoherent relaxation processes that allow the population transfer from state $|2\rangle$ to $|1\rangle$. This scheme results in a off-diagonal peak that represents the process absorbing at ω_{20} and emitting at ω_{10}. In scheme (c), two transitions are coupled via the common ground state. The corresponding 3DFT spectrum has two off-diagonal peaks due to the ground-state bleaching. Moreover, the contributions from the zero-quantum coherence between state $|2\rangle$ to $|1\rangle$ are separated from the off-diagonal peaks and produce two peaks at $\omega_T = \pm\omega_{21}$. In a 2DFT spectrum, the zero-quantum coherences are not separated from the ground-state bleach contribution.

Second, a 3D spectrum contains complete information of the system's third-order optical response and various 2D spectra can be retrived by projecting a 3D spectrum onto a two-dimensional plane. According to the projection-slice theorem for Fourier transforms [108], the projection of a 3D spectrum onto a 2D plane in the frequency domain is equivalent to a 2D slice in the time domain. For example, three different types of 2D spectra can be obtained by projecting a rephasing 3D spectrum onto three 2D planes, as shown in Figs. 8.5 (a-d). The projection onto the bottom plane (ω_τ, ω_t) is a rephasing one-quantum 2D spectrum $S_I(\omega_\tau, \omega_t)$ at $T = 0$. The 2D spectrum at any given T can be calculated by multiplying the projection with a linear phase term $e^{-i\omega_T T}$, according to the Fourier shift theorem [108]. Similarly, zero-quantum 2D spectra at any given τ can be obtained from the projection onto the left-back plane (ω_T, ω_t) with an appropriate phase shift. The projections onto these two planes are plotted in Figs. 8.5(b) and (c). These projections are the spectra that can be acquired by 2D coherent spectroscopy; however, the projection onto the right-back plane (ω_τ, ω_T) is a unique two-dimensional spectrum that is not accessible by conventional 2D spectroscopic techniques based on spectral interferometry. Shown in Fig. 8.5(d), this spectrum is a function of absorption frequency ω_τ and mixing frequency ω_T, and independent of the emission frequency. Furthermore, this spectrum is not simply time integrated, but corresponds to an instantaneous emission time. This projection on the right-back plane can reveal the temporal dynamics of the emission signal in terms of the absorption and mixing frequencies. In general, a 3D spectrum can be projected onto any planes other than the three planes made of three frequency axes. Such skew projections may have special advantages over standard projections.

Third, the most important advantage of 3D coherent spectroscopy is its ability to unravel quantum pathways by isolating their contributions in different spectral peaks. Consider the rephasing 3D spectrum in Fig. 8.5(a), the contributions from different pathways are isolated in six different peaks labeled 3A–3F. These pathways are not completely isolated in the 2D spectra in Figs. 8.5(b–d) in which some peaks consists of the combinations of two 3D peaks. There are eight different quantum pathways, as represented by the double-sided Feynman diagrams in Fig. 8.5(e), that are associated with the rephasing 3D spectrum. In the rephasing 2D spectrum or the bottom plane projection in Fig. 8.5(b), each peak (labeled RA, RB, RCE, and RDF) has contributions from two different pathways. The off-diagonal peak RCE (RDF) is the projection of two 3D peaks 3C and 3E (3D and 3F). The zero-quantum coherence terms can be isolated in the zero-quantum spectrum or the 2D projection in Fig. 8.5(c) as peaks TE and TF. But the diagonal and off-diagonal 3DFT peaks 3A and 3D (3B and 3C) are combined into the peak TAD (TBC). As shown in Fig. 8.5(e), the quantum pathways are most isolated in the 3DFT spectrum where four peaks each represents a single pathway. The remaining two peaks 3A and 3B include two pathways each, which describe equivalent processes in the case of a closed system. Therefore, in a 3D spectrum physical processes are well isolated and the information on a particular process can

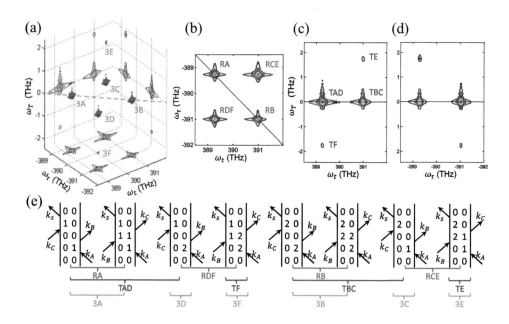

Fig. 8.5 (a) Experimental rephasing 3D spectrum with 2D projections onto three planes. (b) 2D projection onto the bottom plane. (c) 2D projection onto the left-back plane. (d) 2D projection onto the right-back plane. (e) Double-sided Feynman diagrams representing all excitation pathways contributing to the spectra. The pathways are grouped corresponding to specific spectral peaks. Adapted from Ref. 214.

be unambiguously determined by analyzing a single 3D spectral peak corresponding to the process.

For example, as shown in Fig. 8.6, peak 3E can be isolated and analyzed independently of other peaks for the information about the corresponding pathway. The amplitude of peak 3E is

$$|P_{20,E}^{(3)}(\omega_\tau, \omega_T, \omega_t)| \propto$$
$$\mu_{01}^2\mu_{02}^2 \frac{1}{\sqrt{(\omega_\tau - \omega_{01})^2 + \Gamma_{10}^2}} \cdot \frac{1}{\sqrt{(\omega_T - \omega_{21})^2 + \Gamma_{21}^2}} \cdot \frac{1}{\sqrt{(\omega_t - \omega_{20})^2 + \Gamma_{20}^2}}. \quad (8.13)$$

This equation describes the lineshape profile of peak 3E, which is the square root of a Lorentzian along each frequency dimension. One can fit the slices in the three directions, as shown in Fig. 8.6(b), to determine the parameters in this pathway including the resonance frequencies and the relaxation rates.

Overall, the spectral pattern, the strength and the lineshape of each peak in a 3D spectrum can be used to determine the system's energy level scheme and the parameters needed to fully characterize the Hamiltonian, including the transition energy, dipole moment and relaxation rate [214]. This ability to experimentally measure the Hamiltonian can have profound influence on some fields, since quantitative knowledge of a system's Hamiltonian can enable quantitative prediction and control of quantum

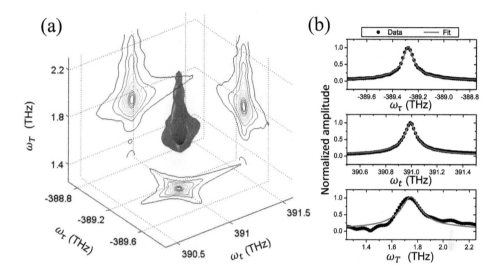

Fig. 8.6 (a) A single 3D spectral peak, labeled 3E in Fig. 8.5(a), with projections on three planes. The peak represents a single-quantum pathway. The dynamics of the associated processes can be understood by analyzing the lineshape and projections. (b) Profiles of peak 3E along three frequency directions. Dotted lines are experimental data and red lines are fits to a square-root Lorentzian function. Linewidths of the profiles determine the relaxation rates. Adapted from Ref. 214.

phenomena. One example is coherent control, where a prior knowledge of the quantum pathways used to construct the Hamiltonian is essential for designing an effective learning algorithm, suitable initial conditions, and even a deterministic control scheme.

Optical 3D coherent spectroscopy is different from simply tracking the amplitude of a single point on a series of 2D spectra taken at different waiting times [66, 106, 185] in the ability to isolate transitions that are overlapping in 2D spectra. By taking 2D spectra at different waiting times, the quantum beats were observed as a function of T due to the interference between two or more quantum pathways, implying that the pathways are not isolated. Furthermore, a series of 2D spectra do not necessarily maintain the correlation in the T dimension, which is key to isolate contributions from pathways. In many systems, the dephasing rates Γ_{10} and Γ_{20} are for optical transitions between potential energy surfaces (in molecules) or bands (in solids). However, Γ_{21} will be the dephasing rate between vibrational levels for one potential surface or for intraband transitions, and thus is likely to be much slower as the fluctuations are often correlated. If this is the case, then transitions that are strongly overlapped in the ω_t and ω_τ dimensions may be well resolved in the ω_T dimension. This resolution allows Γ_{10} and Γ_{20} to be accurately determined from the 3D spectrum, whereas they cannot be from a linear or 2D spectrum.

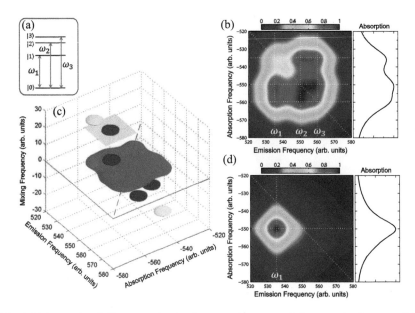

Fig. 8.7 (a) Energy level diagram of a single ground state and three excited states. (b) A 2D spectrum for this system when the linewidths of the optical transitions are larger than the separation between the levels. The projection equivalent to an absorption spectrum is shown on the right panel. (c) A 3D spectrum showing that some peaks become separated in the mixing frequency direction. (d) The 2D slice corresponding to gray plane in (c) and projection showing isolation of $|0\rangle \rightarrow |2\rangle$ transition from which the line width can be measured. Adapted from Ref. 214.

To illustrate this point, consider a system consisting of a single ground state and three excited states as shown in Fig. 8.7(a), where the frequency splitting between the excited states is small compared to the widths of each of their transitions from the ground state. In this case, a one-dimensional spectrum will not resolve the transitions and any attempt to determine their relative strengths and widths will require the use of peak fitting algorithms, which are always problematic. A two-dimensional spectrum actually does not improve the situation very much, as shown in Fig. 8.7(b), the peaks are not well resolved. However, by measuring a 3D spectrum, as shown in Fig. 8.7(c) and then plotting the 2D spectrum corresponding to an appropriate plane of fixed mixing frequency, as shown in Fig. 8.7(d), the transitions are well resolved, allowing the linewidths to easily and uniquely be determined.

Optical 3D coherent spectroscopy can be performed for different energy schemes. Besides the three-level V scheme in K atom, another typical example is the three-level ladder scheme in Rb atom, as shown in Fig. 5.10(a). The access to the doubly excited states require the double-quantum excitation pulse sequence that can also generate a 3D spectrum with three frequency dimensions corresponding to the three time delays. A two-quantum 3D spectrum can be generated by recording the TFWM signal in the frequency domain as both τ and T are scanned and Fourier transforming the signal into a three-dimensional frequency space. Such a double-quantum 3D spectrum

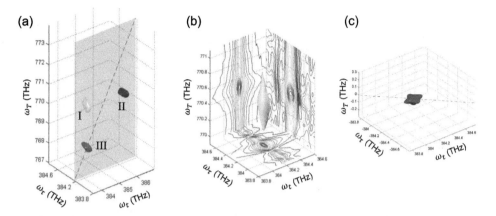

Fig. 8.8 (a) Double-quantum 3D spectrum of Rb. (b) An isolated 3D peak and its 2D projections. (c) Single-quantum 3D spectrum of the D_2 transition of Rb. Adapted from Ref. 214.

for Rb atom is shown in Fig. 8.8(a), where the amplitude of the 3D spectrum is plotted as isosurfaces. There are three spectral peaks. With the maximum amplitude being normalized to one, the red, yellow, and blue peaks are the isosurfaces with the amplitude of 0.35, 0.1, and 0.015, respectively. The gray plane is the plane of $\omega_\tau = 384.2$ THz, and the dashed blue line on it represents $\omega_T = 2\omega_t$.

For this double-quantum 3D spectrum, the peak positions are determined by the energy level scheme in Fig. 5.10(a). All three peaks are located on the gray plane corresponding to the single-quantum resonant frequency during the first time delay, the energy difference between states $|1\rangle$ and $|0\rangle$, i.e., $\omega_\tau = 384.2$ THz. The double-quantum coherence frequency (ω_T) of the yellow and blue peaks is 770.5 THz, the energy difference between states $|2\rangle$ and $|0\rangle$. However, the double-quantum coherence frequency (ω_T) of the red peak is 768.4 THz, double the energy difference between states $|1\rangle$ and $|0\rangle$. The emission frequency (ω_t) of the red and yellow peaks is the D_2 transition frequency (384.2 THz), while the emission frequency (ω_t) of the blue peak is 386.3 THz, the energy difference between states $|2\rangle$ and $|1\rangle$. Therefore, the yellow peak is identified as the spectral contribution from a quantum pathway involving the single-atom doubly excited state 5^2D, corresponding to the single excitation $|0\rangle \rightarrow |1\rangle$, the double excitation $|0\rangle \rightarrow |2\rangle$, and the emission $|1\rangle \rightarrow |0\rangle$. Similarly, the blue peak is also due to a quantum pathway involving single-atom state 5^2D but the signal radiates at the transition $|2\rangle \rightarrow |1\rangle$. The transition dipole moment of $|2\rangle \rightarrow |1\rangle$ is weaker than that of $|1\rangle \rightarrow |0\rangle$. As a result, the amplitude of the blue peak is about an order of magnitude smaller than the yellow peak. With the two double-quantum pathways isolated in two peaks, one can analyze a single 3D peak and extract the information about the corresponding quantum pathway. Fig. 8.8(b) shows the yellow peak isolated from the 3D spectrum in Fig. 8.8(a). The peak can now be projected onto different planes without the interference from other peaks, giving various 2D spectra associated with a single-quantum pathway. The relevant relaxation rates can be revealed by the lineshape analysis of the peak. This analysis gives important information about Hamiltonian including the level energies and transition dipole moments. It also gives full information about all off-diagonal elements of the relaxation matrix Γ. The diagonal elements of Γ corresponding to levels $|0\rangle$ and $|1\rangle$ can be extracted from a single-quantum spectrum, shown in Fig. 8.8(c), as was done for K vapor. The diagonal element of Γ corresponding to $|2\rangle$ cannot be extracted from 3D spectra, however, it will not contribute to a coherent control experiment using third-order (or lower) optical nonlinearities. It can be extracted from a higher-dimensional coherent spectrum. The red peak in the 3D spectrum can not be explained by the energy level scheme of a single Rb atom and is attributed to the collective excited state of two interacting Rb atom, as discussed in Section 5.3. This example demonstrates that 3D coherent spectroscopy can be used to determine the nonlinear optical response of systems with a ladder energy level scheme and reveal information about individual atoms as well as many-body interactions between atoms.

In general, optical 3D coherent spectroscopy can provide sufficient information of the system's nonlinear optical response up to the third order and be used to construct the system's Hamiltonian in many cases. For higher-order nonlinear optical response, the technique in principle can be extended to perform n-dimensional coherent spectroscopy. The abilities of different spectroscopic techniques to determine the Hamiltonian and nonlinear optical response of different systems with increasing complexity are summarized in Table 8.2. In this table, "H" and "Γ" indicates that the Hamiltonian and relaxation matrix can be fully determined, "$\chi^{(n)}$" indicates that the nth-order nonlinear optical response can be fully measured, "$C(\chi^{(i)})$" means that higher-order nonlinear optical response can be calculated, "$\times$" means that this method cannot construct the Hamiltonian or nonlinear optical response of the system.

The simplest system that can be studied using optical methods is probably an isolated two-level system that can only decay by spontaneous emission. The Hamiltonian only consists of a dipole moment and energy splitting as spontaneous emission rate is determined by these two parameters. In this case, the Hamiltonian can be fully determined from a properly calibrated one-dimensional linear spectrum such as an absorption spectrum, or emission spectrum.

There are several ways that the complexity of the system can be increased: 1) inhomogeneous broadening due to Doppler effects in a gas or static disorder in a solid due, for example, to random crystal fields or structural fluctuations, 2) fluctuations of the level energies due to coupling to the environment, which leads to dephasing coherences faster than the population decay, or 3) decay by other (possibly non-radiative) channels. All three of these are considered simultaneously in the second scenario (the third column in the table). The presence of any of these requires that a nonlinear spectroscopic technique be used to determine the Hamiltonian.

The presence of inhomogeneous broadening was one of the primary reasons for developing one-dimensional nonlinear spectroscopic techniques, such as spin or photon echoes in the time domain or hole-burning in the frequency domain. However, time-integrated echo experiments alone do not fully characterize the system as the nature of the broadening, whether it is homogeneous or inhomogeneous must be known. Typically this determination can be made by comparing the nonlinear results with a linear spectrum, thus the third row in the table is for such a combination. Instead of using a

Table 8.2 The ability of different spectroscopic techniques to determine the Hamiltonian and the optical response of different systems with increasing complexity from left to right. The sophistication of the techniques increases from top to bottom.

	Homo. 2 level	InHomo. 2 level	Hetero. 2 level	One manifold $n+1$ level	m manifold $n+1$ level
Linear	H,Γ $\chi^{(1)}, C(\chi^{(i)})$	×	×	×	×
1D Non-lin. + Linear	H,Γ $\chi^{(3)}, C(\chi^{(i)})$	H,Γ $\chi^{(3)}, C(\chi^{(i)})$	×	×	×
2DCS	H,Γ $\chi^{(3)}, C(\chi^{(i)})$	H,Γ $\chi^{(3)}, C(\chi^{(i)})$	H,Γ $\chi^{(3)}, C(\chi^{(i)})$	×	×
3DCS	H,Γ $\chi^{(3)}, C(\chi^{(i)})$	H,Γ $\chi^{(3)}, C(\chi^{(i)})$	H,Γ $\chi^{(3)}, C(\chi^{(i)})$	H,Γ $\chi^{(3)}, C(\chi^{(i)})$	$\chi^{(3)}$
nDCS	H,Γ $\chi^{(N)}, C(\chi^{(i)})$	H,Γ $\chi^{(N)}, C(\chi^{(i)})$	H,Γ $\chi^{(N)}, C(\chi^{(i)})$	H,Γ $\chi^{(N)}, C(\chi^{(i)})$	H,Γ $\chi^{(N)}, C(\chi^{(i)})$

combination of a linear spectrum and a one-dimensional nonlinear spectrum, it is also possible to temporally or spectrally resolve the nonlinear signal, which effectively gives a two-dimensional spectrum, but is not as powerful as full 2D coherent spectroscopy because it relies on interference between different contributions and does not correlate frequencies in the different dimensions.

The next level of complexity is a heterogeneous mixture of two-level systems, and specifically making a distinction with a multi-level system. A one-dimensional non-linear spectrum alone is not sufficient to do so, although it is possible to time- or frequency-resolve the signal. Again using 2D coherent spectroscopy can resolve this ambiguity and completely determine the Hamiltonian.

A system consisting of a single ground state coupled by a dipole transition to a manifold of levels, i.e., a generalization of the classic V-type system, cannot be fully characterized in a 2D spectrum because all peaks in the spectrum will have contributions from multiple quantum pathways. Due to the correlation between frequencies, in a 3D spectrum, peaks corresponding to single pathways can be isolated, and thus all the relevant parameters can be determined. If this is done for $n - 1$ peaks, where n is the number levels in the excited state manifold, then the Hamiltonian can be fully determined. This was demonstrated for the simplest case of $n = 2$. Similar analysis can be done for a system consisting of a manifold of levels for the lower state, i.e., a generalization of a Λ type system, thus a 3D spectrum is sufficient in this case as well.

For a system with multiple manifolds of excited states, with each manifold connected by a dipole-allowed transition, i.e., a generalization of a "ladder" type system, a 3D spectrum can fully determine the third-order nonlinear optical response, however, that is not enough to fully determine the relaxation matrix, Γ. Strictly speaking the Hamiltonian, H, can be fully determined, although the full Hamiltonian really includes both H and Γ. The example of 3D spectroscopy on Rb demonstrates fully determining the third-order optical response for a ladder system ($m = 2$ where m is the number of manifolds) and each manifold containing a single level. The only missing information for this case is the population decay rate of the doubly excited state, Γ_{22}; all other parameters are determined. If $m > 2$, then a 3D spectrum provides no information on manifolds higher than the second one. However, they do not contribute to the third-order nonlinear optical response, and thus are not relevant to a coherent control experiment based on the third-order response. To fully characterize these systems, the dimensionality of the spectrum, N, must be $N \geq 2m + 1$, thus a five dimensional spectrum is needed to fully characterize a system with $m = 2$.

Given full information about the Hamiltonian, it is possible to calculate the higher order nonlinear optical response. This was indicated in the table by "$C(\chi^{(i)})$". This ability means that for most systems, sufficient information is obtained to design a coherent control scheme that uses higher-order optical nonlinearities, even if they are not directly measured.

8.3.2   Three-dimensional spectroscopy of semiconductor quantum wells

Third-order 3D electronic spectroscopy was demonstrated on excitonic resonances in GaAs quantum wells [412]. This measurement used a diffractive pulse shaper to generate four phase locked pulses (three for excitation, one as local oscillator) [413]. Both

S_I and S_{III} spectra were generated. Projections of the 3D spectra were used to show that individual quantum pathways could be isolated, even though they are not isolated in standard 2D spectra. A similar setup was used to study asymmetric double GaAs quantum wells [403]. In addition to using 3D spectra, this work also selects specific pathways by tailoring the spectra of the exciting pulses. The pathway-select spectra reveal coherent interactions between the spatially separated excitons.

Adding a third dimension proved to be quite useful for multidimensional spectroscopy based on phase retrieval [83]. Phase retrieval methods, first demonstrated in two-dimensions [82], do not rely on phase stabilization of the pulse sequence, but rather utilize phase retrieval algorithms, similar to those used in ultrashort pulse measurement method such as frequency resolved optical gating [406], to correct the as-measured phases. Extending this to a third dimension actually improves the method because it eliminates some ambiguities. An example of a 3D spectrum produced by solving the phase retrieval problem is shown in Fig. 8.9 for a GaAs double-quantum well sample, which displays a very rich 2D spectrum [221]. Extending the spectrum into a third dimension helps to isolate the various contributions and give greater insight into the mechanisms responsible for the various peaks. These studies were extended to examine the correlation of the inhomogeneous broadening, which is due to fluctuations of the well widths [143]. The 3D spectra show that the broadening is uncorrelated for excitons localized in different wells, whereas it is correlated for the heavy-hole and light-hole excitons localized in the same well.

When applied to semiconductor nanostructures, 3DCS is an effective tool for resolving ultrafast electronic dynamics and weak optical transitions that are otherwise obscured in a 2D spectrum. In an asymmetric InGaAs DQW nanostructure similar to the one discussed in Section 7.4, 3DCS experiments have revealed new transitions and inter-well coupling mechanisms that are difficult to identify with other optical techniques [53, 83, 143]. Weak optical transitions associated with excitons delocalized within the QW barriers, spatially-indirect excitons with an electron in one QW and the hole in the other, and parity-forbidden excitons have been observed; interactions between them lead to a complex 3D spectrum with at least 5 diagonal peaks and 18 off-diagonal peaks [401]. A quantitative analysis of the 3D spectral lineshapes provides new insight into their coupling mechanisms.

A specific interaction process can be further isolated in a 3D spectrum by selectively exciting and probing specific quantum pathways through spectral shaping of the pulses [403]. Multicolor 3DCS has been demonstrated on the asymmetric InGaAs DQW system to identify, isolate, and analyze weak coherent coupling between excitons in spatially separated QWs. The two-color pulse sequence is shown in Fig. 8.10(a). The spectrum of the first pulse $\mathcal{E}_1$ is filtered so that it is resonant only with the exciton transitions in the NW. Similarly, the second pulse $\mathcal{E}_2$ is resonant only with the WW transition. The interaction of these two pulses with the DQW drives the system into a non-radiative Raman-like coherence between the WW and NW states that evolves during the delay τ_2. The third pulse $\mathcal{E}_3$ has the full spectral bandwidth to excite transitions in both QWs. Using this pulse sequence, the quantum pathways arising from the inter-well stimulated Raman coherence are isolated in the 3D spectrum, as shown in Fig. 8.10(b). The absence of other quantum pathways enhances the signal-to-noise of

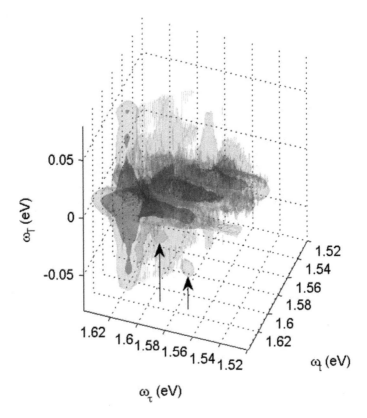

Fig. 8.9 Experimental 3D electronic spectrum produced by solving the phase retrieval problem. The sample is a double GaAs quantum well. Adapted from Ref. 83.

these peaks, enabling quantitative peak lineshape analysis. The ability to isolate specific quantum pathways using the combination of spectral shaping and 3DCS might help to resolve open questions regarding the role of quantum mechanics in the energy transfer processes of photosynthetic complexes and other material systems, including layered semiconductors.

8.3.3   3D coherent spectroscopy of light-harvesting centers

Two-dimensional electronic spectroscopy of light-harvesting centers provided evidence for the involvement of quantum coherence [66, 106]. These results have provoked significant discussion. The evidence was based on monitoring the evolution of peaks in the 2D spectrum as a function of waiting time T, which suggest that a 3D spectrum could provide further insight. Third-order 3D electronic spectroscopy was performed on the Fenna-Matthews-Olson complex by Hayes and Engel [154]. The 3D spectra helped resolve the different excitonic transitions, which are strongly overlapping in 1D and 2D spectra. Analysis of planes of constant ω_T in the 3D spectra improved the determination of the excitonic transition energies. This improved information was used to calculate the Hamiltonian in a site basis.

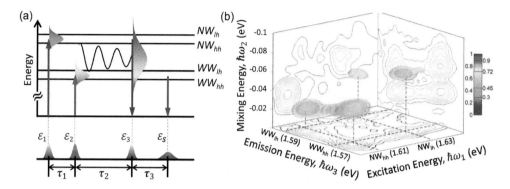

Fig. 8.10 3D rephasing spectrum (amplitude) of an asymmetric double InGaAs quantum well nanostructure at 20 K for collinearly polarized excitation. (a) Shown is the optical excitation sequence with spectrally-shaped pulses to isolate specific off-diagonal peaks corresponding to Raman-type coherence quantum pathways between the WW and NW. (b) 3D spectrum revealing off-diagonal coherence peaks fully isolated in three dimensions allowing for quantitative peak shape analysis. Adapted from Ref. 403.

8.4 3D coherent frequency domain spectroscopy

An alternative approach to multidimensional coherent spectroscopy has been pioneered by Wright [443, 465]. In this approach, the two-dimensional spectrum is produced by tuning two narrow band lasers. A third laser, with a fixed frequency, produces a signal, which can be at a very different wavelength than the tunable lasers. In the first demonstration of producing a 2D spectrum using this method, the tunable lasers were in the infrared and excited vibrations, whereas the third laser was 532 nm, thereby producing a visible signal [465]. The spectra demonstrated the spectral selectivity of this approach, its sensitivity to the interactions causing mode coupling, and ability to spectrally resolve isotopic mixtures.

This method has recently been extended to produce 3D spectra [57]. To produce 3D spectra, the third laser is broadband and a spectrometer is used to measure the signal spectrum. The three dimensions are then the frequencies of the first two lasers and the signal frequency. The goal of this demonstration was to examine gas phase Br_2 with high resolution. The one-dimensional spectrum is very dense and congested. By spreading it out in a higher-dimensional space, it is easier to identify and assign the spectral lines.

9

Two-dimensional spectroscopy of semiconductor quantum dots

Semiconductor quantum dots (QDs) are the zero-dimensional analog of quantum wells, namely, nanocrystals in which the spatial dimensions are reduced to produce exciton quantum confinement effects in not just one spatial dimension, but rather in all three dimensions simultaneously. In turn, this confinement leads to a dramatic flattening of quantum dot dispersion curves and the emergence of well-defined, discretely spaced energy levels. The effects have led to the widespread branding of QDs as "artificial atoms," with tunable absorption and emission lines, coupling interactions, and decoherence rates. In industrial applications, quantum dots play a significant role in devices including LED-based displays, semiconductor lasers, and optical filters. Basic research studies of the fundamental properties of quantum dots are ongoing, with many current efforts driven by the desire to use quantum dot nanostructures and devices in quantum information sciences, such as for single-photon and entangled-photon sources as well a strong optical nonlinearity single-photon-level switching. In the past several years, MDCS has helped elucidate fundamental properties of QD coherence and dephasing, which are relevant to both applications.

Semiconductor QDs are typically produced through one of three mechanisms: (1) they can occur as a localization phenomenon in disordered quantum wells ("interfacial" or "natural" QDs (IFQDs) typically observed in GaAs quantum wells), (2) they can be grown through molecular beam epitaxy on top of an appropriate substrate ("self-assembled" QDs (SAQDs) and droplet-epitaxy QDs), or (3) they they can be grown in solution ("colloidal" QDs (CQDs)). We summarize below the impact that MDCS has had in clarifying the physics behind each QD variety, primarily focusing on characterization and manipulation of the first two classes of QDs, which have been more widely measured in the low-temperature limit, but also touching briefly on the recent progress in understanding the physics of colloidal QDs.

9.1 Optical and electronic properties of quantum dots

Depending on the characteristics of the QD, e.g., epitaxially-grown versus chemically-synthesized, shape, composition, strain, etc., different models are used for the confinement potential Hamiltonian, with the most common being a simple quantum box with infinite or finite barriers, a spherical potential well, and a harmonic oscillator potential. These different models predict different eigenstates and eigenenergies; however, they all point to qualitatively similar electronic and optical properties, in general. Quanti-

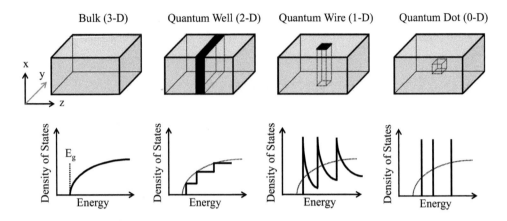

Fig. 9.1 Density of states for particles with three (bulk), two (quantum well), one (quantum wire), and zero (quantum dot) spatial degrees of freedom.

zation of the electron and hole spatial motion has three characteristic effects: 1) the effective band-edge is blue-shifted; 2) confinement forces the electron and hole to overlap, increasing the oscillator strength, and hence the radiative probability; and 3) as shown in Fig. 9.1, the density of states becomes discrete. Figure 9.2 demonstrates the energy level quantization from quantum confinement for a QD embedded in a matrix with a larger bandgap, as depicted in the right-most panel.

If one considers, for example, a square well potential with a barrier height V_0 and a potential width L_B, then the minimum QD size allowed for the existence of a bound exciton state is $L_{min} = \pi\hbar/\sqrt{2\mu_X V_0}$, where μ_X is the exciton reduced mass. For QD sizes smaller than L_{min}, the exciton will be delocalized in the surrounding matrix in which the QD is embedded. As an example, considering the electronic bandstructure of GaAs and a confinement potential of 10 meV (which is typical for GaAs QDs), then $L_{min} \approx 25$ nm. Castella and Wilkins [51] demonstrated that for GaAs IFQDs with lateral dimensions $2-3$ times L_{min}, the PL spectrum is split into two peaks with the lower-energy peak corresponding to excitons weakly localized in the quasi-zero-dimensional QDs and the higher-energy peak corresponding to excitons delocalized in the underlying QW. For the InAs SAQDs, which exhibit a confinement energy ranging from $60-300$ meV, then $L_{min} \approx 8-17$ nm. It is evident that in GaAs QDs, where $L_{min} > a_B$ (a_B is the exciton Bohr radius), excitons are weakly localized, whereas the strong confinement regime is realized in InAs SAQDs, where $L_{min} < a_B$.

Three-dimensional spatial confinement of electrons and holes in semiconductors can be accomplished using several techniques, with the most common relying on epitaxial growth or chemical synthesis of a low-bandgap material embedded in another material with a higher bandgap. Other methods have been developed, including lithographically patterning gate electrodes on a doped QW containing a two-dimensional electron gas or by using genetically-engineered viruses to organize inorganic nanocrystal arrays, although these techniques are not as relevant for ultrafast optical spectroscopy. Colloidal QDs (CQDs) can be chemically synthesized from binary alloys such as CdSe,

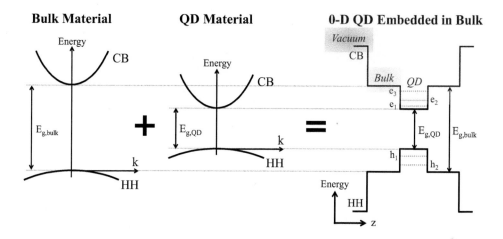

Fig. 9.2 Energy level diagram of the HH and electron conduction band (CB) for bulk material with large bandgap, $E_{g,bulk}$, QD material with small bandgap, $E_{g,QD}$, and the 0-D nanostructure energy diagram with quantized QD states as a result of embedding the QD material into the bulk matrix.

CdS, PbSe, and PbS, or from ternary allows such as CdSeTe. CQDs usually exhibit a complex optical spectrum that is inhomogeneously broadened by up to hundreds of meV from QD size dispersion occurring during synthesis. CQDs present a number of challenges for spectroscopic studies. For example, they typically have resonances at energies outside of the operating range of lasers and detectors that are commonly available. Furthermore, significant inhomogeneity and large energy separation between excited states make studying coherent and incoherent intraband dynamics in CQD ensembles difficult because of limited bandwidth of typical laser systems.

Epitaxially-grown QD ensembles, on the other hand, usually exhibit inhomogeneous line widths less than 100 meV and intraband energy level separation on the order of 10 meV. Epitaxial QDs can be separated into two classes: interfacial fluctuation QDs (IFQDs) that form "naturally" from monolayer changes in the width of a narrow QW, and self-assembled QDs (SAQDs) that form to alleviate lattice-mismatch-induced strain in a narrow QW in which two interfacing layers have dissimilar lattice constants by at least a few percent. The highest quality dots are usually grown by molecular beam epitaxy (MBE), where, for example, GaAs is deposited one atomic layer at a time on AlGaAs. If the GaAs layer is deposited uniformly, and then capped by another AlGaAs layer, the nanostructure would represent a QW. However, under specific growth conditions, GaAs islands can form on top of a thin QW. The sample is then capped with an AlGaAs layer, encasing the islands, resulting in a QW–QD nanostructure, which is depicted in the schematic diagram in Fig. 9.3. The shape of the QD is sensitive to the growth conditions and can take the form of a truncated pyramid, lens, sphere, or box, and due to material composition intermixing, the dots often have a linear composition gradient. For example, in In(Ga)As QDs, due to indium diffusion, the QDs tend to be indium-rich near the top and gallium-rich near

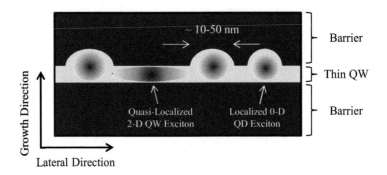

Fig. 9.3 Schematic diagram of epitaxially-grown QDs on a narrow QW. Optical excitation can create 2-D excitons in the QW and 0-D excitons confined in the QDs.

the bottom, resulting in a slight (< 1 nm) displacement of the electron and hole wave functions along the growth direction, inducing a static electric dipole moment [122]. The main limitations of both IFQDs and SAQDs are the cost of fabrication and the lack of control over position of individual QDs. Site-controlled QDs can be fabricated by lithographically-patterning the surface of a semiconductor substrate, such as bulk GaAs, and then growing QDs in the etched nano-imprints. The QD optical properties have so far been diminished compared to SAQDs due to the introduction of additional dephasing due to surface impurities and defects, although significant progress in this area has been made in recent years [404]. Site-controlled growth has also been achieved using droplet epitaxy [327].

The confinement potential for In(Ga)As SAQDs is typically hundreds of meV for the electron and less than 100 meV for the hole. Consequently, the electron-hole pair is bound more by the confinement potential than Coulomb interactions, which act as a perturbation. This is not the case for GaAs interfacial QDs, in which the electron-hole pair is bound by ~ 10 meV relative to 2-D excitons delocalized in the underlying QW. Moreover, despite strong confinement in the SAQDs, electron and hole interactions induce measurable changes of the optical properties, such as exchange-interaction-induced state splitting and bound biexciton effects. The excitonic picture is a useful representation for epitaxially-grown QDs; however, it should be kept in mind that excitons in QDs are different from bulk excitons in several ways. First, a QD exciton cannot dissociate into an electron and hole, unless one of the charges leaves the QD, because the external confinement potential usually dominates over the Coulomb interactions. Second, excited exciton states in a QD are not states of the mutual Coulomb potential of interacting electrons and holes, but instead they are formed by the charge carriers residing in higher-lying single-particle states of the QD potential. Coulomb interactions between electrons and holes then modify the energetics of the few-particle states with respect to the sum of the energies of the individual single particles, and the resulting energy difference is considered the binding energy. For example, the exciton energy is given by $E_X = E_e + E_h - \Delta X$, where E_e and E_h are the single-particle electron and hole energies and ΔX is the binding energy due to renormalization from electron-hole

Coulomb interactions. The biexciton energy would then be $E_B = 2 \cdot E_X - \Delta B$, where ΔB is the biexciton binding energy.

Single-dot spectra reveal rich information about the energetics of various multi-particle states and their radiative decay dynamics. Excitons, bound and anti-bound two-excitons (biexcitons) and negatively or positively charged excitons (trions) are easily identified in PL spectra. In their representative ground states, for which only the lowest-energy electron and hole levels are occupied, these excitonic states would be degenerate if ignoring Coulomb interactions. Experimentally, this is not the case, and instead splitting of several meV are observed. The energetic ordering of the states is sensitive to the QD size, shape and composition; therefore a connection between QD morphology and the recombination energies of the excitonic states is difficult to establish. Experimentally, the energetics can be systematically characterized by recording many single-dot PL spectra for samples prepared using different growth methods, i.e., QDs with various morphology; however, significant dot-to-dot scatter of the data from single-dot experiments and accurate characterization of the dot size, shape, and composition make comparison to theory difficult. Nonetheless, certain fingerprints for specific QD morphology exist, such as the binding energies of the biexciton and trion relative to the exciton ground state and exciton fine-structure splitting arising from the electron-hole exchange interaction [333]. While exchange effects are typically too weak to be observed in higher-dimensional systems, they are enhanced in QDs because confinement increases the electron-hole wave function overlap. A non-zero splitting has important consequences for quantum optical applications, such as entangled-photon pair generation, where the splitting provides energetic "which-path" information to distinguish the photons. MDCS is especially useful for characterizing the fine-structure splitting of each QD in an ensemble, as discussed in detail in [277].

9.2   2D coherent spectroscopy of GaAs quantum dots

In the epitaxial-growth process of QWs, the interfacial surface roughness due to monolyaer well-width fluctuations acts as in-plane disorder potential for excitons. For sufficiently narrow QWs in which the fluctuations represent a significant percentage of the well width, excitons can become localized at the fluctuation site, which acts as a zero-dimensional quantum confinement potential known as a "natural" or interfacial QD with typical confinement energy of 10 meV relative to free excitons in the QW [128]. "Self-assembled" islands can also form in narrow InAs/GaAs QWs due to strain from the large lattice mismatch between indium and gallium. Strong confinement in self-assembled QDs is more reminiscent of an atomic two-level system.

Semiconductor QDs are an exemplary system to demonstrate the capabilities of 2DCS for dissecting disordered ensembles. The stochastic nature of the epitaxial growth process leads to an inhomogeneous distribution of QD sizes that translates into 50-100 meV dispersion of the exciton transition energies for self-assembled InAs/GaAs SAQDs and 1-5 meV for GaAs/AlGaAs IFQDs. The in-plane QD confinement potential tends to be asymmetric with principle axes along the $[110]{\equiv}V$ and $[1\bar{1}0]{\equiv}H$ crystal directions due to several sources, including strain, shape, and piezoelectricity. When considering the electron-hole exchange interaction, anisotropy mixes the elec-

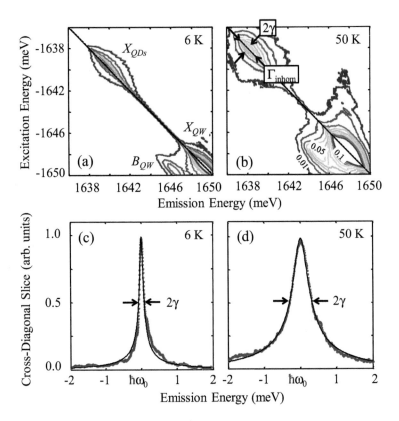

Fig. 9.4 Normalized rephasing single-quantum amplitude spectra of the GaAs IFQD sample are shown for a sample temperature of **(a)** 6 K and **(b)** 50 K. Each spectrum features a QW (X_{QW}) and QD (X_{QD}) exciton peak and a QW biexciton peak (B_{QW}) red-shifted from X_{QW} by the biexciton binding energy. Cross-diagonal slices taken at the maximum amplitude of X_{QD} are shown for **(c)** 6 K and **(d)** 50 K, where the data (points) are fit using a $\sqrt{Lorentzian}$ function (solid line) with a FWHM equal to 2γ. Adapted from Ref. 276.

tronic transitions, resulting in two bright exciton states $|H\rangle$ and $|V\rangle$ that are separated in energy by the so-called fine-structure splitting.

One of the earliest demonstrations of MDCS on QDs was to probe the dephasing mechanisms of excitons confined to GaAs IFQDs [276]. 2D rephasing amplitude single-quantum spectra acquired from the GaAs IFQDs using co-linear excitation and detection are shown in Figs. 9.4(a) and 9.4(b) for low (6 K) and elevated (50 K) temperature, respectively, for a photon excitation density at the sample of 1×10^{12} photons·pulse^{-1}·cm^{-2} corresponding to ≈ 0.1 excitons excited per QD on average. Each spectrum features a QW (X_{QW}) and QD (X_{QD}) exciton peak, and a QW biexciton peak (B_{QW}) appears red-shifted along the emission energy axis from X_{QW} by the biexciton binding energy. A QD biexciton peak can be expected, but it is too weak to be observed using this polarization sequence. The cross-diagonal line shape can be fit using a $\sqrt{Lorentzian}$ function, and the full-width at half-maximum (FWHM) of the fit is equal to twice the homogeneous line width, γ. The homogeneous line shapes (points) and the fits (solid lines) obtained at the maximum amplitude of the QD inhomogeneous distribution are shown in Figs. 9.4(c) and 9.4(d) for temperatures 6 K and 50 K, respectively.

The fact that a $\sqrt{Lorentzian}$ function fits the line shapes well at both low and high temperature indicates that the lineshape is predominantly a measure of the zero-phonon line (ZPL), i.e., the lineshape profile of the exciton-to-ground state transition between exciton and ground states with the same phonon occupation number. Figures 9.4(c) and 9.4(d) clearly show that the ZPL is broader at 50 K than 6 K. An advantage of performing 2DCS on the QDs is that the inhomogeneous and homogeneous line widths are uncoupled, enabling the extraction of the homogeneous line width dependence on the QD resonance energy by taking cross-diagonal slices along the diagonal dashed line. The ZPL width dependence on the QD resonance energy is shown in Fig. 9.5 for a set of temperatures ranging from 6 K to 50 K. The line center of the inhomogeneous distribution at each temperature is indicated by the solid circles. The ZPL width increases with energy (decreasing QD size) across the inhomogeneous distribution for all temperatures, with the effect being more dramatic at elevated temperatures. With increasing temperature, exciton-phonon interactions dominate the lineshape broadening. The absence of any off-diagonal peaks suggests that these interactions are an elastic pure dephasing process and do not arise from inelastic phonon-assisted excitation of excitons.

GaAs IFQDs are also a model system to study population transfer between localized and delocalized states, such as in light-harvesting complexes [398] and colloidal nanocrystals designed for multi-exciton generation [193]. Because the QW–QD energy separation is smaller than the excitation laser spectral bandwidth, the nonlinear optical response of both the QW and QDs can be simultaneously probed. Moreover, the population transfer and decay rates are slow compared to the duration of the pulses, i.e., the change in population is negligible during the light-matter interaction, but fast compared to the maximum achievable pulse delays in the 2DCS experiment, enabling complete characterization of the population relaxation dynamics [275]. Figure 9.6(a) shows a normalized rephasing single-quantum spectrum for a sample temperature of 35 K and for a delay between pulses B and C of $T = 5$ ps. Quantum pathways associated with the system being in a ground or excited state population after the first two pulses do not accumulate phase during the delay T, but instead decay through interband recombination or non-radiative relaxation. Since the phase does not evolve during T for these processes, population transfer dynamics can be studied by recording 2D rephasing spectra for increasing delay T. Normalized spectra are shown in Figs. 9.6(b)–(c) for T increasing from 20 ps to 100 ps, also acquired for a sample temperature of 35 K. The maximum amplitudes of the spectra relative to the spectrum taken at $T = 5$ ps are shown on the respective color bars. At short times, the QW and QD peaks dominate the spectrum and are inhomogeneously broadened along the diagonal. With increasing delay T, radiative recombination and population transfer processes decrease the amplitudes of these peaks. Moreover, the peak shapes become more symmetric, which indicates loss of correlation between the exciton excitation and emission energies during T. The appearance of cross peaks at long T reveals incoherent exciton

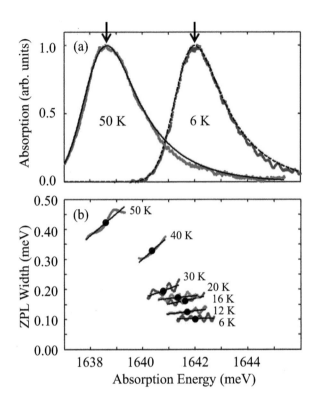

Fig. 9.5 **(a)** The asymmetric absorption line shape, obtained by projecting the 2D amplitude rephasing spectrum onto the vertical axis. Line center of the inhomogeneous distribution is marked by the vertical arrows. **(b)** The ZPL widths (points) within the FWHM of the inhomogeneous distribution for different sample temperatures. Line center of the inhomogeneous distribution at each temperature is marked by the larger solid circle. Linear fits (solid lines) are performed at each temperature. Adapted from Ref. [276].

population transfer between the QW and QDs, which would otherwise be concealed in one-dimensional linear and nonlinear spectroscopies. A relaxation peak (RP) appears at the excitation energy of the QW and emission energy of the QDs, revealing incoherent QW → QD population relaxation and localization. Thermal activation of QD excitons is observed at longer T as an excitation peak (EP) at the excitation energy of the QDs and emission energy of the QW.

A detailed rate-equation analysis of the recombination and transfer dynamics allows several conclusions to be drawn regarding the primary decay and QW–QD coupling mechanisms. It is well-established that QW exciton-bound hole spin relaxation occurs on a picosecond timescale due to confinement-enhanced electron-hole exchange [250], and this mechanism is attributed to the fast decay of the QW population. Excitons initially in the $|\pm 1\rangle$ bright states can hole-spin-flip to the $|\mp 2\rangle$ dark states, after which they can non-radiatively decay or spin flip back to the bright states. The onset of the slow decay occurs after formation of a quasi-equilibrium between excitons in

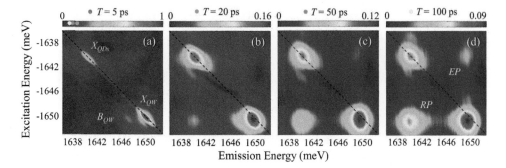

Fig. 9.6 QW (X_{QW}) and QD (X_{QDs}) excitons, QW biexciton (B_{QW}), QW → QD relaxation (RP) and QD → QW excitation (EP) peaks are observed in the normalized 2D rephasing one-quantum spectra for increasing delay T. The sample temperature is 35 K for all spectra. The maximum amplitude of each spectrum relative to the $T = 5$ ps spectrum is shown on the color bar. Adapted from Ref. [275].

the bright and dark states. Exciton population relaxation in QDs is expected to occur more slowly than in QWs because the discrete density of states suppresses many of the effective scattering mechanisms [183, 386]. However, the mechanisms responsible for fast spin relaxation in QWs have been found to be effective for excitons weakly localized in GaAs IFQDs [372]. Coupling between the exciton spin states through electron or exciton spin flips can contribute to population decay, but these relaxation processes occur on a much slower time scale. The similarity between the QW and QD fast population decay suggests that the same exciton-bound hole-spin-flip mechanisms in the QW are prevalent in the IFQDs as well.

The most likely QW → QD exciton transfer path is directly between the bright $|\pm 1\rangle$ states, which requires acoustic phonon emission for energy and momentum conservation. A possible exciton-phonon coupling mechanism is the following: zero-momentum excitons initially in the QW acquire momentum by scattering from the phonon population, after which they relax to the QD states while emitting one or multiple phonons with the necessary energy and momentum. Because the momentum distribution of excitons in the QDs is centered about zero with a non-zero width $\Delta\mathbf{k}$, QW excitons with zero initial momentum can also lose 10 meV of energy and relax into the QD states via phonon emission. Additionally, the 10 meV energy separation between the QW and QDs can be overcome through a cascaded process involving dark excited states in the QDs [294, 330].

Excitation from the QDs to the QW states involves similar paths as the QW relaxation processes; however, the mechanisms are quite different. Since excitons in the QDs must overcome the 10 meV confinement potential, the EP is only expected to appear at higher temperature where phonon-assisted activation processes can occur. The QD → QW activation process is observed only for sample temperatures at 35 K and above, at which significant acoustic phonon population with energy ≥ 10 meV exists. While multi-phonon activation is possible, the absence of the EP at lower temperatures indicates that QD → QW excitation arises from a single-phonon process.

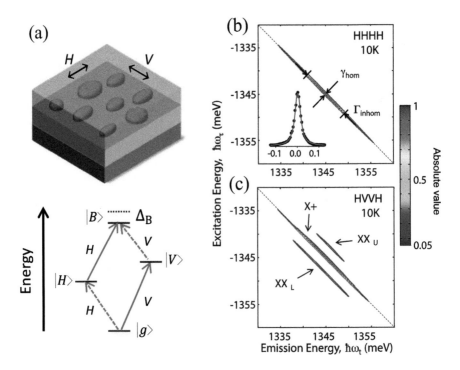

Fig. 9.7 (a) Illustration of a semiconductor self-assembled quantum dot ensemble (top). The lowest-energy exciton ($|H\rangle$ and $|V\rangle$) and biexciton ($|B\rangle$) transitions are represented by a four-level diamond energy diagram. Rephasing one-quantum spectra of a self-assembled InAs/GaAs ensemble measured at 10 K for (b) co-linearly (HHHH) and (c) cross-linearly (HVVH) polarized optical excitation and detection. Figures (b) and (c) are adapted from Ref. 272.

9.3   2D spectroscopy of self-assembled In(Ga)As quantum dots

To measure the exciton homogeneous linewidth (γ) and fine-structure splitting of SAQD, optical studies originally required isolating single QDs due to inhomogeneous broadening. Using photoluminescence spectroscopy, the exciton linewidth and emission polarization properties have been characterized, and multi-particle states associated with charged excitons and biexcitons have been identified [54, 86, 127]. In single-dot studies, characterizing how statistical fluctuations within the ensemble affect the optical properties requires repeating multiple measurements on different QDs, and details of inter-dot interactions are not accessible. Alternatively, ensemble measurements are possible with nonlinear spectroscopy techniques [35, 210]. Transient FWM studies have been useful for characterizing the exciton dephasing time ($T_2 = \hbar/\gamma$) and radiative lifetime (T_1). Oscillations of the FWM signal have also been identified as coherent quantum beats between $|H\rangle$ and $|V\rangle$ and $|H\rangle$ and $|B\rangle$). Despite the rich dynamics identified by these studies, transient FWM is not sensitive to dot-to-dot fluctuations of the homogeneous properties because it provides an ensemble-averaged response.

Optical 2DCS offers the specificity of single-dot studies while probing the entire ensemble [277–279]. This capability is illustrated by the rephasing spectrum shown in Fig. 9.7(b) for an InAs QD ensemble at 10 K, where the emission energies are plotted as negative for this pulse sequence as explained in Section 2.2 [277, 278]. The spectrum is shown for collinearly polarized excitation and detection aligned along H (denoted as an HHHH polarization scheme corresponding to the polarization of pulses $\mathcal{E}_1, \mathcal{E}_2, \mathcal{E}_3$ and the signal). The single peak on the diagonal dashed line is attributed to GSB and ESE of the $|H\rangle$ exciton transition. The large linewidth along the diagonal compared to the anti-diagonal direction indicates that material is strongly inhomogeneously broadened ($\sigma \gg \gamma$). Anti-diagonal slices as a function of exciton resonance energy throughout the inhomogeneous distribution reveals a relatively uniform homogeneous linewidth $\gamma \approx 10\ \mu\text{eV}$. Ensemble measurements of InAs quantum dots have also demonstrated the ability of MDCS to characterize energy-dependent linewidths in an inhomogeneous ensemble, as discussed in relation to p-shell excitons in Ref. 382.

For an HVVH polarization scheme, the GSB and ESE quantum pathways for the exciton state are no longer accessible. Cross-linear polarization emphasizes the biexcitonic interaction pathway and suppresses the excitonic pathway, making it possible to more clearly identify charged trion resonances on the diagonal, biexcitonic resonances below the diagonal as a third-order response, and even a biexcitonic fifth-order response (due to six-wave mixing), above the diagonal. A rephasing 2D spectrum for the cross-linear polarizations (HVVH) is shown in Fig. 9.7(c). In this spectrum, the diagonal peak X_+ is associated with positively charged excitons (trions) in charged QDs within the ensemble, and the off-diagonal peaks (XX_U and XX_L) correspond to the biexciton state $|B\rangle$ in neutral QDs.

Another contribution that MDCS has made in understanding the physics of quantum dots is in quantifying biexciton binding energies, and particularly the dependence of the binding energy on quantum dot frequency. The shift of off-diagonal peak XX_L to lower energy along the emission frequency axis provides a measure the biexciton binding energy Δ_B relative to $E_H + E_V$. An analysis of these effects in both InAs self-assembled quantum dots and in GaAs interfacial quantum dots [278] found that biexciton binding energies exhibit different dependences on emission frequency in the two different types of materials. Whereas the biexciton binding energy increases with emission energy in interfacial quantum dots, it remains an almost perfect constant in self-assembled InAs dots, which is attributed to thermal annealing of the sample that homogenizes the QD properties. The latter finding is striking because complementary studies on single InAs quantum dots had found this binding energy to vary [210, 321, 333]. The results may indicate that the dependence of biexciton binding energy is obscured by local environment modifications resulting from the necessary etching of mesas or patterning of masks that is required to isolate single dots.

When just a few QDs are probed within the ensemble, evidence of coherent coupling between excitons at different energies is observed. Figure 9.8 is a rephasing spectrum from a GaAs/AlGaAs "natural" QD nanostructure, where only a few QDs are probed in this case using a sparse QD sample and collinear 2DCS geometry with 600 nm spatial resolution. Coupling between excitons at energy 1.691 eV and 1.692 eV is evident by the off-diagonal peaks between them. An analysis of the spectral phase

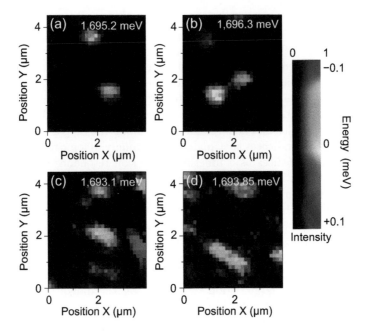

Fig. 9.8 Hyperspectral imaging of the four-wave mixing signal from a GaAs interfacial quantum dot sample. (a–b) Images from low-density portions of the sample. (c–d) Images from high-density portions of the sample. Courtesy of J. Kasprzak and W. Langbein.

reveals that the nature of their coupling arises from biexcitonic interactions between two excitons. Hyperspectral spatial images of the FWM signal elucidate the spatial extent across which excitons are coupled. By correlating the FWM signal energies of coupled resonances with their spatial position, the authors find that long-range coupling of extended exciton states is also important.

One aspect that eludes ensemble measurements taken even with MDCS is the prospect of measuring quantum mechanical coupling between dots in different locations. Because interfacial and self-assembled quantum dot ensembles grow randomly, this coupling gets washed out in ensemble measurements even as the homogeneous linewidth is preserved. Coupling has been observed and characterized, however, using frequency-based MDCS in which excitation beams are arranged to impinge upon the sample in collinear geometry [87, 178, 206, 207, 257, 266, 302, 303].

Noteworthy work in this area is displayed in Fig. 9.8 [178]. By combining parallelized collection of four-wave mixing with a collinear series of excitation pulses, it was shown possible to perform hyperspectral imaging, with sub-micron spatial resolution, of an ensemble of individually identifiable interfacial GaAs quantum dots with results including the observation of individual excitons and biexcitons in individual quantum dots and on coherent coupling between excitons in relatively distant quantum dots. Somewhat surprisingly, the authors found that coupling persisted out to a large inter-exciton distance of almost 1 μm. They attributed the coupling mechanism to

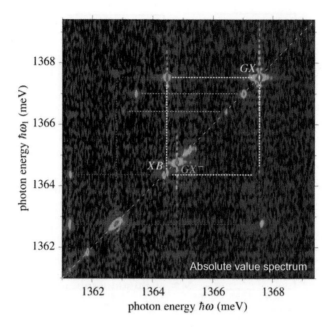

Fig. 9.9 MDCS rephasing spectrum, showing biexcitons and coupling effects pertaining to individual InAs self-assembled quantum dots. The excitation axis is shown on a positive frequency scale, such that the diagonal runs from lower left to upper right. Adapted from Ref. 266.

a binding interaction mediated by spatially extended states, for example, the oblong green resonance displayed in the lower portion of Fig. 9.8(d).

Although the dipoles are much weaker, subsequent work has demonstrated that similar measurements can also be performed on small ensembles of InAs quantum dots. Mermillod *et al.* used the hyperspectral imaging technique to identify biexcitons and inter-dot coupling effects using an MDCS rephasing pulse sequence [266], and Delmonte *et al.* have followed up on this experiment with a study comparing rephasing and double-quantum MDCS pulse sequences [87]. An example of the kinds of observable coupling signatures is displayed in Fig. 9.9. Coupling between excitons at different energies is evident by the off-diagonal peaks between them. An analysis of the spectral phase reveals that the nature of their coupling arises from biexcitonic interactions between two excitons. Hyperspectral spatial images of the FWM signal elucidate the spatial extent across which excitons are coupled. By correlating the FWM signal energies of coupled resonances with their spatial position, the authors find that long-range coupling of extended exciton states is also important.

9.4 Coherent control of quantum dots

The identification of MDCS resonances in self-assembled and interfacial quantum dots has opened the door toward being able to manipulate coherent effects between dots and within ensembles of dots in a controlled manner. Among the most prominent

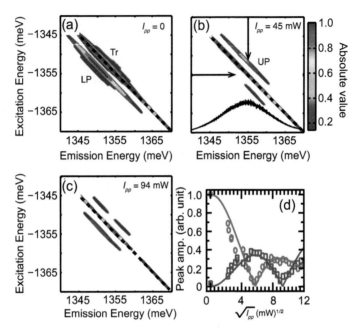

Fig. 9.10 Coherent control of an exciton-biexciton system in an InAs self-assembled quantum dot ensemble. (a) Rephasing spectrum in the absense of a pre-pulse. (b) Rephasing spectrum with a pre-pulse of 45 mW. (c) Rephasing spectrum with a pre-pulse of 94 mW. (d) Rabi oscillations of the lower and upper peaks from (a)–(c). The red circles correspond to the amplitude of the lower left peak (LP). The blue squares correspond to the amplitude of the upper right peak (UP). Adapted from Ref. 380.

recent advances in this direction has been a result in which pulse sequences were used to generate four-wave-mixing and six-wave-mixing effects in a single dot to be able to manipulate the intrinsic coherence of an InAs quantum dot dipole to be able to engineer its coherent emission [119].

9.4.1   Coherent control within an ensemble of quantum dots

The possibility of exerting coherent control of quantum dots can also be scaled up to larger ensembles. The ability of 2DCS to resolve different frequency groups within the ensemble also allows the technique to serve as a sensitive probe of quantum states in coherent control experiments of ensemble. For example, by preceding an MDCS experiment with a resonant pre-pulse, Rabi flopping with greater contrast than that visible using transient absorption spectroscopy [36, 202] was observed in an ensemble of InAs quantum dots containing as many as 10 million distinct emitters [380, 381]. Figure 9.10 shows an example of how this Rabi flopping manifests. In the absence of a pre-pulse, the rephasing spectrum for an ensemble of InAs quantum dots consists of a trion peak along the diagonal, and a biexciton peak to the lower left of the diagonal [Fig. 9.10(a)]. Applying a pre-pulse of 45 mW drives the ensemble to emit with a biexciton resonance that has moved to the upper right of the diagonal

[Fig. 9.10(b)]. Increasing the pre-pulse power up to 94 mW drives the ensemble back toward its original state [Fig. 9.10(c)]. The amplitudes of the lower left and upper right peak can be quantified and understood within the context of Rabi oscillations of the exciton-biexciton system, as depicted in Fig. 9.10(d). The results clearly display Rabi oscillations (experiment and theory indicated by the points and curves, respectively), which usually require isolation of single QDs to avoid the detrimental effects of inhomogeneity [33, 58, 175, 219, 373].

9.4.2   Coherent control of interactions between individual quantum dots

Beyond inducing Rabi oscillations, it is possible to use a pre-pulse to controllably mediate inter-dot interactions in an ensemble of interfacial quantum dots [257]. Characterizing the sample using double-quantum spectroscopy with excitation pulses of narrow bandwidth, showed that inter-dot interactions are normally absent [Fig. 9.11(c)]. However, interactions could be turned on by applying a pre-pulse resonant with the quantum well wetting layer [Figs. 9.11(d) and 9.11(e)].

A potential puzzle in comparing the work of Kasprzak *et al.* [178] to the work of Martin and Cundiff [257] is the question of why interactions were observed in absence of a pre-pulse by the authors of the former paper and yet they were not observed in the absence of a pre-pulse by the authors of the latter paper. Sample variability is one possible explanation. MDCS excitation bandwidth, which was much larger in the work of Kasprzak *et al.*, is an alternate possibility. Further study is required to settle the question more definitively.

9.5   Two-dimensional spectroscopy of colloidal quantum dots

In comparison to their interfacial and self-assembled epitaxial counterparts, colloidal quantum dots are more tantalizing in their technological potential, but also more experimentally challenging to measure. On one hand, the growth process of colloidal dots in solution is more efficient than the process used to manufacture self-assembled and interfacial dots, and the ability to disperse colloidal dots in solution opens up a number of possibilities for colloidal quantum dot applications. On the other hand, colloidal quantum dots typically have a larger inhomogeneous broadening than self-assembled or interfacial quantum dots, obscuring features (such as electron-phonon coupling effects) with energy scales smaller than that of the inhomogeneous broadening. Beyond this, the increased importance of the material surface in colloidal materials can often lead to "blinking" effects, in which electronic trap states at the dot surface lead to Auger recombination that suppresses individual quantum dot emission for even seconds at a time [104].

Room-temperature colloidal quantum dot MDCS studies were the first set of experiments thus far undertaken, and have already been able to cut through some of the difficulties in understanding physical processes [49]. Among the more prominent systems of study has been CdSe. In 2011, for example, Turner *et al.* used MDCS to report the existence of an electronic zero-quantum coherence between the two lowest lying excitonic states in an ensemble of CdSe quantum dots, lasting about 15 fs [410]. Shortly thereafter, Griffin *et al.* reported that hole relaxation occurs with a timescale

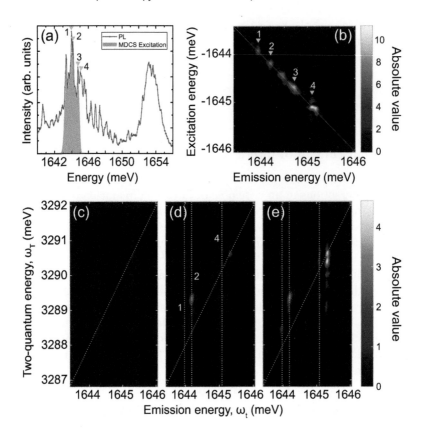

Fig. 9.11 Optical control of interactions in a small ensemble of interfacial quantum dots. **(a)** Photoluminescence measurement of a small ensemble of GaAs interfacial quantum dots, demonstrating the existence of isolated dots and illustrating the bandwidth of the excitation pulses used to generate (b)–(d). **(b)** MDCS rephasing spectrum, depicting four individually identifiable quantum dots. **(c)–(e)** MDCS two-quantum spectra, displayed as a function of increasing pre-pulse power, with the pre-pulse being tuned to the quantum well wetting layer. Adapted from Ref. 257.

very similar to that of the decay reported by Turner [139], offering an alternate explanation for the initial results. Subsequent measurements by Caram *et al.* put the claim of electronic coherences in CdSe on firmer footing by demonstrating the existence of an electronic coherence between the second and third excited states of CdSe, lasting 80 fs (Fig. 9.12) [48]. More recently, Cassette *et al.* measured room-temperature electronic coherences in a related system of CdSe/CdZnS core/shell nanoplatelets, identifying an unambiguous electronic coherence in the system with a dephasing time of 10–20 fs [50]. The identification was possible because the nanoplatelet system has a cleaner spectrum than the spectrum of CdSe quantum dots (i.e., vibrational coherences and ensemble dephasing do not interfere with the observation of electronic coherences). In all of these systems, the observation of room-temperature electronic coherences is

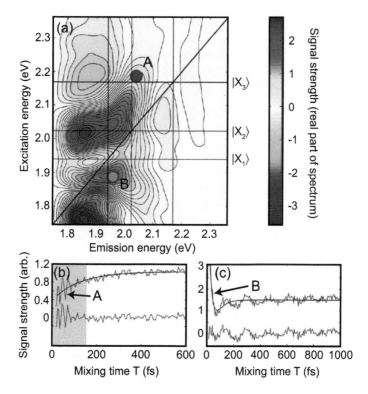

Fig. 9.12 Room-temperature measurements of coherent coupling in CdSe quantum dots. (a) One-quantum absorptive spectrum (real part of the sum of rephasing and nonrephasing spectra). The spectrum encompasses three excitonic resonances, respectively labeled $|X_1\rangle$, $|X_2\rangle$, and $|X_3\rangle$. (b)–(c) Line-outs as a function of the mixing time T at points A and B from (a) reveal an electronic coherence [shaded pink region of (b)], which can be distinguished from both phonon/solvent relaxation [unshaded portion of (b)] and phonon oscillations [(c)]. Adapted from Ref. 48.

important not only because it informs the physics of quantum dot light-matter inter-actions, but also because of its potential relevance to more complicated systems such as photosynthetic light-harvesting complexes [66, 106].

Performing low-temperature studies on colloidal quantum dots is more challenging than epitaxial dots since they are grown in solution. However, the advantages of low-temperature studies, namely reducing or removing the spectral broadening due to phonons, thereby revealing underlying processes, and, similarly, being able to vary the density of phonons, thereby quantifying the electron-phonon coupling, have provided motivation to overcome these challenges. To reach low temperatures, colloidal quantum dots can be drop- or spin-cast on a substrate to form a film. However, such films typically result in strong scattering that can overwhelm the TFWM signal in MDCS experiments relying on wave-vector selection to separate the signal from the excitation pulses. Another consideration for low-temperature experiments is the inability to flow

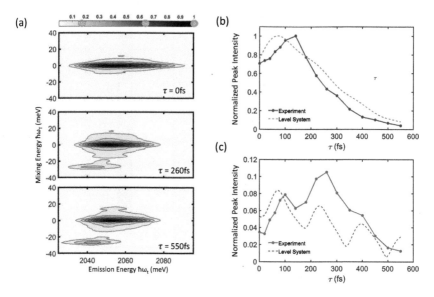

Fig. 9.13 Low-temperature multidimensional coherent spectra of CdSe/CdZnS core-shell colloidal quantum dots demonstrating non-Markovian spectal diffusion. (a) Zero-quantum spectra for 3 value of τ, the delay between the first and second pulses. (b) Comparison of the measured integrated strength of the peak at zero mixing energy to a level model incorporating Markovian dephasing. (c) Comparison of the measured integrated strength of the peak at the negative mixing energy corresponding to the LO phonon to a level model incorporating Markovian dephasing. Adapted from Ref. 224.

the sample. Flowing samples are often used to avoid photo-damage, or the effects of long-lived states.

One route to performing low-temperature experiments on solution-phase samples is to use a glass-forming liquid as the solvent. This method has successfully been used to realize low-temperature MDCS experiments on colloidal both CdSe [224, 225] and perovskite nanocrystals [226, 229] by using heptomethylnonane as a solvent, which forms a glass at temperature below 100 K.

Two-dimensional coherent spectroscopy on CdSe/CdZnS core-shell colloidal quantum dots at cryogenic temperatures revealed non-Markovian exciton-phonon interactions on a femtosecond timescale [224]. In the two-dimensional spectra, sidebands due to electronic coupling with CdSe lattice LO-phonon modes had non-exponential dephasing, which deviated from the expected exponential decay resulting from Markovian spectral diffusion, which is instantaneous and memoryless. As shown in Fig. 9.13, comparison to simulations provided evidence that LO-phonon coupling induces energy-gap fluctuations on the finite timescales of nuclear motion. The femtosecond resolution of the technique probes exciton dynamics directly on the timescales of phonon coupling in nanocrystals.

Coupling to phonon modes is a primary mechanism of excitonic dephasing and energy loss in semiconductors. However, low-energy phonons in colloidal quantum dots and their coupling to excitons are poorly understood because their experimental signatures are weak and usually obscured by the unavoidable inhomogeneous broadening of colloidal dot ensembles. Multidimensional coherent spectroscopy at cryogenic temperatures was used to extract the homogeneous nonlinear optical response of excitons in a CdSe/CdZnS core/shell colloidal quantum dot ensemble [225]. A comparison to the simulation provides evidence that the observed lineshapes arise from the coexistence of confined and delocalized vibrational modes, both of which couple strongly to excitons in CdSe/CdZnS colloidal quantum dots.

10

Two-dimensional spectroscopy of atomically thin 2D materials

Among the most exciting semiconductor discoveries of the past decade has been the discovery of a direct bandgap optical transition in single-layer versions of the transition metal dichalcogenides (TMDs) [251, 371]. Single-layer TMDs exhibit a number of optical properties not shared by their many-layer parent compounds, including spin-valley coupling effects in the lowest energy excitations, exceptionally large optical interaction dipole moments, and deeply bound excitons with binding energies on the order of several hundred meV. TMD structures and heterostructures hold potential in a number of opto-mechanical applications, including atomically thin transistors and sensors [63] and valleytronic nanolasers and LEDs [331]. Because excitonic interactions are of paramount importance in TMDs [313], MDCS is uniquely poised to elucidate interactions in a way that neither linear measurements (such as photoluminescence) nor one-dimensional nonlinear measurements (such as pump-probe spectroscopy) have been able to achieve.

10.1 Introduction to 2D materials

Transition metal dichalcogenides (TMDs) exhibit unique electronic and optical properties at the ultimate two-dimensional limit [251, 371, 445]. TMDs have a hexagonal crystal lattice structure with chemical formula MX_2 (M = Mo, W; X = S, Se, Te), where the M and X atoms reside at the A and B lattice sites, as shown in Fig. 10.1(a). Similar to graphene and other 2D materials, a single atomically thin monolayer can be isolated through mechanical exfoliation or grown with vapor deposition techniques. Monolayer TMDs are a direct gap semiconductor, where the minimum of the conduction band and maximum of the valence band reside at the two degenerate +K and -K (K') points (dubbed "valleys") in the first Brillouin zone, as shown in Fig. 10.1(b). The K and K' valleys are related to each other through time-reversal symmetry; combined with large spin-orbit coupling and broken spatial inversion symmetry in monolayers, the symmetry leads to opposite electronic spin, orbital magnetic moment, and Berry curvature at the two valleys [448]. Consequently, the dipole-allowed optical transitions are also valley dependent: left- (right-) circularly polarized light excites an electron-hole pair in the K (K') valley.

   One consequence of the atomic thickness of monolayer TMDs is a significant enhancement of the Coulomb force due to reduced dielectric screening in two dimensions. Combined with the large carrier effective masses, reduced screening leads to

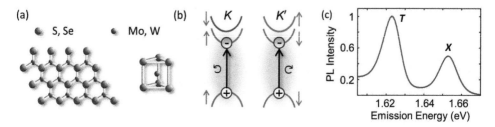

Fig. 10.1 (a) Monolayer MoSe$_2$ has a honeycomb crystal lattice structure with the transition metal sandwiched between two layers of chalogen atoms. (b) Band-edge direct optical transitions occur at the +K and -K (K') points, or valleys, in momentum space. Transition dipole selection rules dictate that electron-hole pairs are photo-excited at the K and K' valleys with left- (right-) circularly polarized optical excitation, respectively. (c) Photoluminescence spectrum featuring two peaks near 1.65 eV and 1.62 eV associated with the neutral exciton (X) and charged exciton (trion, T).

pronounced excitonic effects with neutral and charged exciton binding energies on the order of 500 meV and 30 meV, respectively—an order of magnitude larger than conventional semiconductors such as GaAs-based nanostructures [60]. At room temperature, excitons dominate the optical response of TMDs and absorb up to $\sim 10\%$ of the incident light. A typical photoluminescence spectrum from MoSe$_2$ on a sapphire substrate at 10 K is shown in Fig. 10.1(c). The two peaks correspond to the neutral exciton (X) and charged exciton (T) transitions (common to monolayer TMDs is unintentional doping from the substrate that is responsible for the trion peak).

Although monolayer TMDs are relatively new to the large family of layered semiconductors, a large body of work already exists examining the origin, dynamics, transport, and interactions of excitons [44, 134, 274, 331, 431]. Recent 2DCS experiments have added to this knowledge-base through measurements of the coherent exciton dynamics and many-body interactions.

10.2   Homogeneous linewidth in 2D materials

Among the first MDCS measurements on the TMDs were 2D rephasing spectra aimed at extracting the homogeneous linewidth of WSe$_2$ [273]. Measuring both the temperature and excitation density dependence of a CVD-grown sample of WSe$_2$ on sapphire demonstrated that the exciton resonance in WSe$_2$ exhibits a significant degree of inhomogeneity [Fig. 10.2(a)], notably hiding the intrinsic homogeneous linewidth in simpler measurements such as photoluminescence.

As is the case with the excitons in more established materials such as GaAs and ZnSe, the authors observed a strong excitation induced dephasing (EID) in WSe$_2$ [Fig. 10.2(b)], which they interpreted as an indication that many-body effects play a significant role in the nonlinear optical signature of the TMDs. Notably, when EID effects are normalized according to inter-exciton separation and exciton Bohr radius, the interaction broadening for WSe$_2$ turns out to be significantly larger than it is for either GaAs or ZnSe in either bulk or quantum well form. Such strong interactions

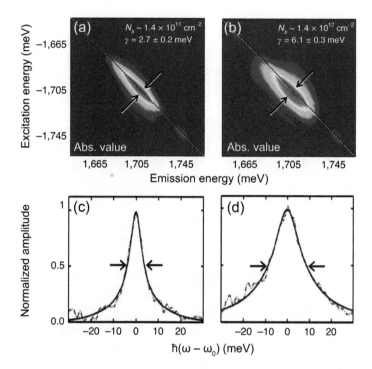

Fig. 10.2 MDCS spectra of the bright valley exciton in WSe$_2$ (single-quantum spectrum, rephasing pulse sequence). (a) Low excitation density. (b) High excitation density. (c) Cross–diagonal amplitude profile from (a). (d) Cross-diagonal amplitude profile from (b). Adapted from Ref. 273.

might be explained by a reduced degree of dielectric screening in two-dimensional materials as opposed to three-dimensional materials.

Quantitative measurements of the homogeneous linewidth [see Figs. 10.2(c) and 10.2(d)] in the limit of both low excitation density and temperature led to an extrapolated linewidth of 1.6 ± 0.3 meV, leading to a coherence time of $T_2 = 410 \pm 5$ fs. This value is almost exactly twice the population relaxation time, which was measured at $T_1 \approx 200$ fs. Such a relationship between T_1 and T_2 is expected for a system with population-decay-limited dephasing (known as the "coherent limit"), but markedly different from traditional semiconductors. In GaAs quantum wells, for example, typical population times are significantly longer than their corresponding dephasing times [110].

The $T_2 = 2T_1$ relationship between coherence and population decay times has more recently been corroborated by Jakubczyk *et al.* [165], where the authors probed the dynamics of an exfoliated monolayer sample of MoSe$_2$ on a Si/SiO$_2$ substrate using a collinear MDCS experiment with excitation pulses exhibiting a greatly reduced spot size of 700 nm. Importantly, the reduction in spot size allowed the authors to confine their excitation spot to within less than the spatial extent of a single flake of exfoliated monolayer material, and to be able to distinguish portions of the flake that

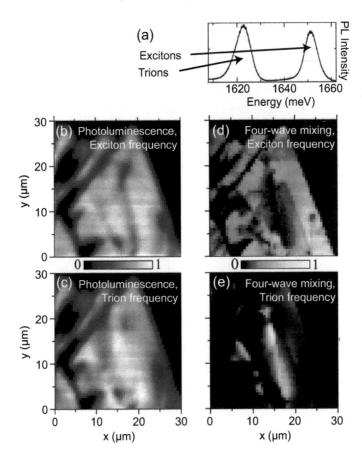

Fig. 10.3 Hyperspectral imaging of an MoSe$_2$ monolayer. (a) Photoluminescence measurements reveal two resonances, attributable to neutral bound excitons and charged trions. (b)–(c) Spatially resolved images of this photoluminescence at the resonant frequencies of the exciton [1650 meV, panel (b)] and of the trion [1625 meV, panel (c)] reveal charged and neutral regions of the sample in real space. (d)–(e) The contrast is more dramatic, and the spatial resolution sharper, for spatially resolved images of resonant four-wave mixing at the same frequencies. Adapted from Ref. 165.

were predominantly charged and predominantly neutral (Fig. 10.3), contributing to a significantly reduced degree of spectral inhomogeneity than that reported by Moody *et al.*, which was obtained using a spot size of 30 μm. More recent measurements on single-layer exfoliated WS$_2$ performed by the same group [166] have yielded similar results and conclusions.

One challenge of early measurements on TMD monolayers, particularly exfoliated flakes, was the inconsistency of the measurements, both from flake-to-flake, and due to apparent aging of samples. The discovery that encapsulating the TMD flakes in hexagonal boron nitride (hBN) improved their quality also turned out to provide more reproducible results. The claims for improvement in quality were based on the decrease on optical linewidth of the exciton resonances, however, MDCS measurements showed that both the homogeneous and inhomogeneous linewidths decreased, such that samples still exhibited weak inhomogeneous broadening at low temperature [259]. The same study showed that the homogeneous linewidth could be permanently modified by both temperature cycling and application of too much laser power in unencapsulated samples, whereas this was not the case for encapsulated samples.

The microscopic origin of dephasing and decoherence in the TMDs is of importance for their practical usefulness as quantum materials. One might optimistically hope, for example, that improvements in sample quality might reduce the scattering contributions to the dephasing rate, thereby increasing T_2 and by extension creating a more advantageous set of material properties for coherent information processing. The current measurements indicating that T_1 and T_2 are already related by a factor of 2 suggest that there may not be much more room for improvement in this area. Beyond this, microscopic calculations [273] indicate that radiative decay with a residual T_2 time of 500 fs becomes the dominant dephasing process for completely delocalized excitons in an ideal single-layer WSe$_2$ crystal. In addition, spectrally integrated photon-echo measurements on a variety of TMD samples in both bulk and monolayer form [93] found little degradation of TMD dephasing rates in monolayer samples as compared to bulk. The authors of the photon-echo study compared their results to a first-principles model of electron-phonon coupling and concluded that electron-phonon coupling, rather than defect or impurity driven phenomena, are the driving mechanisms in the decoherence rates in TMD monolayers. Ultimately, the issue remains an open question as chemical vapor deposition (CVD) is known to produce materials that are far from the ideal limit, and the low quantum efficiency of single-layer TMD photoluminescence [9] guarantees that radiative decay processes are still a long way from serving as the dominant decay mechanism. Studies on exfoliated samples that are encapsulated with boron nitride may provide some clarification.

10.3   Valley coherence and coupling in 2D materials

Perhaps the most interesting contributions that MDCS has been able to make in advancing the TMD field have been connected to the technique's ability to reveal signatures of coherent coupling between different resonances. A number of recent studies have investigated the relationship between neutral excitons and charged trions in the materials, most prominently in MoSe$_2$ [150, 151, 165, 355, 356], and have found evidence of coherent coupling between the two types of excitations. In MDCS such coupling emerges as a cross peak between diagonal exciton and trion resonances [Fig. 10.4(a)] that oscillates in intensity as the intermediate delay time T is varied. Wider bandwidth studies using a multiresonant variant of MDCS have also demonstrated coupling between excitons originating from different sub-bands, and between excitons and the continuum in a sample of few-layer MoS$_2$ [77].

One MDCS measurement of the exciton-trion coupling is shown in Fig. 10.4. A rephasing single-quantum spectrum of monolayer MoSe$_2$ at 10 K on a sapphire substrate is shown in Fig. 10.4(a) for cocircular excitation and detection [150]. Several important details are immediately evident from this spectrum. First, the exciton (X) and trion (T) resonances exhibit moderate inhomogeneous broadening due to dis-

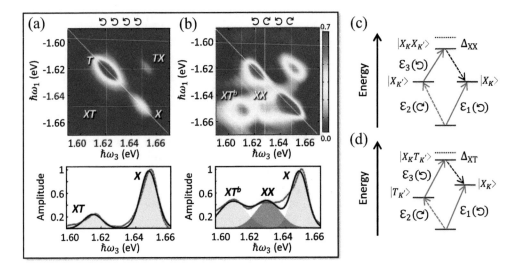

Fig. 10.4 Polarization-resolved 2D rephasing single-quantum spectroscopy of doped (n-type) monolayer transition metal dichalcogenide $MoSe_2$ at 10 K. (a) The rephasing single-quantum spectrum for co-circularly polarized excitation and detection features exciton and trion peaks on the diagonal and their off-diagonal coupling peaks XT and TX. A horizontal slice near the exciton excitation energy $\hbar\omega_1 = -1.65$ eV is shown in the bottom panel with Gaussian fits. (b) Similar to (a) but for alternating circular polarization of the excitation pulses and detection. Two additional peaks XX and XT^b are attributed to the neutral and charged biexcitons forming between two excitons or an exciton and trion in opposite valleys, respectively. The quantum pathways corresponding to XX and XT^b are shown in (c) and (d), respectively. Adapted from Ref. 150.

order potentials possibly from impurities, defects, and strain in the system. At low temperature and density, the homogeneous linewidths are on the order of 1 meV. A comparison to the population lifetime reveals that the exciton coherence time is lifetime limited ($T_2 = 2T_1$), whereas trion decoherence is significantly faster than its lifetime ($T_2 \ll T_1$). This observation might be attributed to a greater susceptibility for trion interactions with the Fermi sea or charged surface adsorbates and defects in the monolayer. A similar behavior of the trion resonance has also been reported in a MDCS measurement of the intrinsic homogeneous linewidth for the trion resonance in a CVD-grown $MoSe_2$ monolayer [396]. Second, an analysis of the off-diagonal peaks reveals coherent coupling and quantum beats between X and T that arise from EIS between excitons and trions [151, 165]. This coupling is seen as oscillations of the LCP and HCP in Fig. 10.5 at short delays, $T < 1$ ps, and exponential growth of the relative peak amplitudes compared to the exciton and trion amplitudes a long delays $T > 1$ ps.

The rephasing spectrum is quite different for cross-circularly polarized excitation and detection in which excitons and trions in both valleys are now accessible, as shown in Fig. 10.4(b). The key feature is the observation of a new off-diagonal peak (XX) that has been assigned to the bound biexciton with a binding energy $\Delta B = \hbar\omega_X - \hbar\omega_B = 20$ meV along the emission energy axis $\hbar\omega_3$. Additionally, the below-diagonal peak

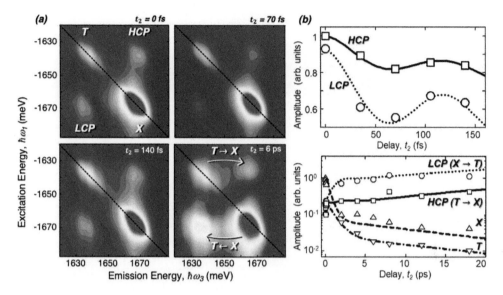

Fig. 10.5 (a) Normalized 2D rephasing spectrum (amplitude) of monolayer MoSe$_2$ for co–circular polarization at 10 K versus delay T equal to 0 fs, 70 fs, 140 fs, and 6 ps, as labeled. At short delays (T less than a few hundred femtoseconds), oscillations of the off-diagonal coupling peaks HCP and LCP indicate coherent coupling between excitons (X) and trions (T). At longer delays ($T > 1$ ps), an increase in the relative magnitude of LCP and HCP compared to X and T indicate incoherent energy and charge transfer between excitons and trions due to trion formation (LCP) and dissociation (HCP). (b) Oscillations of the cross peaks HCP and LCP versus delay T arise from quantum beats between the exciton and trion. The data are modeled with the optical Bloch equations for a four-level diamond system, revealing the dephasing time $\tau_c = 250$ fs and the oscillation period $\tau_{XT} = 130$ fs (corresponding to the exciton-trion energy difference of 31 meV). (b) On longer timescales, the exciton (X) and trion (T) peaks exhibit biexponential relaxation dynamics due to radiative and nonradiative recombination and exciton-trion energy transfer. The increase in relative magnitude of the LCP and HCP (compared to X and T) indicate exciton-to-trion and trion-to-exciton energy transfer, respectively. Adapted from Ref. 151.

(XT^b) shifts to lower emission energy by $\Delta XT = \hbar\omega_T - \hbar\omega_{XT^b}) = 5$ meV relative to the co-circularly polarized spectrum, which is consistent with a bound charged biexciton state. Both binding energies extracted from the spectra are in reasonable agreement with current theoretical models [186, 203]. Furthermore, these states only appear for cross-circular polarization, indicating that they originate from intervalley coupling between excitons and trions. The quantum pathways for these two peaks are shown in Figs. 10.4(c) and 10.4(d), respectively. It is worth noting that the spectral proximity of XX and XT^b emission with T makes them difficult to access in one-dimensional spectroscopy techniques, whereas they are clearly separated in the 2D spectra as shown by the horizontal slices taken along $\hbar\omega_3$ at the exciton excitation energy $\hbar\omega_1 = 1.648$ eV.

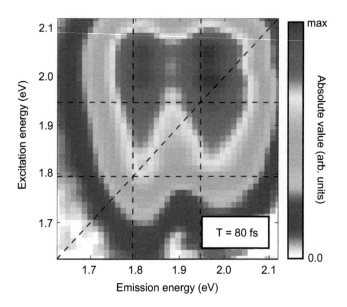

Fig. 10.6 Multiresonant MDCS measurement of exciton transition in few-layer MoS$_2$, depicting coupling effects between the A exciton at 1.79 eV and the B exciton at 1.94 eV. Adapted from Ref. 77.

MDCS also opens the possibility of probing the zero-quantum coherences between nearly degenerate excited states, offering a powerful test of the degree to which the excitons with opposite pseudospins and in opposite valleys are able to maintain valley coherence. For example, a study of exciton dynamics in WSe$_2$ [149] recently measured the zero-quantum coherence between excitons in opposite valleys (accessed using MDCS beams with cross-circular polarization) to have a linewidth of 6.9 meV. This value is almost twice as broad as the exciton population inverse lifetime of 3.4 meV, measured in the same manner but using co-circular polarization. The results place an important bound on the intervalley coherence time of TMDs.

Finally, MDCS studies have facilitated new ways of accessing higher-order states in TMDs such as biexcitons and excitons formed between excited conduction and valence band states (so-called A and B excitons, as shown in Fig. 10.6). A number of studies have reported evidence for biexciton resonances in TMD materials recently [249, 308, 346, 352, 458] but there has been some disagreement in the field between theoretical and reported experimental values of biexciton binding energies. A recent MDCS study revealed the presence and binding energy in the biexciton in a much more unambiguous fashion than would be possible using other types of spectroscopy [150]. By illuminating a sample of MoSe$_2$ with cross-circularly polarized light ($\sigma^+\sigma^-\sigma^-\sigma^+\sigma^-$), spectroscopic signatures of the biexciton in MoSe$_2$ emerge that are not visible in a co-circular arrangement of pulses that reveals only exciton resonances, trion resonances, and exciton-trion coupling. As shown in Fig. 10.4(b), among the most prominent of these is a cross peak, labeled XX, that has no analog in a spectrum derived from co-circular excitation pulses [Fig. 10.4(a)]. Comparisons to theory reveal that this peak is a direct signature of the neutral biexciton. Further comparison to theory also demonstrates that the peaks labeled TX^b and XT^b are multi-exciton states corresponding to charged biexciton states [150]. Further analysis indicates that all of these states are intervalley in nature, which—due to the spin-valley coupling inherent in TMDs—may exhibit a number of interesting quantum mechanical properties including entanglement between the pair of valley pseudospins [352].

2DCS experiments of other TMD materials (WSe_2 and MoS_2) have revealed rich exciton dynamics arising from nonradiative coherences between opposite valley states [149], exciton-phonon and exciton-exciton interactions [93, 273], and coupling between the exciton and electron-hole continuum [77]. 2DCS has also helped identify the mechanism of exciton dephasing in layered InSe [91] and energy transfer between singlet and triplet states in layered GaSe [92]. The ability to isolate and stack different 2D materials with precision, transfer them to nanopatterned substrates, and incorporate them with photonic microcavities and plasmonic nanostructures provide unprecedented possibilities for tailoring their physical properties for novel opto-electronic, photonic, and coherent valley/spintronic applications. Optical 2DCS will likely play a role in understanding and characterizing TMD-based heterostructures and nanophotonics in the same way that it has impacted our understanding of optical phenomena in monolayers.

11

Other applications of multi-dimensional coherent spectroscopy in Physics

As the technique of MDCS has become more widespread, more versatile and easier to implement, it has been applied to an ever broader set of materials. Previous chapters have highlighted some of the material system where it has been used extensively and provided unique insight. Indeed in some cases, MDCS has supplanted traditional one-dimensional or incoherent techniques and become the primary spectroscopic method.

Given the constant evolution of the field, we cannot capture all the applications of MDCS, and even if we did, it would quickly be out of date. However, in this chapter, we try to highlight a handful of the other material systems where MDCS has been used relatively recently to obtain new insight. As we note in the Preface, in this book we are focusing samples of interest to researchers in the fields of Physics and Material Science. There is an extensive literature on using MDCS to study molecules, however, those results are well covered in other books [62, 147].

11.1   Semiconducting carbon nanotubes

Single-walled carbon nanotubes (SWNTs) possess many unique electrical and optical properties that make them appealing for next-generation opto-electronics and photonics [422]. The photophysics of quasi one-dimensional excitons confined along the circumferential direction are highly tunable via the configuration and diameter of the SWNT [425]. The structure can be conceptualized by wrapping a monolayer sheet of graphene into cylindrical form along an axis Z defined by a pair of chiral indices (n, m) (see Fig. 11.1(a)) that correspond to the number of unit vectors $\vec{a}_1$ and $\vec{a}_2$ in the real-space honeycomb crystal lattice. The properties of a SWNT are highly dependent on its chirality. In general, for $|m - n| = 3k$, where k is an integer, the SWNT is metallic; otherwise, the SWNT is a semiconductor, although there are exceptions to this rule [102]. The optical spectrum of a single semiconductor SWNT contains a series of sharp peaks (denoted E_{11}, E_{22}, etc.) whose energy scales inversely with the tube diameter $d = (|\vec{a}|/\pi)\sqrt{n^2 + nm + m^2}$, which is on the order of ~ 1 nm. The ability to control the optical bandgap and electrical behavior makes SWNTs a model system for studying one-dimensional exciton and charge transfer physics.

Films containing a mesoscale network of interwoven SWNTs are readily synthesized with solution-based processing techniques [290]. A purification processing step enables

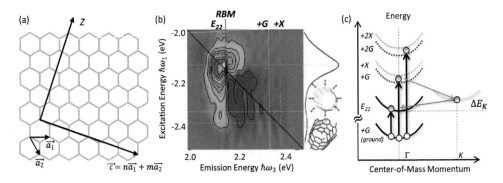

Fig. 11.1 (a) Hexagonal lattice structure of a nanotube with real-space unit vectors $\vec{a}_1$ and $\vec{a}_2$. The vector $\vec{c}$ denotes the chirality of the nanotube, where Z is the tube axis. (b) Real part of a 2D rephasing spectrum from a (6,5) single-walled carbon nanotube under ambient conditions for $\tau_2 = 100$ fs. Population transfer between the diagonal exciton (E_{22}) and phonon sideband states (G and X) is revealed by the off-diagonal cross peaks. (b) Energy level diagram depicting absorption quantum pathways of pulse E_3 from the ground-state vibrational coherence of the G-band mode induced by stimulated Raman-type processes from the first two pulses. Excitation from the G-band (ground) to the excited-state phonon sidebands is accompanied by phonon-mediated relaxation to the E_{22} exciton state. Adapted from Ref. 137.

sorting of semiconducting nanotubes from metallic ones that detrimentally quench exciton emission. The energetic manifold of a SWNT film, which comprises an ensemble with many different chiralities, is extremely complex, since there are multiple discrete transitions for each structure type. Steady-state photoluminescence excitation (PLE) spectroscopy has been the predominant method for resolving the different chiralities and energy transfer between them; however, without any time resolution, PLE cannot distinguish between different energy transfer pathways that lead to emission at the same energy, even for an isolated SWNT [263].

Several variants of broadband optical 2DCS have been developed to study these dynamics. The results from one such experiment is shown in the single-quantum rephasing spectrum in Fig. 11.1(b), which corresponds to excitation and emission near the E_{22} transition of an aqueous suspension of (6,5)-enriched SWNTs. The strongest diagonal feature corresponds to the ESE and GSB pathways of the E_{22} exciton transition. The off-diagonal peaks explicitly show that excitons also couple to several prominent phonon modes, including the radial breathing mode (RBM at 296 cm^{-1}, 37 meV) and the G-band (1590 cm^{-1}, 197 meV). An analysis of the dynamics of the off-diagonal peaks during delay τ_2 further elucidates their origin. Oscillations of the peak amplitudes with periods of 21 fs and 110 fs indicate that impulsive optical excitation creates a coherent wavepacket corresponding to a vibrational superposition of the exciton with the G-band (21 fs) and the RBM (110 fs) through stimulated Raman-type processes. The off-diagonal peaks are also enhanced by multiphonon scattering involving higher-energy X-band modes and the K-point state. This pathway is depicted in Fig. 11.1(c).

2DCS experiments have also found utility for understanding energy transfer in mixed ensembles with multiple chiralities. Energy transfer from higher-bandgap to lower-bandgap SWNTs ranging from tens of femtoseconds between bare nanotubes to several picoseconds for nanotubes wrapped in a conjugated polymer coating has been identified [264, 265]. Packing SWNTs with synthetic heterodimers also serves as a model system for emulating natural light-harvesting complexes; observations of quantum beating in the 2D spectra from these SWNTs have helped elucidate the origin of energy transfer in photosynthesis [430].

11.2   Color centers in diamond

Color centers in diamond/indexdiamond are combinations of elemental impurities and carbon vacancies that absorb and emit light at visible or near-infrared frequencies within an otherwise pristine and non-absorbing diamond host lattice. In recent times, these defects have drawn a significant amount of attention in the Material Science and Quantum Optics communities in conjunction with potential applications in sensing and the construction of quantum networks. To date, the most commonly studied color-center systems within this material classification are the negatively charged nitrogen-vacancy (NV^-) centers and negatively charged silicon-vacancy (SiV^-) centers, and MDCS has been used to investigate both of these.

Among the most heavily utilized advantages of using the technique to study these systems has been its ability to isolate and characterize electron-phonon coupling interactions—which are a relevant coherence-limiting and linewidth-limiting factor in applications involving quantum networks. The ability to extract homogeneous electronic T_2 times in the midst of inhomogeneous ensembles has also proved useful, particularly in cases involving densely packed color centers subject to large amounts of strain or exhibiting inter-particle quantum mechanical coupling effects.

The first application of MDCS to study color centers in diamond was published in 2013 [160], in a work where the authors reported amplitude spectra corresponding to a series of combined rephasing-plus-nonrephasing measurements [with spectra combined according to $S_I(\kappa_t, -\kappa_\tau) + S_{II}(\kappa_t, \kappa_\tau)$; see Chapter 3.9] of the coherent properties of an NV^- center ensemble at room temperature. Among the notable aspects of the measurement is that coupling between the NV^- center zero-phonon line at 15,700 cm^{-1} (equivalently, at 1.95 eV or 637 nm) and a prominent vibrational sideband at 16,180 cm^{-1} can be observed in the form of a cross peak in the 2DCS data with a maximum at $(\bar{\nu}_t, \bar{\nu}_\tau) = (15{,}700 \text{ cm}^{-1}, 16{,}180 \text{ cm}^{-1})$. The work remains an benchmark in the field because of its demonstration that MDCS studies of color centers in diamond are feasible.

More recently, researchers have used narrower-bandwidth pulses (15 meV full-width at half-maximum, as compared to Ref. 160, where the bandwidth was 1,600 cm^{-1} or 400 meV) to study the properties of the zero-phonon line in greater detail, measuring the line's temperature dependence as a sample of NV^- centers in bulk diamond was tuned across the phonon excitation energy threshold [228]. By combining a temperature-dependent measurement on the sample with the power of MDCS to isolate homogeneous linewidths, the authors identified thermal dephasing effects resulting from interactions with quasi-localized vibrational modes, and they observed

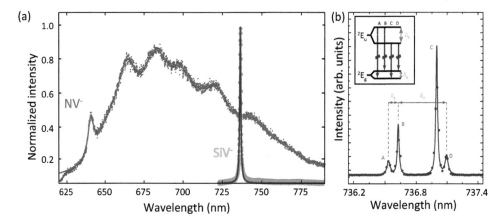

Fig. 11.2 (a) Comparison between the room-temperature photoluminescence spectrum of NV⁻ centers in diamond (gray dots) and SiV⁻ centers in diamond (blue circles). Provided by J. Becker. (b) Low-temperature (4 K) photoluminescence spectrum of the SiV⁻ center zero-phonon line, split into its characteristic four-peak structure. The inset shows the system's accompanying energy-level diagram. Adapted from Ref. 23.

ultrafast spectral diffusion on the picosecond timescale. The authors also extrapolated cross-diagonal 2DCS linewidths down to the zero-temperature limit and found the intrinsic, ensemble-averaged homogeneous linewidth of their sample to be in the range of tens of GHz.

Among the exciting new developments in the spectroscopy of color centers beyond NV⁻ centers in recent times has been the application of MDCS techniques to the silicon-vacancy (SiV⁻) center system, which has been demonstrated to exhibit a significantly more pronounced zero-phonon line than that of the NV⁻ center system as well as less spectral diffusion as illustrated in Fig. 11.2, making it desirable for applications in quantum networks.

While the measurements displayed in Fig. 11.2 were conducted through photoluminescence (and indeed photoluminescence measurements constitute the most commonly utilized means of accessing color-center optical properties), the ability of MDCS to incorporate other forms of optical detection schemes in a background-free manner has recently led to the discovery that not all SiV⁻ centers are created alike. Figure 11.3 shows a comparison, using collinear MDCS, between two such complementary schemes, with Figs. 11.3(a) and 11.3(b) showing an MDCS experiment where the signal was collected in the form of a photoluminescence beatnote, and Figs. 11.3(c) and 11.3(d) showing an MDCS experiment where the signal was collected through heterodyne detection (see Section 4.4 for a discussion of these methods). While the photoluminescence-detected signal bears a strong resemblance in many ways to the signal that would be expected in the sample based on standard linear photoluminescence techniques, the heterodyne-detected signal reveals a striking new population of "hidden" silicon-vacancy centers with more than 60 times as much spectral inhomogeneity than the photoluminescence-detected bright centers.

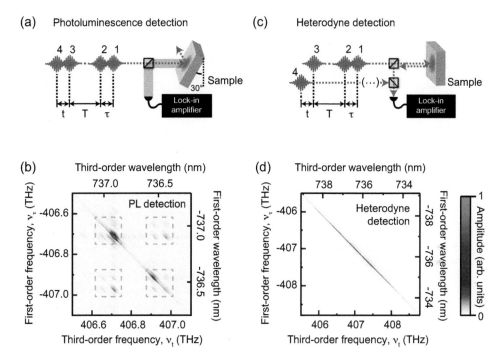

Fig. 11.3 MDCS characterization of hidden resonance features in a high-density sample of SiV$^-$ centers in diamond. (a)–(b) Rephasing plot of the sample as measured with collinear excitation pulses and photoluminescence detection. (c)–(d) Rephasing plot of the sample as measured with collinear excitation pulses and heterodyne detection. Adapted from Ref. 364.

Perhaps more interesting yet is the discovery, facilitated by the ability of MDCS to independently characterize the electronic T_2 times of these different populations of centers by analyzing cross-diagonal linewidths, that the hidden SiV$^-$ center population exhibits significantly enhanced T_2 times in comparison to the population of PL-emitting centers. The findings may be useful in the development of quantum devices where strain is intentionally applied to silicon-vacancy centers embedded (for example) in nanowires to controllably tune color-center brightness or optical coherence properties.

Flipping the relationship around, recent studies have also demonstrated that the sensitivity of SiV$^-$ centers in diamond to strain effects can be used in concert with the ability of MDCS to see through congested spectra to make inroads into the development of improved methods for quantifying strain in the first place [19]. Figure 11.4 shows a position-dependent measurement of photoluminescence-detected two-pulse correlation measurements collected on the same sample as was used to collect the data for Fig. 11.3. By comparing the peaks in these measurements to 2DCS rephasing plots and then mapping the peak positions onto the different ways that the strain tensor has been understood to shift around optical resonance features, it was shown

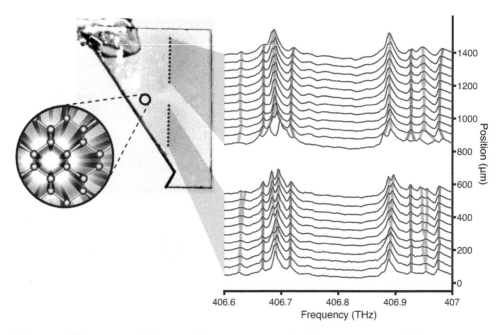

Fig. 11.4 Illustration of the use of resonances in a high-density sample of SiV⁻ centers in diamond to map out the strain tensor in the sample. Adapted from Ref. 19.

possible to map out the sample's position-dependent strain tensor in a contact-free manner that could potentially be applied in many other types of systems.

Beyond their sensitivity to strain, SiV⁻ systems are expected to experience electron-phonon coupling effects, taking on a similar form to the mechanisms observed in NV⁻ centers but on a much weaker scale. In a recent theoretical investigation into the subject [227] it was argued that MDCS may prove uniquely suited among spectroscopic techniques to be able to characterize these effects. To date, experimental verification of this prediction remains lacking, but the possibility of such verification remains an intriguing future prospect in the field.

11.3 Perovskite materials

Hybrid organic-inorganic perovskite materials, especially the organometal halide perovskites, have attracted tremendous attention in the past few years due to their potential for solar cells [138] and other optoelectronic applications [88, 174, 388, 446]. In organometal halide perovskite structure (ABX_3), A represents organic cation (MA, FA), inorganic cation (Cs) or their mixture; B denotes metal cation (Pb or Sn); and X refers to halide anion. As shown in Fig.11.5, BX_6 octahedra are corner-shared by cations (MA, FA, or Cs) in 3D perovskites. Perovskite materials can be synthesized in the form of thin films, single crystals, and nanocrystals. By tuning the cation, metal halide, and structure of building block for thin films, single crystals, and nanocrystals, it is possible to have different photophysical behaviors and facilitate various applications.

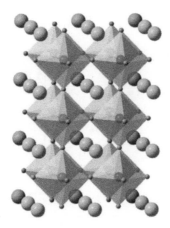

Fig. 11.5 Schematic representation of the perovskite structure.

Perovskite materials have unique properties such as long carrier lifetimes, tolerance to defects, high charge-carrier mobility, a tunable absorption edge, ease of fabrication by solution process, etc. These properties make them promising candidate materials for high-performance optoelectronics in applications such as solar cells, field-effect transistors, light-emitting diodes, lasers, X-ray detectors, and photodetectors. Perovskite materials also have potential applications in spintronics due to strong spin-orbit coupling tied to the heavy metal Pb. This property opens new opportunities to explore quantum control of spin coherence to advance the field of spintronics, which aims to use the superposition of opposite spins to process information. To realize these applications, it is essential to understand the optical properties, the photoexcitation landscape, and the quantum dynamics of photoexcited carriers, phonons, and spins in various forms of perovskite materials.

Ultrafast spectroscopy, especially, transient absorption and reflection spectroscopy, has been used to study the carrier and phonon dynamics of perovskites. Through linearly polarized transient absorption spectra, hot carrier relaxation has been observed to last up to a few hundred picoseconds [120, 142, 310, 407, 451, 455]. The existence of large polaron [267, 268, 467] and hot phonon bottleneck [113, 270, 300, 429] could be the possible mechanisms for these long-lived hot charge carriers. Carrier mobility was measured using ultrafast microscopy and found to be dominated by quasi-ballistic transfer dynamics [142]. The coherent phonon due to the angular distortion of Pb-I-Pb with optical character has been observed at both low temperature [113, 429] and room temperature [270, 300]. Temperature-dependent time-resolved photoluminescence and microwave conductance experiments also suggest phonon-assisted electron-hole recombination, indicating the direct-indirect character of the band gap [159]. This might explain why the electron-hole recombination is slow [376, 447], which plays a significant role in achieving decent solar cell efficiency.

Optical 2DCS has also been applied to various perovskite materials to study coherent phonon interactions [270], ultrafast dissociation of excitons into charge carriers [169], initial thermal carrier distribution due to carrier interactions [318], existence

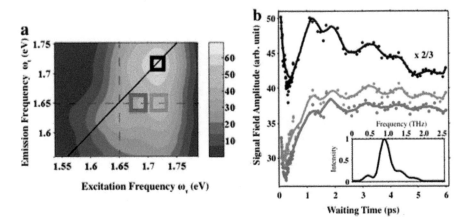

Fig. 11.6 (a) 2D amplitude spectrum with $T = 10$ ps for a MAPI film at room temperature. The boxed areas of interest correspond to the main diagonal peak (black) above the band edge and two lower cross-peak areas (orange and red). The dashed lines mark the band edge at 1.65 eV. (b) The averaged amplitudes over different boxed regions in (a) are plotted as a function of the waiting time T. The inset is the Fourier power spectrum of the orange line from 2 to 6 ps. Adapted from Ref. 270.

and properties of biexcitons [393], vibrational coupling [291], carrier dynamics and coupling of the dual emission from orthorhombic phase at 115 K [395], and homogeneous linewidth in perovskite nanoplatelets [229]. These studies have utilized unique advantages of optical 2DCS such as the abilities to separate and measure homogeneous and inhomogeneous linewidths, reveal coupling and interactions, and probe quantum coherence. Here we discuss a few examples of optical 2DCS experiments in perovskites.

One of the first experiments was performed in methylammonium lead iodide (MAPI) films at room temperature [270]. With the rephasing excitation pulse sequence, single-quantum 2D electronic spectra were obtained in the energy region near the band edge. The excited carrier density is $n = 1.3 \times 10^{18}$ cm^{-3}. A 2D amplitude spectrum at waiting time $T = 10$ ps is shown in Fig. 11.6(a). The spectrum consists of a main diagonal peak above the band edge and off-diagonal signals below the diagonal line. A series of 2D spectra were measured at different waiting times. To track the spectral change over the waiting time, the average amplitudes within the three boxed areas in Fig. 11.6(a) are plotted as a function of the waiting time, as shown in Fig. 11.6(b). The cross-peak amplitudes, shown as the red and orange traces, grow quickly due to carrier cooling with an exponential rise time of about 300 fs. The diagonal-peak amplitude, represented by the black trace, decays slowly, indicating that the majority of the carriers stay above the band edge for over 6 ps. The traces in Fig. 11.6(b) also show clear amplitude oscillations that have the same frequency and phase. The Fourier power spectrum of the orange trace from 2 to 6 ps, as shown in the inset, has a strong peak at 0.9 THz. These oscillations can be attributed to the evolution of vibrational wavepackets (phonons) due to an indirect band gap excitation and interacting with the charge carriers. The phonon modes in the 0.9-1 THz frequency range have been

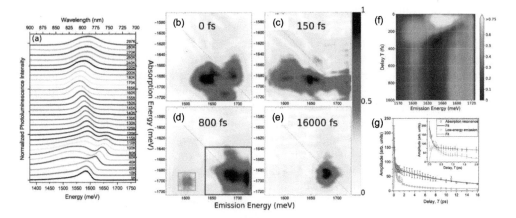

Fig. 11.7 (a) PL spectra of a MAPI thin film at different temperatures. (b)−(e) Representative 2D spectra at different waiting times with the sample at 115 K. (f) Projection of 2D spectra onto the emission axis for different waiting times up to 1 ps. (g) Decay of the peak amplitude within the highlighted regions in (d). Adapted from Ref. 395.

assigned to a set of Pb-I-Pb angular distortions with mixed transverse-longitudinal optical character [209]. Coherent phonon oscillations at this frequency were observed in pump-probe experiments for MAPI films at 77 K but not at room temperature [429]. The 2DCS measurement is background free and more sensitive in detecting these oscillations compared to transient absorption experiments. However, with the limited waiting time range, this experiment does not have the sufficient resolution to resolve the actual linewidth of the Fourier peak. The observed strong coupling between the electronic transition and the 0.9 THz phonon modes likely cause a phonon bottleneck that is responsible for slow hot carrier cooling in MAPI films.

In another example, optical 2DCS was used to study the coupling between two resonances in MAPI films. Perovskite MAPI has three phases including tetragonal, orthorhombic, and cubic phase depending on the temperature. The corresponding PL spectra at different temperatures display three characteristic regions [81, 111, 198, 446], as shown in Fig. 11.7(a). From room temperature to 160 K, the sample is in the tetragonal phase and the PL speak shifts to red and becomes narrower as the temperature decreases. The tetragonal phase transitions into the orthohombic phase around 160 K. The PL peak blue shifts from 160 to 120 K and a new peak emerges at a higher energy around 125 K. Below 120 K, both PL peaks red shift and the higher energy peak becomes dominant as the temperature decreases. The dual emissions below 125 K have been of interest and several underlying physical mechanisms have been proposed to explain the molecular origin of the dual-emission feature [81, 111, 198, 298, 434, 444, 446]. These studies mainly focused on the PL measurements with either time-integrated PL or time-resolved PL at ns or longer time scales. Optical 2DCS has also been used to study the carrier dynamics and coupling of the dual emissions in a MAPI film at 115 K.

Single-quantum 2D spectra were measured for the dual-emission resonances at different waiting times T ranging from 0 to 16 ps. Four representative 2D spectra are shown in Fig. 11.7(b−e). The 2D spectrum at $T = 0$ shows that the excited carriers already started to redistribute into lower energy states within the 35-fs pulse duration. The 2D spectra at later T reveal that the carriers move into an intermediate resonance at 1650 meV and finally settle into a lower energy resonance at 1600 meV after 250 fs. The 2D spectrum at $T = 800$ fs, as shown in Fig. 11.7(d), has a peak on the diagonal line while the off-diagonal signal has only one peak at the low-energy emission resonance (1600 meV). The high-energy emission resonance (1650 meV) is short lived and its signal disappears after 250 fs. As the waiting time increases up to 16 ps, the diagonal peak and the low-energy off-diagonal peak persist but the signal amplitudes decrease. This carrier redistribution process can be better visualized in Fig. 11.7(f) which plots the projection of 2D spectra onto the emission axis at different waiting time T. It is equivalent to a time-resolved population measurement with a time resolution of 35 fs. A short-lived intermediate state within 250 fs is clearly shown and the carrier population in the low-energy emission resonance and the absorption resonance have a longer lifetime. The integrated signal amplitude of these two resonances, as indicated by green and blue squares in Fig. 11.7(d), is plotted as a function of the waiting time in Fig. 11.7(g). The off-diagonal peak (green curve) displays clear oscillations in amplitude, indicating coherent phonon interactions with the low-energy emission resonance. In this work, a set of 2D spectra revealed the ultrafast dynamics of the carrier redistribution into a short-lived transient state and subsequently into a lower energy emission state. The time evolution of the cross-diagonal peak also indicated coherent phonon interactions in the system.

The ability of 2DCS to separate homogeneous and inhomogeneous linewidths in an inhomogeneously broadened system allows measurements of intrinsic linewidth and line-broadening mechanisms in perovskite nanocrystals. Lead-halide perovskite nanocrystals were first synthesized in 2015 [312] and attracted much interest due to their unique light absorption and emission properties [5, 345]. Perovskite nanoscrystals can be sythesized as nanocubes or nanoplatelets [26, 405].

Perovskite nanocubes attracted attention due to their high luminescence quantum efficiency, as compared to more traditional colloidal nanoparticles synthesize from II−VI or III−V semiconductors. It was suggested that the high quantum efficiency was due to the position of the dark state relatively to the optically active triplet states. In traditional nanocrystals, the dark state is energetically below the triplet states, thus electron-hole pairs that relax into the dark state usually undergo nonradiative recombination and thus do not luminesce. In contrast, calculations suggest that the dark state is energetically above the triplet states in perovskite nanocubes, thereby reducing the favoring radiative recombinaton [24]. MDCS measurements on peroviskite nanocrystals at low temperature provided information on the structure of the triplets states, including energetic spacing, dipole orientation, and dephasing of the coherences between the triplet states [226]. In addition, these results suggested that the dark state may actually lie between the triplet states, with its exact position depending on nanocube size.

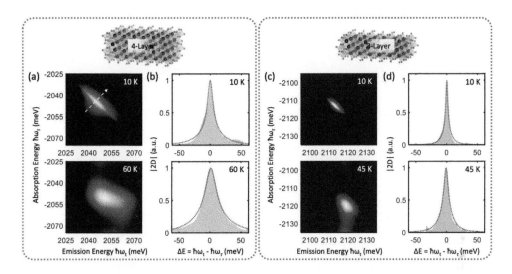

Fig. 11.8 Single-quantum 2D amplitude spectra of (a) four-layer and (c) three-layer nanoplatelets synthesized at a reaction temperature of 170 °C. The waiting time is set to $T = 1$ ps. Cross-diagonal slices of (b) four-layer and (d) three-layer 2D spectra centered at 2045 meV and 2113 meV respectively. Adapted from Ref. 229.

Besides sharing the efficient and tunable photophysics of their nanocube counterparts, nanoplatelets' planar geometry allows for directional light emission and absorption [173], efficient energy transfer in stacked superlattices [323], and remarkable homogeneity in the dominant out-of-plane quantum confinement by precisely controlling the polyhedral layer thickness [435]. Due to size and shape dispersion, spectra of colloidal nanoplatelets usually have inhomogeneous broadening. Optical 2DCS was performed on four-layer and three-layer $CsPbI_3$ perovskite nanoplatelet ensembles to determine the intrinsic homogeneous linewidth and elucidate the underlying homogeneous broadening mechanisms. Both four-layer and three-layer thick nanoplatelets were synthesized at a reaction temperature of 170 °C. Single-quantum 2D amplitude spectra are shown in Figs. 11.8(a) and (c) for four-layer and three-layer samples respectively. The diagonal elongation of the spectra indicates inhomogeneous broadening due to varying confinement of lateral exciton center-of-mass motion, despite dominant out-of-plane quantum confinement. The degree of inhomogeneous broadening depends strongly on layer thickness as shown in 2D spectra for four-layer and three-layer nanoplatelets. The homogeneous linewidth γ is extracted by fitting the cross-diagonal slices at 2045 meV and 2113 meV in Figs. 11.8(b) and (d), respectively.

As the temperature increases, the 2D spectra broaden in the cross-diagonal direction, which is characteristic of thermal dephasing due to elastic exciton-phonon scattering [273, 357]. The temperature dependence of the homogeneous linewidth follows a linear dependence as [324]

$$\gamma(T, N_x) = \gamma_0(N_x) + AT, \tag{11.1}$$

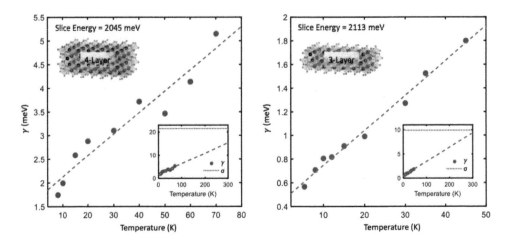

Fig. 11.9 The dependence on temperature of the homogeneous linewidths retrieved from the 2D spectra for 4-layer (left) and 3-layer (right) nanoplatelets. The inhomogeneous linewidths are also plotted as horizontal dotted lines in the insets. Adapted from Ref. 229.

where $\gamma_0(N_x)$ is the zero-temperature linewidth at an excitation density N_x, T is the temperature, and A represents the strength of the coupling to low-energy acoustic phonon modes. A series of 2D spectra were obtained at different temperatures to extract the temperature-dependent homogeneous linewidths, which are plotted in Fig. 11.9 for four-layer (left) and three-layer (right) nanoplatelets. By extrapolating to zero temperature, the intrinsic linewidths are found to be $\gamma_0 = 0.45$ meV and 1.67 meV for three-layer and four-layer nanoplatelets, respectively. The intrinsic linewidth is fundamentally limited by the population relaxation rate. The dominant population relaxation channel in colloidal nanocrystals is a spin-flip process between an optically-active bright-exciton manifold and an optically-inactive dark state [3, 260]. As the nanoplatelet thickness decreases, the bright-dark energy splitting increases [351], resulting in a slower spin-flip relaxation rate. This explains that the intrinsic linewidth γ_0 is smaller for three-layer nanoplatelets. The thermal dephasing parameters extracted from the fits are $A = 0.032$ meV/K (three-layer) and $A = 0.045$ meV/K (four-layer), indicating relatively weaker acoustic phonon coupling in three-layer nanoplatelets.

12
New trends in multidimensional coherent spectroscopy

Optical multidimensional coherent spectroscopy, encompassing the infrared and visible regions of the electromagnetic spectrum, was first demonstrated in the late 1990s. As is typical for the first realizations of a technique, the initial results were often sub-optimal in terms of both results and approach, nevertheless they demonstrated the potential of the technique and thereby inspired both refinements of the initial approaches and the development of entirely new approaches.

While MDCS has matured significantly at this point, becoming a more easily accessible technique, and perhaps the preferred technique in ultrafast spectroscopy, MDCS continues to undergo improvement, broadening and refinement. In this chapter, we briefly describe some of the current trends in the development of the technique, which we think will be significant in the near future.

12.1 Broadening the spectral range: from terahertz to x-rays

As noted in Chapter 1, the concepts of MDCS were originally developed in NMR spectroscopy. Implementing them in the optical regions of the electromagnetic spectrum constituted opening a new spectral region for these methods. Generally, the "optical" spectral region is defined as the visible through mid-infrared, where it is possible to use familiar optical elements such as lens and mirrors and where detectors measure the intensity of the light, perhaps at the single-photon level. The success in the optical regions inspired efforts to further the spectral coverage of MDCS.

One direction was the development of MDCS methods in the "terahertz" (THz) spectral region, i.e, the spectral range that has lower frequency (longer wavelength) than the optical region, but at high enough frequency that typical electronics technologies no longer work. In this region, some familiar optical elements still work, such as lenses and mirrors, but typically the detection methods directly measure the electric field of the electromagnetic radiation, rather than the intensity. In addition, the longer wavelength often means that the sample is not necessarily large compared to the wavelength, making beams of well-defined wavevector less useful. In this sense, the THz region is more similar to the radio-frequency regime where MDCS techniques were originally developed.

In the opposite direction, implementing MDCS using x-ray or extreme ultraviolet (XUV) pulses, most likely produced by an x-ray free electron laser, has been proposed, although not yet realized. MDCS with x-rays is most interesting for molecular systems as it allows atomic core states to be accessed. The high photon energy makes single-photon detection easier than the optical spectral regime, however, the short wavelength makes phase-resolved measurements challenging, which is exacerbated by the lack of ordinary optical elements.

12.1.1 THz MDCS

THz pulses are typically generated from ultrashort optical pulses using methods such as optical rectification, difference frequency generation or biased photoconductive switches [114]. Conversely, detection of THz pulses is realized using ultrashort optical pulses as gate or probe pulses to measure the electric field of the THz signal using electro-optic sampling or photoconductive switches. Since these methods directly measure the electric field of the THz pulse, heterodyne methods employing a THz reference pulse are not needed. Manipulating THz beams is typically more challenging than optical beams, due to the combination of greater divergence and difficulty in detection, thus as much manipulation as possible, for example generation of time delays, is done for the optical beams prior to THz generation.

The large size of propagating THz beams and the large spot size of focused beams, both a consequence of the long wavelength (hundreds of microns to 1 millimeter), make it difficult to implement the noncollinear geometries often used for MDCS in the optical part of the spectrum. Instead, fully collinear geometries are typically used. The use of a fully collinear geometry requires a method to separate the excitation pulses from the signal. Unlike to the optical spectral region where phase-cycling methods are typically used, in the THz separation is typically achieved by taking the difference between measurements with all excitation pulses and an appropriate combination of measurements where individual excitation pulses are absent [105]. This process can separate the nonlinear signal from the excitation pulses, but does not separate different coherence pathways from one another, thus the resulting time-domain signal typically includes all pathways. For example, coherent photon-echo signals for various pulse orderings as well as transient absorption signals. The time-domain data are typically a function of the delay between two of the THz pulses and the delay of the gate pulse used for detection of the THz signal. The different pathways become separate in the frequency domain by taking a two-dimensional Fourier transform. Because the THz pulses are only a few cycles long, and the durations can be comparable to the relevant relaxation times, the scans often have regions of significant pulse overlap, thus multiple possible time-orderings have to be considered. Again, these can typically be separated in the two-dimensional frequency spectrum.

The first experimental demonstration of 2D THz spectroscopy studied intersubband transitions in semiconductor quantum wells [199, 200]. These transitions can have extremely large dipole moments, making them good candidates for nonlinear spectroscopy. Indeed it is possible to achieve Rabi oscillations using THz pulses. Including nonperturbative contributions was critical to understanding the resulting 2D THz spectra [200].

By adjusting the quantum well width, the intersubband transition energy can be adjusted into the frequency range of the optical phonons that occur in polar semiconductors such as GaAs. This results in coupling of intersubband and lattice excitations, which can be studied using 2D THz correlation spectra [201]. The 2D THz spectra reveal a highly complex nonlinear signal with many diagonal and off-diagonal components of both positive and negative signs, i.e., corresponding to both induced transmission and induced absorption.

Coupling of phonon excitations to electronic interband resonances can also occur in bulk narrow-gap semiconductors due to the deformation potential and polar optical excitations. These effects are manifest in the second-order Raman spectrum of such materials, which can be ascribed to double-quantum excitations that are revealed by the anharmonicity of the associated phonons. These effects have been studied using double-quantum 2D THz spectroscopy [367, 368].

12.2   Improving the spatial resolution

Nonlinear spectroscopy, especially optical 2DCS, has been mainly used to study large ensembles, leading to important limitations in interpreting data in connection with underlying microscopic mechanisms. Ensemble measurements only provide an average of properties of individuals within an ensemble, which can include two or more species, heterogeneity in one species, and other variations from individual to individual. Although optical 2DCS can isolate the homogeneous linewidth even in the presence of inhomogeneous broadening, measurements of many properties are still not distinguishable, especially when the resonances are energetically degenerate. The average of an ensemble property may not be an accurate representation of individuals. Moreover, the ensemble measurement of a property does not provide information about its spatial distribution. This information is critical to understand the role of heterogeneities (both static and dynamic) due to, for example, grain boundaries, defects, disorder, etc. Optical 2DCS measurements with a high spatial resolution are especially useful to reveal the dynamics of spatial variations. Finally, ensemble measurements may not be able to detect rare events or small sub-groups in the ensemble, such as defect-induced quantum dots in 2D materials and polarons in perovskites. The identification of rare events associated with particular individuals or sub-groups is critical to understand the underlying causes and therefore to engineer materials of desired properties.

Optical 2DCS with a sub-diffraction spatial resolution can benefit many studies of sample heterogeneity, single-molecule systems, and more. For instance, electronic dynamics in perovskite thin films show spatial variations across grain boundaries [335], the energy transfer time scale in light-harvesting proteins varies across individual proteins [156], the charge transfer and transport in single molecular junctions are sensitive to sub-molecule changes, the local structure and environment play important roles in determining carrier dynamics in semiconductor nanostructures, and the localized 2DCS can be used to construct the two-exciton wave function of coupled quantum emitters [334].

The spatial resolution of most optical 2DCS techniques is determined by the size of beam spot on the sample. In the box geometry, the beams can be focused to the size of ~ 10 μm by an optical lens that converges the excitation beams. A better spatial resolution can be achieved in the collinear geometry, where an objective lens can be used to focus the beams to a tight spot. However, the spatial resolution is ultimately

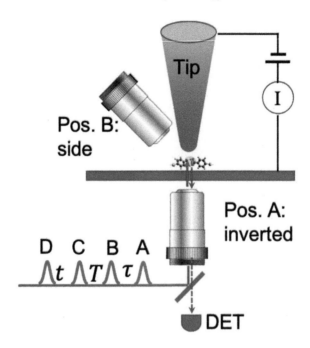

Fig. 12.1 Schematic showing collinear optical 2DCS on a tip. The co-propagating excitation pulses are focused on the tip by an objective. The objective can be either inverted (Pos. A) or coupled from the side (Pos. B). The generated fluorescence signal is collected by the same objective and detected by a detector (DET).

limited by the diffraction limit. It is still a challenge for optical 2DCS to achieve a high spatial resolution beyond the diffraction limit.

One approach to achieve a sub-diffraction spatial resolution in optical 2DCS is to use nonoptical detection. The resolution of nonoptical detection is not limited by the diffraction limit of optical fields. Optical 2DCS with a 50-nm spatial resolution has been implemented by using photoemission electron microscopy as detection [4]. In this experiment, even though the excitation beams have a diffraction-limited spot ($\sim$ 400 nm) on the sample, the final state after a four-pulse excitation is detected via photoemission electron microscopy, which has 50-nm spatial resolution. The method was demonstrated [4] on a corrugated silver surface to image nanoscale coherence. The results showed lineshape variations in 2D spectra at the subwavelength scale.

An all optical approach can also achieve a sub-diffraction spatial resolution by using a metal tip surface, such as a Scanning Tunneling Microscope (STM) tip or a gold nanoelectrode (GNE), to enhance and confine a light field to sub-10 nm spatial dimensions. Ultrafast surface-enhanced Raman spectroscopy (SERS) has been developed in recent years [140]. The application of ultrafast Raman spectroscopic technique on the plasmonic surface can provide high temporal and spatial resolution and high sensitivity while minimizing the sample damage problem. Compared with the traditional nonlinear optical measurements, the number of probed molecules can be greatly reduced, even reaching single-molecule resolution [222, 450]. The success of tip-enhanced nonlinear spectroscopy shed light onto the feasibility to integrate optical 2DCS and tip-based single-entity electrochemical techniques [432] to achieve sub-diffraction spatial resolution. This can be accomplished by coupling the excitation beam of a collinear 2DCS apparatus to a GNE or STM tip through a high NA objective.

A schematic overview is shown in Fig. 12.1. Four co-propagating excitation pulses derived from a collinear optical 2DCS are used. They are focused to a diffraction-limited spot onto the apex of a tip by an objective. There are two different configurations. The objective can be either inverted (Pos. A) or coupled from the side (Pos. B). The laser field is enhanced by the tip to form a hotspot with a small volume to achieve a sub-diffraction spatial resolution. The excitation pulses generate a high-order fluorescence signal which can be collected by the same objective and measured by a detector. A single-photon detector can be used to provide sufficient sensitivity for single-molecule measurements. The recorded signal is then processed by a lock-in amplifier to generate 2D spectra.

Although both optical 2DCS and STM single-entity study have been well established separately, there are critical challenges to integrate the two techniques. First, a direct consequence of high spatial resolution is a smaller sample volume and thus weaker signals. The experiment requires a higher detection sensitivity than the ensemble measurements. For example, the nonlinear signal from single molecules is weaker compared to that from ensembles. The detection sensitivity can be improved by using a single-photon counting module as the detector, capable of detecting signals at the single-photon level. The nonresonance background due to parametric processes can be reduced by avoiding zero time delays between the excitation pulses. Second, a tightly focused femtosecond laser beam may cause damage to the plasmonic substrates and the sample. To prevent the damage, one can use a femtosecond laser with a high repetition rate to provide low pulse energy while maintaining a good signal-to-noise ratio. It is also possible to add a protective layer on the surface of tips and samples. For example, in single-molecule studies, the target molecule can be mixed with alkanethiol molecules to form an ordered self-assembled monolayer on the gold surfaces. For some experiments, a thin silica layer can be coated on the gold surfaces. Third, molecules in contact with a metal surface can have a strong quenching effect, which limits the molecules' ability to emit fluorescence. To avoid this problem, optical 2DCS experiment can be performed by measuring the nonlinear TFWM signal instead of the fluorescence, in which case, the fourth pulse is used a reference pulse.

12.3   Multidimensional spectroscopy with quantum light

Throughout this book a semiclassical treatment of the light-matter interaction has been used, namely the matter has been treated quantum mechanically while the light has been treated as a classical electromagnetic field. This approach is generally sufficient for spectroscopy, where the goal is to determine the properties of the matter, in which case the quantum nature of the light is generally not relevant. In addition, the quantum nature of light is typically important at very low intensities, such that the discrete nature of photons becomes relevant, whereas nonlinear spectroscopies typically require high intensities. Nevertheless, it has been suggested that using nonclassical light, also known as quantum light, could provide advantages.

Initial theoretical investigations considered using entangled photons in double-quantum MDCS of excitonic states [315, 319]. The calculations suggested that the use of entangle photons could provide enhancements in temporal/spectral resolution as well as a linear scaling of the signal, rather than the quadratic scaling typically associated with two-photon absorption. However, realizing a linear scaling requires very low photon numbers, which is contradictory to the need to achieve high intensities to realize a nonlinear response. Current analysis suggest that linear response regime may not be achievable [301, 314].

Currently, MDCS with nonclassical light has not been demonstrated. Based on the apparent challenges of achieving a quantum-advantage and doubts as to whether or not it is even achievable, it is not clear if effort will be sufficiently rewarding.

12.4   Photoemission-detected MDCS

The ability to extract multidimensional coherent spectroscopy information from incoherent signals such as photoluminescence and photocurrent using phase-cycling techniques (see Chapter 4) has invited the possibility of wedding the MDCS technique to sophisticated material probes like photoemission electron microscopy (PEEM) and angle-resolved photoemission spectroscopy (ARPES). The two probes have tantalizing potential. PEEM, for example, holds promise for leveraging the short wavelength of electrons to achieve spatial resolution improvements, similar to the way that plasmonic interactions improve the spatial resolution of tip-based MDCS techniques. ARPES holds promise for disentangling the coherent dynamics of metals and superconductors, where resonances that would otherwise be smeared out in the energy domain all by itself can be sharply separated when unfolded into simultaneous energy and momentum dimensions.

The technical challenges associated with incorporating MDCS techniques into interrogation techniques of this sort are substantial, as they require both a deep understanding of and ability to manipulate ultrashort pulses of light to make the MDCS part of the experiment work, and an equally deep level of familiarity with the ultrahigh vacuum techniques associated with any photoemission-based technique's inherent surface sensitivity. On top of all this, in order to generate the photoelectrons that ultimately end up being measured in PEEM and ARPES techniques, the material sample's work function needs to be overcome, and so one either needs to incorporate ultraviolet light into the last stage of the experiment or employ light in sufficient intensity at the final stage to generate multiphoton photoemission.

In spite of these difficulties, a few inroads have successfully been made. An important benchmark example is a study published in 2011 using an MDCS/PEEM setup capable of measuring the properties of a corrugated silver surface with 50-nm spatial resolution, fully an order of magnitude better than the maximum spatial resolution achievable using optics alone [4]. Figure 12.2 shows a cartoon depiction of a PEEM-detected MDCS apparatus experimental geometry. In order to generate photoelectrons with enough energy to overcome the work function of their sample, the authors of Ref. 4 had to increase the power of their MDCS pulses to the point where the system

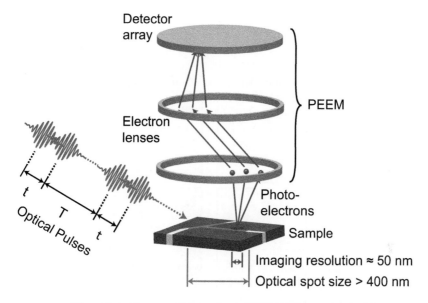

Fig. 12.2 Cartoon illustration of PEEM-detected MDCS.

departs from the perturbative regime, somewhat hindering the measurements' ability to be interpreted. Future measurements incorporating weak MDCS pulses and a true ultraviolet source may yet prove an even more powerful tool.

The incorporation of ARPES probes into MDCS experiments has proved a more elusive endeavor than the merger of MDCS and PEEM, but advances along this direction are beginning to be seen here as well. In a recent study on Ag(111), it was shown possible to use Fourier-transform techniques, coherent pulses of light, and multiphoton photoemission techniques to probe the Shockley surface state of the material as a simultaneous function of both photoelectron energy and the photon energy of the intermediate nonlinear interactions that proceeded the photoemission event [316]. Figure 12.3 shows an example illustration. More recently, the same group has demonstrated the application of a similar array of techniques to study dressed states of the Cu(111) surface, this time incorporating momentum resolution into the measurements and thereby coming closer to utilizing the full power of the ARPES technique on their samples.

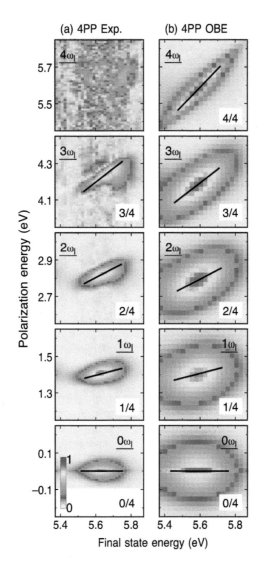

Fig. 12.3 (a) Experimental measurements and (b) theoretical simulations of a set of photoemission-detected MDCS study of the Shockley surface state of Ag(111). The "polarization energy" (y-axis) corresponds to the Fourier transform of optical excitation pulses. The "final state energy" corresponds to photoelectron energy. Adapted from Ref. 316

Figure Credits

Figure 3.2 is adapted with permission from Ref. 354. Copyright (2010) The Optical Society.

Figures 3.12 and 3.12 are adapted with permission from Ref. 362. Copyright (2017) The Optical Society.

Figure 4.4(b) is adapted with permission from Ref. 215. Copyright (2013) The Optical Society.

Figures 4.6 and 4.7 are adapted with permission from Ref. 285 . Copyright (2013) The Optical Society.

Figure 5.4 is reprinted with permission from Ref. 78. Copyright (2010) by the American Physical Society.

Figure 5.6 is adapted with permission from Ref. 217. Copyright (2013) American Chemical Society.

Figure 5.7 is adapted with permission from Ref. 370. Copyright (2013) American Chemical Society.

Figures 5.8(a), (b), 5.9(a), (b), and (d~g) are reprinted with permission from Ref. 79. Copyright (2012) by the American Physical Society.

Figure 5.10 is adapted with permission from Ref. 129. Copyright (2016) The Optical Society.

Figure 5.11 is adapted with permission from Ref. 459. Copyright (2019) The Optical Society.

Figures 5.12, 5.13, and 5.14 are adapted with permission from Ref. 460. Copyright (2019) The Optical Society.

Figures 6.3, 6.4, 6.5, and 6.6 are adapted with permission from Ref. 231. Copyright (2017) The Optical Society.

Figure 7.3 is adapted with permission from Ref. 462. Copyright (2007) National Academy of Sciences.

Figure 7.4 is reprinted with permission from Ref. 42. Copyright (2009) by the American Physical Society.

Figure 7.5 is reprinted with permission from Ref. 414. Copyright (2012) by the American Physical Society.

Figure 7.6 is reprinted with permission from Ref. 359. Copyright (2017) by the American Physical Society.

Figures 7.8 and 7.9 are reprinted with permission from Ref. 176. Copyright (2010) by the American Physical Society.

Figure 7.12 is reprinted with permission from Ref. 401. Copyright (2016) by the American Physical Society.

References

[1] Abella, ID, Kurnit, NA, and Hartmann, SR (1966). Photon echoes. *Phys. Rev.*, **141**, 391–406.

[2] Abramavicius, Darius, Palmieri, Benoit, Voronine, Dmitri V., Šanda, František, and Mukamel, Shaul (2009). Coherent multidimensional optical spectroscopy of excitons in molecular aggregates; quasiparticle versus supermolecule perspectives. *Chem. Rev.*, **109**, 2350–2408.

[3] Accanto, Nicolò, Masia, Francesco, Moreels, Iwan, Hens, Zeger, Langbein, Wolfgang, and Borri, Paola (2012). Engineering the spin–flip limited exciton dephasing in colloidal CdSe/CdS quantum dots. *ACS Nano*, **6**, 5227–5233.

[4] Aeschlimann, Martin, Brixner, Tobias, Fischer, Alexander, Kramer, Christian, Melchior, Pascal, Pfeiffer, Walter, Schneider, Christian, Strüber, Christian, Tuchscherer, Philip, and Voronine, Dmitri V. (2011). Coherent two-dimensional nanoscopy. *Science*, **333**, 1723–1726.

[5] Akkerman, Quinten A, Rainò, Gabriele, Kovalenko, Maksym V, and Manna, Liberato (2018). Genesis, challenges and opportunities for colloidal lead halide perovskite nanocrystals. *Nat. Mater.*, **17**, 394–405.

[6] Albrecht, T. F., Bott, K., Meier, T., Schulze, A., Koch, M., Cundiff, S. T., Feldmann, J., Stolz, W., Thomas, P., Koch, S. W., and Göbel, E. O. (1996). Disorder mediated biexcitonic beats in semiconductor quantum wells. *Phys. Rev. B*, **54**, 4436–4439.

[7] Ali, A. W. and Griem, H. R. (1965). Theory of resonance broadening of spectral lines by atom-atom impacts. *Phys. Rev.*, **140**, A1044–A1049.

[8] Allen, L. and Eberly, J.H. (2012). *Optical Resonance and Two-Level Atoms.* Dover Publications.

[9] Amani, Matin, Taheri, Peyman, Addou, Rafik, Ahn, Geun Ho, Kiriya, Daisuke, Lien, Der-Hsien, Ager, Joel W., Wallace, Robert M., and Javey, Ali (2016). Recombination kinetics and effects of superacid treatment in sulfur- and selenium-based transition metal dichalcogenides. *Nano Lett.*, **16**, 2786–2791.

[10] Anderson, P. W. (1949). Pressure broadening in the mircrowave and infra-red regions. *Phys. Rev.*, **76**, 647–661.

[11] Anderson, P. W. (1972). More is different. *Science*, **177**, 393–396.

[12] Asahara, Akifumi and Minoshima, Kaoru (2017). Development of ultrafast time-resolved dual-comb spectroscopy. *APL Photonics*, **2**, 041301.

[13] Autry, Travis M., Moody, Galan, Fraser, James, McDonald, Corey, Mirin, R. P., and Silverman, Kevin (2019). Single-scan acquisition of multiple multidimensional spectra. *Optica*, **6**, 735.

[14] Autry, Travis M., Nardin, Gaël, Smallwood, Christopher L., Silverman, Kevin, Bajoni, Daniele, Lemaître, Aristide, Bouchoule, Sophie, Bloch, Jacqueline, and Cundiff, Steven (2020). Excitation ladder of cavity polaritons. *Phys. Rev. Lett.*, **125**, 067403.

[15] Autry, T. M., Smallwood, C. L., Singh, R., and Cundiff, S. T. Theory of degenerate wave mixing in a collective gas. p. to be published.

[16] Axt, V. M. and Stahl, A. (1994). A dynamics-controlled truncation scheme for the hierarchy of density matrices in semiconductor optics. *Z. Phys. B*, **93**, 195–204.

[17] Axt, V. M. and Stahl, A. (1994). The role of the biexciton in a dynamic density matrix theory of the semiconductor band edge. *Z. Phys. B*, **93**, 205–211.

[18] Barends, R., Kelly, J., Megrant, A., Veitia, A., Sank, D., Jeffrey, E., White, T. C., Mutus, J., Fowler, A. G., Campbell, B., Chen, Y., Chen, Z., Chiaro, B., Dunsworth, A., Neill, C., O'Malley, P., Roushan, P., Vainsencher, A., Wenner, J., Korotkov, A. N., Cleland, A. N., and Martinis, John M. (2014). Superconducting quantum circuits at the surface code threshold for fault tolerance. *Nature*, **508**, 500–503.

[19] Bates, Kelsey M., Day, Matthew W., Smallwood, Christopher L., Owen, Rachel C., Schröder, Tim, Bielejec, Edward, Ulbricht, Ronald, and Cundiff, Steven T. (2021). Using silicon-vacancy centers in diamond to probe the full strain tensor. *J. Appl. Phys.*, **130**, 024301.

[20] Batsch, M., Meier, T., Thomas, P., Lindberg, M., Koch, S. W., and Shah, J. (1993). Dipole-dipole coupling of excitons in double quantum wells. *Phys. Rev. B*, **48**, 11817–11826.

[21] Bauer, M., Marienfeld, A., and Aeschlimann, M. (2015). Hot electron lifetimes in metals probed by time-resolved two-photon photoemission. *Prog. Surf. Sci.*, **90**, 319–376.

[22] Baumann, Kristian, Guerlin, Christine, Brennecke, Ferdinand, and Esslinger, Tilman (2010). Dicke quantum phase transition with a superfluid gas in an optical cavity. *Nature*, **464**, 1301–1306.

[23] Becker, Jonas Nils, Görlitz, Johannes, Arend, Carsten, Markham, Matthew, and Becher, Christoph (2016). Ultrafast all-optical coherent control of single silicon vacancy colour centres in diamond. *Nat. Commun.*, **7**, 13512.

[24] Becker, Michael A., Vaxenburg, Roman, Nedelcu, Georgian, Sercel, Peter C., Shabaev, Andrew, Mehl, Michael J., Michopoulos, John G., Lambrakos, Samuel G., Bernstein, Noam, Lyons, John L., Stöferle, Thilo, Mahrt, Rainer F., Kovalenko, Maksym V., Norris, David J., Rainò, Gabriele, and Efros, Alexander L. (2018). Bright triplet excitons in caesium lead halide perovskites. *Nature*, **553**, 189–193.

[25] Becker, P. C., Fragnito, H. L., Cruz, C. H. B., Fork, R. L., Cunningham, J. E., Henry, J. E., and Shank, C. V. (1988). Femtosecond photon-echoes from band-to-band transitions in GaAs. *Phys. Rev. Lett.*, **61**, 1647–1649.

[26] Bekenstein, Yehonadav, Koscher, Brent A, Eaton, Samuel W, Yang, Peidong, and Alivisatos, A Paul (2015). Highly luminescent colloidal nanoplates of perovskite

cesium lead halide and their oriented assemblies. *J. Am. Chem. Soc.*, **137**, 16008–16011.

[27] Belabas, Nadia and Jonas, David M (2005). Three-dimensional view of signal propagation in femtosecond four-wave mixing with application to the boxcars geometry. *J. Opt. Soc. Am. B*, **22**, 655–674.

[28] Bell, Joshua D., Conrad, Rebecca, and Siemens, Mark E. (2015). Analytical calculation of two-dimensional spectra. *Opt. Lett.*, **40**, 1157–1160.

[29] Bennett, Kochise, Rouxel, Jeremy R., and Mukamel, Shaul (2017). Linear and nonlinear frequency- and time-domain spectroscopy with multiple frequency combs. *J. Chem. Phys.*, **147**, 094304.

[30] Bernien, Hannes, Schwartz, Sylvain, Keesling, Alexander, Levine, Harry, Omran, Ahmed, Pichler, Hannes, Choi, Soonwon, Zibrov, Alexander S., Endres, Manuel, Greiner, Markus, Vuletic, Vladan, and Lukin, Mikhail D. (2017). Probing many-body dynamics on a 51-atom quantum simulator. *Nature*, **551**, 579–584.

[31] Birkedal, D., Singh, J., Lyssenko, V. G., Erland, J., and Hvam, J. M. (1996). Binding of quasi-two-dimensional biexcitons. *Phys. Rev. Lett.*, **76**, 672–675.

[32] Blank, David A., Kaufman, Laura J., and Fleming, Graham R. (1999). Fifth-order two-dimensional raman spectra of cs$_2$ are dominated by third-order cascades. *J. Chem. Phys.*, **111**, 3105.

[33] Bonadeo, N. H., Erland, J., Gammon, D., Park, D., Katzer, D. S., and Steel, D. G. (1998). Coherent optical control of the quantum state of a single quantum dot. *Science*, **282**, 1473–1476.

[34] Borca, C. N., Zhang, T., Li, X., and Cundiff, S. T. (2005). Optical two-dimensional fourier transform spectroscopy of semiconductors. *Chem. Phys. Lett.*, **416**, 311.

[35] Borri, P., Langbein, W., Schneider, S., Woggon, U., Sellin, R., Ouyang, D., and Bimberg, D. (2001). Ultralong dephasing time in InGaAs quantum dots. *Phys. Rev. Lett.*, **87**, 157401.

[36] Borri, P., Langbein, W., Schneider, S., Woggon, U., Sellin, R. L., Ouyang, D., and Bimberg, D. (2002). Rabi oscillations in the excitonic ground-state transition of ingaas quantum dots. *Phys. Rev. B*, **66**, 081306.

[37] Borri, Paola, Langbein, Wolfgang, Woggon, Ulrike, Esser, Axel, Jensen, Jacob R, and rn M Hvam, J (2003). Biexcitons in semiconductor microcavities. *Semicond. Sci. Technol.*, **18**, S351–S360.

[38] Bott, K., Heller, O., Bennhardt, D., Cundiff, S. T., Thomas, P., Mayer, E. J., Smith, G. O., Eccleston, R., Kuhl, J., and Ploog, K. (1993). Influence of exciton-exciton interactions on the coherent optical response in gaas quantum wells. *Phys. Rev. B*, **48**, 17418–17426.

[39] Boudreau, Sylvain, Levasseur, Simon, Perilla, Carlos, Roy, Simon, and Genest, Jérôme (2013). Chemical detection with hyperspectral lidar using dual frequency combs. *Opt. Express*, **21**, 7411.

[40] Boyd, Robert W. (2009). *Nonlinear Optics* (3 edn). Elsevier, Waltham, MA.

[41] Bristow, Alan D., Karaiskaj, Denis, Dai, Xingcan, and Cundiff, Steven T. (2008). All-optical retrieval of the global phase for two-dimensional fourier-transform spectroscopy. *Opt. Express*, **16**, 18017.

[42] Bristow, Alan D., Karaiskaj, Denis, Dai, Xingcan, Mirin, Richard P., and Cundiff, Steven T. (2009). Polarization dependence of semiconductor exciton and biexciton contributions to phase-resolved optical two-dimensional Fourier-transform spectra. *Phys. Rev. B*, **79**, 161305.

[43] Bristow, A D, Karaiskaj, D, Dai, X, Zhang, T, Carlsson, C, Hagen, K R, Jimenez, R, and Cundiff, S T (2009). A versatile ultrastable platform for optical multidimensional Fourier-transform spectroscopy. *Rev. Sci. Instrum.*, **80**, 073108.

[44] Britnell, L., Ribeiro, R. M., Eckmann, A., Jalil, R., Belle, B. D., Mishchenko, A., Kim, Y.-J., Gorbachev, R. V., Georgiou, T., Morozov, S. V., Grigorenko, A. N., Geim, A. K., Casiraghi, C., Neto, A. H. C., and Novoselov, K. S. (2013). Strong light-matter interactions in heterostructures of atomically thin films. *Science*, **340**, 1311–1314.

[45] Brixner, T, Mancal, T, Stiopkin, IV, and Fleming, GR (2004). Phase-stabilized two-dimensional electronic spectroscopy. *J. Chem. Phys.*, **121**, 4221–4236.

[46] Brixner, T, Stiopkin, I V, and Fleming, G R (2004). Tunable two-dimensional femtosecond spectroscopy. *Opt. Lett.*, **29**, 884.

[47] Bruder, Lukas, Eisfeld, Alexander, Bangert, Ulrich, Binz, Marcel, Jakob, Max, Uhl, Daniel, Schulz-Weiling, Markus, Grant, Edward R., and Stienkemeier, Frank (2019). Delocalized excitons and interaction effects in extremely dilute thermal ensembles. *Phys. Chem. Chem. Phys.*, **21**, 2276–2282.

[48] Caram, Justin R., Zheng, Haibin, Dahlberg, Peter D., Rolczynski, Brian S., Griffin, Graham B., Fidler, Andrew F., Dolzhnikov, Dmitriy S., Talapin, Dmitri V., and Engel, Gregory S. (2014). Persistent interexcitonic quantum coherence in cdse quantum dots. *J. Phys. Chem. Lett.*, **5**, 196–204.

[49] Cassette, Elsa, Dean, Jacob C., and Scholes, Gregory D. (2016). Two-dimensional visible spectroscopy for studying colloidal semiconductor nanocrystals. *Small*, **12**, 2234–2244.

[50] Cassette, Elsa, Pensack, Ryan D., Mahler, Benoît, and Scholes, Gregory D. (2015). Room-temperature exciton coherence and dephasing in two-dimensional nanostructures. *Nat. Commun.*, **6**, 6086.

[51] Castella, H. and Wilkins, J. W. (1998). Splitting of the excitonic peak in quantum wells with interfacial roughness. *Phys. Rev. B*, **58**, 16186.

[52] Chemla, D S and Shah, J (2001). Many-body and correlation effects in semiconductors. *Nature*, **411**, 549–557.

[53] Chen, Aixi (2014). Coherent manipulation of spontaneous emission spectra in coupled semiconductor quantum well structures. *Opt. Express*, **22**, 26991.

[54] Chen, Gang, Stievater, T. H., Batteh, E. T., Li, Xiaoqin, Steel, D. G., Gammon, D., Katzer, D. S., Park, D., and Sham, L. J. (2002). Biexciton quantum coherence in a single quantum dot. *Phys. Rev. Lett.*, **88**, 117901.

[55] Chen, Peter C. (2010). High resolution coherent 2D spectroscopy. *J. Phys. Chem. A*, **114**, 11365–11375.

[56] Chen, Peter C. and Gomes, Marcia (2008). Two-dimensional coherent double resonance electronic spectroscopy. *J. Phys. Chem. A*, **112**, 2999–3001.

[57] Chen, Peter C, Wells, Thresa A, and Strangfeld, Benjamin R (2013). High-resolution coherent three-dimensional spectroscopy of Br_2. *J. Phys. Chem. A*, **117**, 5981–5986.

[58] Cheng, Guang-Ling, Zhong, Wen-Xue, and Chen, Ai-Xi (2015). Phonon induced phase grating in quantum dot system. *Opt. Express*, **23**, 9870.

[59] Cheng, Yuan-Chung and Fleming, Graham R. (2008). Coherence quantum beats in two-dimensional electronic spectroscopy. *J. Phys. Chem. A*, **112**, 4254–4260.

[60] Chernikov, Alexey, Berkelbach, Timothy C., Hill, Heather M., Rigosi, Albert, Li, Yilei, Aslan, Ozgur Burak, Reichman, David R., Hybertsen, Mark S., and Heinz, Tony F. (2014). Exciton binding energy and nonhydrogenic rydberg series in monolayer WS_2. *Phys. Rev. Lett.*, **113**, 076802.

[61] Cho, Minhaeng (2001). Nonlinear response functions for the three-dimensional spectroscopies. *J. Chem. Phys.*, **115**, 4424–4437.

[62] Cho, Minhaeng (ed.) (2019). *Coherent Multidimensional Spectroscopy*. Springer Singapore.

[63] Choi, Wonbong, Choudhary, Nitin, Han, Gang Hee, Park, Juhong, Akinwande, Deji, and Lee, Young Hee (2017). Recent development of two-dimensional transition metal dichalcogenides and their applications. *Mater. Today*, **20**, 116–130.

[64] Coddington, Ian, Newbury, Nathan, and Swann, William (2016). Dual-comb spectroscopy. *Optica*, **3**, 414–426.

[65] Coddington, Ian, Swann, William C., and Newbury, Nathan R. (2008). Coherent multiheterodyne spectroscopy using stabilized optical frequency combs. *Phys. Rev. Lett.*, **100**, 013902.

[66] Collini, Elisabetta, Wong, Cathy Y., Wilk, Krystyna E., Curmi, Paul M. G., Brumer, Paul, and Scholes, Gregory D. (2010). Coherently wired light-harvesting in photosynthetic marine algae at ambient temperature. *Nature*, **463**, 644–647.

[67] Cooley, James W. and Tukey, John W. (1965). An algorithm for the machine calculation of complex fourier series. *Math. Comp.*, **19**, 297–297.

[68] Cowan, M. L., Ogilvie, J. P., and Miller, R. J D (2004). Two-dimensional spectroscopy using diffractive optics based phased-locked photon echoes. *Chem. Phys. Lett.*, **386**, 184–189.

[69] Cundiff, S.T. (2002). Phase stabilization of ultrashort optical pulses. *J. Phys. D*, **35**, R43–R59.

[70] Cundiff, ST, Ye, J, and Hall, JL (2001). Optical frequency synthesis based on mode-locked lasers. *Rev. Sci. Instr.*, **72**, 3749–3771.

[71] Cundiff, S. T. (1994). Effects of correlation between inhomogeneously broadened transitions on quantum beats in transient four-wave mixing. *Phys. Rev. A*, **49**, 3114–3118.

[72] Cundiff, S T (2002). Time domain observation of the Lorentz-local field. *Laser Phys.*, **12**, 1073–1078.

[73] Cundiff, Steven T. (2008). Coherent spectroscopy of semiconductors. *Opt. Express*, **16**, 4639–4664.

[74] Cundiff, Steven T. (2014). Optical three dimensional coherent spectroscopy. *Phys. Chem. Chem. Phys.*, **16**, 8193–8200.

[75] Cundiff, S. T. and Ye, J (2003). Colloquium: Femtosecond optical frequency combs. *Rev. Mod. Phys.*, **75**, 325–342.

[76] Cundiff, Steven T., Zhang, Tianhao, Bristow, Alan D., Karaiskaj, Denis, and Dai, Xingcan (2009). Optical two-dimensional fourier transform spectroscopy of semiconductor quantum wells. *Acc. Chem. Res.*, **42**, 1423–1432.

[77] Czech, Kyle J., Thompson, Blaise J., Kain, Schuyler, Ding, Qi, Shearer, Melinda J., Hamers, Robert J., Jin, Song, and Wright, John C. (2015). Measurement of ultrafast excitonic dynamics of few-layer MoS_2 using state-selective coherent multidimensional spectroscopy. *ACS Nano*, **9**, 12146–12157.

[78] Dai, Xingcan, Bristow, Alan D, Karaiskaj, Denis, and Cundiff, Steven T (2010). Two-dimensional fourier-transform spectroscopy of potassium vapor. *Phys. Rev. A*, **82**, 052503.

[79] Dai, Xingcan, Richter, Marten, Li, Hebin, Bristow, Alan D., Falvo, Cyril, Mukamel, Shaul, and Cundiff, Steven T. (2012). Two-dimensional double-quantum spectra reveal collective resonances in an atomic vapor. *Phys. Rev. Lett.*, **108**, 193201.

[80] Dai, Xingcan, Richter, Marten, Li, Hebin, Bristow, Alan D., Falvo, Cyril, Mukamel, Shaul, and Cundiff, Steven T. (2012, May). Two-dimensional double-quantum spectra reveal collective resonances in an atomic vapor. *Phys. Rev. Lett.*, **108**, 193201.

[81] Dar, M. Ibrahim, Jacopin, Gwénolé, Meloni, Simone, Mattoni, Alessandro, Arora, Neha, Boziki, Ariadni, Zakeeruddin, Shaik Mohammed, Rothlisberger, Ursula, and Grätzel, Michael (2016). Origin of unusual bandgap shift and dual emission in organic-inorganic lead halide perovskites. *Sci. Adv.*, **2**, e1601156.

[82] Davis, J. A., Dao, L .V., Do, M. T., Hannaford, P., Nugent, K. A., and Quiney, H. M. (2008). Noninterferometric two-dimensional fourier-transform spectroscopy of multilevel systems. *Phys. Rev. Lett.*, **100**, 227401.

[83] Davis, J. A., Hall, C. R., Dao, L. V., Nugent, K. A., Quiney, H. M., Tan, H. H., and Jagadish, C. (2011). Three-dimensional electronic spectroscopy of excitons in asymmetric double quantum wells. *J. Chem. Phys.*, **135**, 044510.

[84] Day, Matthew W., Dong, Mark, Smith, Bradley C., Owen, Rachel C., Kerber, Grace C., Ma, Taigao, Winful, Herbert G., and Cundiff, Steven T. (2020). Simple single-section diode frequency combs. *APL Photonics*, **5**, 121303.

[85] de Boeij, Wim P., Pshenichnikov, Maxim S., and Wiersma, Douwe A. (1996). On the relation between the echo-peak shift and Brownian-oscillator correlation function. *Chem. Phys. Lett.*, **253**, 53–60.

[86] Dekel, E., Gershoni, D., Ehrenfreund, E., Spektor, D., Garcia, J. M., and Petroff, P. M. (1998). Multiexciton spectroscopy of a single self-assembled quantum dot. *Phys. Rev. Lett.*, **80**, 4991–4994.

[87] Delmonte, Valentin, Specht, Judith F., Jakubczyk, Tomasz, Höfling, Sven, Kamp, Martin, Schneider, Christian, Langbein, Wolfgang, Nogues, Gilles, Richter, Marten, and Kasprzak, Jacek (2017). Coherent coupling of individual quantum dots measured with phase-referenced two-dimensional spectroscopy: Photon echo versus double quantum coherence. *Phys. Rev. B*, **96**, 041124.

[88] Deschler, Felix, Price, Michael, Pathak, Sandeep, Klintberg, Lina E., Jarausch, David-Dominik, Higler, Ruben, Hüttner, Sven, Leijtens, Tomas, Stranks, Samuel D., Snaith, Henry J., Atatüre, Mete, Phillips, Richard T., and Friend, Richard H. (2014). High photoluminescence efficiency and optically pumped lasing in solution-processed mixed halide perovskite semiconductors. *J. Phys. Chem. Lett.*, **5**, 1421–1426.

[89] Deveaud, B., Chomette, A., Clerot, F., Auvray, P., Regreny, A., Ferreira, R., and Bastard, G. (1990). Subpicosecond luminescence study of tunneling and relaxation in coupled quantum wells. *Phys. Rev. B*, **42**, 7021–7032.

[90] Deveaud, B., Clérot, F., Roy, N., Satzke, K., Sermage, B., and Katzer, D. S. (1991). Enhanced radiative recombination of free excitons in GaAs quantum wells. *Phys. Rev. Lett.*, **67**, 2355–2358.

[91] Dey, P., Paul, J., Glikin, N., Kovalyuk, Z. D., Kudrynskyi, Z. R., Romero, A. H., and Karaiskaj, D. (2014). Mechanism of excitonic dephasing in layered InSe crystals. *Phys. Rev. B*, **89**, 125128.

[92] Dey, P., Paul, J., Moody, G., Stevens, C. E., Glikin, N., Kovalyuk, Z. D., Kudrynskyi, Z. R., Romero, A. H., Cantarero, A., Hilton, D. J., and Karaiskaj, D. (2015). Biexciton formation and exciton coherent coupling in layered GaSe. *J. Chem. Phys.*, **142**, 212422.

[93] Dey, P., Paul, J., Wang, Z., Stevens, C. E., Liu, C., Romero, A. H., Shan, J., Hilton, D. J., and Karaiskaj, D. (2016). Optical coherence in atomic-monolayer transition-metal dichalcogenides limited by electron-phonon interactions. *Phys. Rev. Lett.*, **116**, 127402.

[94] Dicke, Robert H (1954). Coherence in spontaneous radiation processes. *Phys. Rev.*, **93**, 99–110.

[95] Diederich, Geoffrey M. and Siemens, Mark E. (2019). Absolute phase calibration in phase-modulated multidimensional coherent spectroscopy. *Opt. Lett.*, **44**, 3054.

[96] Ding, Feng and Zanni, Martin T. (2007). Heterodyned 3D IR spectroscopy. *Chem. Phys.*, **341**, 95–105.

[97] Do, Thanh Nhut, Chen, Lipeng, Belyaev, Andrey K., Tan, Howe-Siang, and Gelin, Maxim F. (2018). Pulse-shape effects in fifth-order multidimensional optical spectroscopy. *Chemical Physics*, **515**, 119–128.

[98] Do, Thanh Nhut, Gelin, Maxim F., and Tan, Howe-Siang (2017). Simplified expressions that incorporate finite pulse effects into coherent two-dimensional optical spectra. *The Journal of Chemical Physics*, **147**, 144103.

[99] Dong, Mark, Day, Matthew W., Winful, Herbert G., and Cundiff, Steven T. (2020). Quantum-well laser diodes for frequency comb spectroscopy. *Opt. Express*, **28**, 21825.

[100] Dorfman, Konstantin E., Fingerhut, Benjamin P., and Mukamel, Shaul (2013). Broadband infrared and raman probes of excited-state vibrational molecular dynamics: simulation protocols based on loop diagrams. *Phys. Chem. Chem. Phys.*, **15**, 12348.

[101] Draeger, Simon, Roeding, Sebastian, and Brixner, Tobias (2017). Rapid-scan coherent 2D fluorescence spectroscopy. *Opt. Express*, **25**, 3259–3267.

[102] Dresselhaus, Mildred S., Dresselhaus, Gene, and Avouris, Phaedon (ed.) (2001). *Carbon Nanotubes: Synthesis, Structure, Properties, and Applications*. Springer.

[103] Eden, J.G., Ricconi, B.J., Xiao, Y., Shen, F., and Senin, A.A. (2008). Interactions between thermal ground or excited atoms in the vapor phase: Many-body dipole–dipole effects, molecular dissociation, and photoassociation probed by laser spectroscopy. In *Adv. At. Mol. Opt. Phys.*, Volume 56, pp. 49–118.

[104] Efros, Alexander L. and Nesbitt, David J. (2016). Origin and control of blinking in quantum dots. *Nat. Nanotechnol.*, **11**, 661.

[105] Elsaesser, Thomas, Riemann, Klaus, and Woerner, Michael (2019). *Concepts and Applications of Nonlinear Terahertz Spectroscopy*. IOP Publishing.

[106] Engel, Gregory S, Calhoun, Tessa R, Read, Elizabeth L, Ahn, Tae-Kyu, Mančal, Tomáš, Cheng, Yuan-Chung, Blankenship, Robert E, and Fleming, Graham R (2007). Evidence for wavelike energy transfer through quantum coherence in photosynthetic systems. *Nature*, **446**, 782–786.

[107] Enomoto, K., Kitagawa, M., Kasa, K., Tojo, S., and Takahashi, Y. (2007). Determination of the s-wave scattering length and the c_6 van der Waals coefficient of ^{174}yb via photoassociation spectroscopy. *Phys. Rev. Lett.*, **98**, 203201.

[108] Ernst, R, Bodenhausen, G, and Wokaun, A (1987). *Principles of Nuclear Magnetic Resonance in One and Two Dimensions*. Oxford Science Publications, Oxford, U.K.

[109] Falvo, Cyril, Palmieri, Benoit, and Mukamel, Shaul (2009). Coherent infrared multidimensional spectra of the OH stretching band in liquid water simulated by direct nonlinear exciton propagation. *J. Chem. Phys.*, **130**, 184501.

[110] Fan, Xudong, Takagahara, T., Cunningham, J.E., and Wang, Hailin (1998). Pure dephasing induced by exciton-phonon interactions in narrow GaAs quantum wells. *Solid State Commun.*, **108**, 857–861.

[111] Fang, Hong-Hua, Raissa, Raissa, Abdu-Aguye, Mustapha, Adjokatse, Sampson, Blake, Graeme R., Even, Jacky, and Loi, Maria Antonietta (2015). Photophysics of organic-inorganic hybrid lead iodide perovskite single crystals. *Adv. Funct. Mater.*, **25**, 2378–2385.

[112] Farrow, Darcie A., Yu, Anchi, and Jonas, David M. (2003). Spectral relaxation in pump–probe transients. *J. Chem. Phys.*, **118**, 9348–9356.

[113] Fei, C., Sarmiento, J. S., and Wang, H. (2018). Generation of coherent optical phonons in methylammonium lead iodide thin films. *J. Phys. Chem. C*, **122**, 17035–17041.

[114] Ferguson, Bradley and Zhang, Xi-Cheng (2002, sep). Materials for terahertz science and technology. *Nature Materials*, **1**, 26–33.

[115] Ferrio, K. B. and Steel, D. G. (1996). Observation of the ultrafast two-photon coherent biexciton oscillation in a GaAs/al$_x$ga$_{1-x}$As multiple quantum well. *Phys. Rev. B*, **54**, R5231–R5234.

[116] Feynman, R. P., Leighton, R. B., and Sands, M. (2011). *The Feynman Lectures on Physics, Vol. I: The New Millennium Edition: Mainly Mechanics, Radiation, and Heat*. The Feynman Lectures on Physics. Basic Books.

[117] Fidler, Andrew F., Harel, Elad, and Engel, Gregory S. (2010). Dissecting hidden couplings using fifth-order three-dimensional electronic spectroscopy. *J. Phys. Chem. Lett.*, **1**, 2876–2880.

[118] Fox, A. M. (2010). *Optical Properties of Solids*. Oxford University Press.

[119] Fras, F., Mermillod, Q., Nogues, G., Hoarau, C., Schneider, C., Kamp, M., Höfling, S., Langbein, W., and Kasprzak, J. (2016). Multi-wave coherent control of a solid-state single emitter. *Nat. Photonics*, **10**, 155–158.

[120] Frost, J. M., Whalley, L. D., and Walsh, A. (2017). Slow cooling of hot polarons in halide perovskite solar cells. *ACS Energy Lett.*, **2**, 2647–2652.

[121] Frostig, Hadas, Bayer, Tim, Dudovich, Nirit, Eldar, Yonina C., and Silberberg, Yaron (2015). Single-beam spectrally controlled two-dimensional Raman spectroscopy. *Nat. Photonics*, **9**, 339–343.

[122] Fry, P. W., Itskevich, I. E., Mowbray, D. J., Skolnick, M. S., Finley, J. J., Barker, J. A., O'Reilly, E. P., Wilson, L. R., Larkin, I. A., Maksym, P. A., Hopkinson, M., Al-Khafaji, M., David, J. P. R., Cullis, A. G., Hill, G., and Clarck, J. C. (2000). Inverted electron-hole alignment in InAs-GaAs self-assembled quantum dots. *Phys. Rev. Lett.*, **84**, 733.

[123] Fuller, Franklin D. and Ogilvie, Jennifer P. (2015). Experimental implementations of two-dimensional Fourier transform electronic spectroscopy. *Annu. Rev. Phys. Chem.*, **66**, 667–690.

[124] Fuller, Franklin D., Wilcox, Daniel E., and Ogilvie, Jennifer P. (2014). Pulse shaping based two-dimensional electronic spectroscopy in a background free geometry. *Opt. Express*, **22**, 1018–1027.

[125] Fulmer, Eric C., Ding, Feng, Mukherjee, Prabuddha, and Zanni, Martin T. (2005). Vibrational dynamics of ions in glass from fffifth-order two-dimensional infrared spectroscopy. *Phys. Rev. Lett.*, **94**, 067402.

[126] Gaëtan, Alpha, Miroshnychenko, Yevhen, Wilk, Tatjana, Chotia, Amodsen, Viteau, Matthieu, Comparat, Daniel, Pillet, Pierre, Browaeys, Antoine, and Grangier, Philippe (2009). Observation of collective excitation of two individual atoms in the Rydberg blockade regime. *Nat. Phys.*, **5**, 115–118.

[127] Gammon, D., Snow, E. S., and Katzer, D. S. (1995). Excited state spectroscopy of excitons in single quantum dots. *Appl. Phys. Lett.*, **67**, 2391–2393.

[128] Gammon, D., Snow, E. S., Shanabrook, B. V., Katzer, D. S., and Park, D. (1996). Homogeneous linewidths in the optical spectrum of a single gallium arsenide quantum dot. *Science*, **273**, 87–90.

[129] Gao, Feng, Cundiff, Steven T, and Li, Hebin (2016). Probing dipole–dipole interaction in a rubidium gas via double-quantum 2D spectroscopy. *Opt. Lett.*, **41**, 2954–2957.

[130] Garrett-Roe, S and Hamm, P (2009). Purely absorptive three-dimensional infrared spectroscopy. *J. Chem. Phys.*, **130**, 164510.

[131] Garrett-Roe, S and Hamm, P (2009). What can we learn from three-dimensional infrared spectroscopy? *Acc. Chem. Res.*, **42**, 1412–1422.

[132] Garrett-Roe, Sean, Perakis, Fivos, Rao, Francesco, and Hamm, Peter (2011). Three-dimensional infrared spectroscopy of isotope-substituted liquid water reveals heterogeneous dynamics. *J. Phys. Chem. B*, **115**, 6976–6984.

[133] Ge, Nien-Hui, Zanni, Martin T., and Hochstrasser, Robin M. (2002). Effects of vibrational frequency correlations on two-dimensional infrared spectra. *J. Phys. Chem. A*, **106**, 962–972.

[134] Geim, A. K. and Grigorieva, I. V. (2013). Van der Waals heterostructures. *Nature*, **499**, 419–425.

[135] Godbout, Martin, Deschênes, Jean-Daniel, and Genest, Jerômè (2010). Spectrally resolved laser ranging with frequency combs. *Opt. Express*, **18**, 15981.

[136] Golonzka, O., Khalil, M., Demirdöven, N., and Tokmakoff, A. (2001). Vibrational anharmonicities revealed by coherent two-dimensional infrared spectroscopy. *Phys. Rev. Lett.*, **86**, 2154–2157.

[137] Graham, M. W., Calhoun, T. R., Green, A. A., Hersam, M. C., and Fleming, G. R. (2012). Two-dimensional electronic spectroscopy reveals the dynamics of phonon-mediated excitation pathways in semiconducting single-walled carbon nanotubes. *Nano Lett.*, **12**, 813–819.

[138] Green, Martin A., Ho-Baillie, Anita, and Snaith, Henry J. (2014). The emergence of perovskite solar cells. *Nat. Photonics*, **8**, 506–514.

[139] Griffin, Graham B., Ithurria, Sandrine, Dolzhnikov, Dmitriy S., Linkin, Alexander, Talapin, Dmitri V., and Engel, Gregory S. (2013). Two-dimensional electronic spectroscopy of CdSe nanoparticles at very low pulse power. *J. Chem. Phys.*, **138**, 014705.

[140] Gruenke, Natalie L., Cardinal, M. Fernanda, McAnally, Michael O., Frontiera, Renee R., Schatz, George C., and Van Duyne, Richard P. (2016). Ultrafast and nonlinear surface-enhanced Raman spectroscopy. *Chem. Soc. Rev.*, **45**, 2263–2290.

[141] Grumstrup, Erik M, Shim, Sang-Hee, Montgomery, Matthew a, Damrauer, Niels H, and Zanni, Martin T (2007). Facile collection of two-dimensional electronic spectra using femtosecond pulse-shaping Technology. *Opt. Express*, **15**, 16681–16689.

[142] Guo, Z., Wan, Y., Yang, M., Snaider, J., Zhu, K., and Huang, L. (2017). Long-range hot-carrier transport in hybrid perovskites visualized by ultrafast microscopy. *Science*, **356**, 59–62.

[143] Hall, C. R., Tollerud, J. O., Quiney, H. M., and Davis, J. A. (2013). Three-dimensional electronic spectroscopy of excitons in asymmetric double quantum wells. *New J. Phys.*, **15**, 045028.

[144] Hall, K. L., Lenz, G., Ippen, E. P., and Raybon, G. (1992, jun). Heterodyne pump–probe technique for time-domain studies of optical nonlinearities in waveguides. *Opt. Lett.*, **17**, 874.

[145] Hamm, Peter (2006). Three-dimensional-ir spectroscopy: Beyond the two-point frequency fluctuation correlation function. *J. Chem. Phys.*, **124**, 124506.

[146] Hamm, P, Lim, M, DeGrado, W F, and Hochstrasser, R M (1999). The two-dimensional IR nonlinear spectroscopy of a cyclic penta-peptide in relation to its three-dimensional structure. *Proc. Natl. Acad. Sci. U.S.A*, **96**, 2036–2041.

[147] Hamm, P. and Zanni, M. (2011). *Concepts and Methods of 2D Infrared Spectroscopy*. Cambridge University Press.

[148] Hamner, Chris, Qu, Chunlei, Zhang, Yongping, Chang, JiaJia, Gong, Ming, Zhang, Chuanwei, and Engels, Peter (2014). Dicke-type phase transition in a spin-orbit-coupled Bose-Einstein condensate. *Nat. Commun.*, **5**, 4023.

[149] Hao, Kai, Moody, Galan, Wu, Fengcheng, Dass, Chandriker Kavir, Xu, Lixiang, Chen, Chang-Hsiao, Sun, Liuyang, Li, Ming-Yang, Li, Lain-Jong, MacDonald, Allan H., and Li, Xiaoqin (2016). Direct measurement of exciton valley coherence in monolayer WSe_2. *Nat. Phys.*, **12**, 677–682.

[150] Hao, Kai, Specht, Judith F., Nagler, Philipp, Xu, Lixiang, Tran, Kha, Singh, Akshay, Dass, Chandriker Kavir, Schüller, Christian, Korn, Tobias, Richter, Marten, Knorr, Andreas, Li, Xiaoqin, and Moody, Galan (2017). Neutral and charged inter-valley biexcitons in monolayer $MoSe_2$. *Nat. Commun.*, **8**, 15552.

[151] Hao, Kai, Xu, Lixiang, Nagler, Philipp, Singh, Akshay, Tran, Kha, Dass, Chandriker Kavir, Schüller, Christian, Korn, Tobias, Li, Xiaoqin, and Moody, Galan (2016). Coherent and incoherent coupling dynamics between neutral and charged excitons in monolayer $mose_2$. *Nano Lett.*, **16**, 5109–5113.

[152] Harel, Elad, Fidler, Andrew F., and Engel, Gregory S. (2010). Real-time mapping of electronic structure with single-shot two-dimensional electronic spectroscopy. *Proc. Natl. Acad. Sci. U.S.A*, **107**, 16444–16447.

[153] Haug, Hartmut and Koch, Stephan W. (1993). *Quantum Theory of the Optical and Electronic Properties of Semiconductors* (2nd edn). World Scientific, River Edge, NJ.

[154] Hayes, Dugan and Engel, Gregory S (2011). Extracting the excitonic Hamiltonian of the Fenna-Matthews-Olson complex using three-dimensional third-order electronic spectroscopy. *Biophys. J.*, **100**, 2043–2052.

[155] Hettich, C, Schmitt, C, Zitzmann, J, Kühn, S, Gerhardt, I, and Sandoghdar, V (2002). Nanometer resolution and coherent optical dipole coupling of two individual molecules. *Science*, **298**, 385.

[156] Hildner, Richard, Brinks, Daan, Nieder, Jana B., Cogdell, Richard J., and van Hulst, N. F. (2013). Quantum coherent energy transfer over varying pathways in single light-harvesting complexes. *Science*, **340**, 1448–1451.

[157] Hipke, Arthur, Meek, Samuel A., Ideguchi, Takuro, Hänsch, Theodor W., and Picqué, Nathalie (2014). Broadband doppler-limited two-photon and stepwise excitation spectroscopy with laser frequency combs. *Phys. Rev. A*, **90**, 011805(R).

[158] Hu, Y. Z., Binder, R., Koch, S. W., Cundiff, S. T., Wang, H., and Steel, D. G. (1994). Excitation and polarization effects in semiconductor four-wave-mixing spectroscopy. *Phys. Rev. B*, **49**, 14382–14386.

[159] Hutter, E. M., Gélvez-Rueda, M. C., Osherov, A., Bulović, V., Grozema, F. C., Stranks, S. D., and Savenije, T. J. (2017). Direct–indirect character of the bandgap in methylammonium lead iodide perovskite. *Nat. Mater.*, **16**, 115.

[160] Huxter, V. M., Oliver, T. A. A., Budker, D., and Fleming, G. R. (2013). Vibrational and electronic dynamics of nitrogen-vacancy centres in diamond revealed by two-dimensional ultrafast spectroscopy. *Nat. Phys.*, **9**, 744.

[161] Hybl, JD, Ferro, AA, and Jonas, DM (2001). Two-dimensional Fourier transform electronic spectroscopy. *J. Chem. Phys.*, **115**, 6606–6622.

[162] Hybl, John D., Albrecht, Allison W., Gallagher Faeder, Sarah M., and Jonas, David M. (1998). Two-dimensional electronic spectroscopy. *Chem. Phys. Lett.*, **297**, 307–313.

[163] Ideguchi, T., Bernhardt, B., Guelachvili, G., Hänsch, T. W., and Picqué, N. (2012). Raman-induced kerr-effect dual-comb spectroscopy. *Opt. Lett.*, **37**, 4498.

[164] Ideguchi, Takuro, Holzner, Simon, Bernhardt, Birgitta, Guelachvili, Guy, Picqué, Nathalie, and Hänsch, Theodor W. (2013). Coherent raman spectro-imaging with laser frequency combs. *Nature*, **502**, 355–358.

[165] Jakubczyk, Tomasz, Delmonte, Valentin, Koperski, Maciej, Nogajewski, Karol, Faugeras, Clément, Langbein, Wolfgang, Potemski, Marek, and Kasprzak, Jacek (2016). Radiatively limited dephasing and exciton dynamics in mose$_2$ monolayers revealed with four-wave mixing microscopy. *Nano Lett.*, **16**, 5333–5339.

[166] Jakubczyk, Tomasz, Nogajewski, Karol, Molas, Maciej R, Bartos, Miroslav, Langbein, Wolfgang, Potemski, Marek, and Kasprzak, Jacek (2018). Impact of environment on dynamics of exciton complexes in a WS$_2$ monolayer. *2D Mater.*, **5**, 031007.

[167] Jang, H. U., Lomsadze, B., Trachy, M. L., Veshapidze, G., Fehrenbach, C. W., and DePaola, B. D. (2010). Interaction of a finite train of short pulses with an atomic ladder system. *Phys. Rev. A*, **82**, 043424.

[168] Jeon, Jonggu, Kim, JunWoo, Yoon, Tai Hyun, and Cho, Minhaeng (2019, Feb). Dual frequency comb photon echo spectroscopy. *J. Opt. Soc. Am. B*, **36**, 223–234.

[169] Jha, A., Duan, H.-G., Tiwari, V., Nayak, P. K., Snaith, H. J., Thorwart, M., and Miller, R. J. D. (2018). Direct observation of ultrafast exciton dissociation in lead iodide perovskite by 2D electronic spectroscopy. *ACS Photonics*, **5**, 852–860.

[170] Jones, K M, Julienne, P S, Lett, P D, Phillips, W D, Tiesinga, E, and Williams, C J (1996). Measurement of the atomic Na(3P) lifetime and of retardation in the interaction between two atoms bound in a molecule. *EPL (Europhysics Letters)*, **35**, 85.

[171] Jones, K. M., Tiesinga, E., Lett, P. D., and Julienne, P. S. (2006). Ultracold photoassociation spectroscopy: Long-range molecules and atomic scattering. *Rev. Mod. Phys.*, **78**, 483–535.

[172] Joo, Taiha, Jia, Yiwei, Yu, Jae-Young, Lang, Matthew J., and Fleming, Graham R. (1996). Third-order nonlinear time domain probes of solvation dynamics. *J. Chem. Phys.*, **104**, 6089–6108.

[173] Jurow, Matthew J., Morgenstern, Thomas, Eisler, Carissa, Kang, Jun, Penzo, Erika, Do, Mai, Engelmayer, Manuel, Osowiecki, Wojciech T., Bekenstein, Yehonadav, Tassone, Christopher, Wang, Lin-Wang, Alivisatos, A Paul, Brütting, Wolfgang, and Liu, Yi (2019). Manipulating the transition dipole moment of

CsPbBr$_3$ perovskite nanocrystals for superior optical properties. *Nano Lett.*, **19**, 2489–2496.

[174] Kagan, C. R. (1999). Organic-inorganic hybrid materials as semiconducting channels in thin-film field-effect transistors. *Science*, **286**, 945–947.

[175] Kamada, H., Gotoh, H., Temmyo, J., Takagahara, T., and Ando, H. (2001). Exciton rabi oscillation in a single quantum dot. *Phys. Rev. Lett.*, **87**, 246401.

[176] Karaiskaj, Denis, Bristow, Alan D., Yang, Lijun, Dai, Xingcan, Mirin, Richard P., Mukamel, Shaul, and Cundiff, Steven T. (2010). Two-quantum many-body coherences in two-dimensional Fourier-transform spectra of exciton resonances in semiconductor quantum wells. *Phys. Rev. Lett.*, **104**, 117401.

[177] Karki, Khadga J., Widom, Julia R., Seibt, Joachim, Moody, Ian, Lonergan, Mark C., Pullerits, Tõnu, and Marcus, Andrew H. (2014). Coherent two-dimensional photocurrent spectroscopy in a PbS quantum dot photocell. *Nat. Commun.*, **5**, 5869.

[178] Kasprzak, J., Patton, B., Savona, V., and Langbein, W. (2011). Coherent coupling between distant excitons revealed by two-dimensional nonlinear hyperspectral imaging. *Nat. Photon.*, **5**, 57–63.

[179] Kasprzak, J., Richard, M., Kundermann, S., Baas, A., Jeambrun, P., Keeling, J. M. J., Marchetti, F. M., Szymańska, M. H., André, R., Staehli, J. L., Savona, V., Littlewood, P. B., Deveaud, B., and Dang, Le Si (2006). Bose–einstein condensation of exciton polaritons. *Nature*, **443**, 409–414.

[180] Keilmann, Fritz, Gohle, Christoph, and Holzwarth, Ronald (2004). Time-domain mid-infrared frequency-comb spectrometer. *Opt. Lett.*, **29**, 1542.

[181] Keusters, D and Warren, W S (2003). Effect of pulse propagation on the two-dimensional photon echo spectrum of multilevel systems. *J. Chem. Phys.*, **119**, 4478–4489.

[182] Keusters, D and Warren, W S (2004). Propagation effects on the peak profile in two-dimensional optical photon echo spectroscopy. *Chem. Phys. Lett.*, **383**, 21–24.

[183] Khaetskii, A. V. and Nazarov, Yu. V. (2000). Spin-dephasing processes in semiconductor quantum dots. *Physica E*, **6**, 470.

[184] Khalil, M., Demirdöven, N., and Tokmakoff, A. (2003). Obtaining absorptive line shapes in two-dimensional infrared vibrational correlation spectra. *Phys. Rev. Lett.*, **90**, 047401.

[185] Khalil, M., Demirdöven, N., and Tokmakoff, A. (2004). Vibrational coherence transfer characterized with Fourier-transform 2D IR spectroscopy. *J. Chem. Phys.*, **121**, 362.

[186] Kidd, Daniel W., Zhang, David K., and Varga, Kálmán (2016). Binding energies and structures of two-dimensional excitonic complexes in transition metal dichalcogenides. *Phys. Rev. B*, **93**, 125423.

[187] Kiesel, N., Schmid, C., Tóth, G., Solano, E., and Weinfurter, H. (2007). Experimental observation of four-photon entangled Dicke state with high fidelity. *Phys. Rev. Lett.*, **98**, 063604.

[188] Kim, D. S., Ko, H. S., Kim, Y. M., Rhee, S. J., Hohng, S. C., Yee, Y. H., Kim, W. S., Woo, J. C., Choi, H. J., Ihm, J., Woo, D. H., and Kang, K. N. (1996). Percolation of carriers through low potential channels in thick $al_x ga_{1-x} as$ (x<0.35) barriers. *Phys. Rev. B*, **54**, 14580–14588.

[189] King, G. W. and van Vleck, J. H. (1939). Dipole-dipole resonance forces. *Phys. Rev.*, **55**, 1165–1172.

[190] King, John T., Anna, Jessica M., and Kubarych, Kevin J. (2011). Solvent-hindered intramolecular vibrational redistribution. *Phys. Chem. Chem. Phys.*, **13**, 5579–5583.

[191] Kippenberg, T. J., Holzwarth, R., and Diddams, S. A. (2011). Microresonator-based optical frequency combs. *Science*, **332**, 555–559.

[192] Kira, M. and Koch, S.W. (2011). *Semiconductor Quantum Optics*. Cambridge University Press.

[193] Klimov, V. I. (2007). Spectral and dynamical properties of multiexcitons in semiconductor nanocrystals. *Annu. Rev. Phys. Chem.*, **58**, 635.

[194] Kner, P., Schäfer, W., Lövenich, R., and Chemla, D. S. (1998). Coherence of four-particle correlations in semiconductors. *Phys. Rev. Lett.*, **81**, 5386–5389.

[195] Koch, Martin, Shah, Jagdeep, and Meier, Torsten (1998). Coupled absorber-cavity system: Observation of a characteristic nonlinear response. *Phys. Rev. B*, **57**, R2049–R2052.

[196] Kohler, Daniel D., Thompson, Blaise J., and Wright, John C. (2017). Frequency-domain coherent multidimensional spectroscopy when dephasing rivals pulsewidth: Disentangling material and instrument response. *The Journal of Chemical Physics*, **147**, 084202.

[197] Kohnle, V., Léger, Y., Wouters, M., Richard, M., Portella-Oberli, M. T., and Deveaud, B. (2012). Four-wave mixing excitations in a dissipative polariton quantum fluid. *Phys. Rev. B*, **86**, 064508.

[198] Kong, Weiguang, Ye, Zhenyu, Qi, Zhen, Zhang, Bingpo, Wang, Miao, Rahimi-Iman, Arash, and Wu, Huizhen (2015). Characterization of an abnormal photoluminescence behavior upon crystal-phase transition of perovskite $CH_3NH_3PbI_3$. *Phys. Chem. Chem. Phys.*, **17**, 16405–16411.

[199] Kuehn, W., Reimann, K., Woerner, M., and Elsaesser, T. (2009). Phase-resolved two-dimensional spectroscopy based on collinear n-wave mixing in the ultrafast time domain. *J. Chem. Phys.*, **130**, 164503.

[200] Kuehn, W., Reimann, K., Woerner, M., Elsaesser, T., and Hey, R. (2011). Two-dimensional terahertz correlation spectra of electronic excitations in semiconductor quantum wells. *J. Phys. Chem. B*, **115**, 5448–5455.

[201] Kuehn, W., Reimann, K., Woerner, M., Elsaesser, T., Hey, R., and Schade, U. (2011). Strong correlation of electronic and lattice excitations in GaAs / AlGaAs semiconductor quantum wells revealed by two-dimensional terahertz spectroscopy. *Phys. Rev. Lett.*, **107**, 067401.

[202] Kujiraoka, Mamiko, Ishi-Hayase, Junko, Akahane, Kouichi, Yamamoto, Naokatsu, Ema, Kazuhiro, and Sasaki, Masahide (2010). Optical rabi oscillations in a quantum dot ensemble. *App. Phys. Express*, **3**, 092801.

[203] Kylänpää, Ilkka and Komsa, Hannu-Pekka (2015). Binding energies of exciton complexes in transition metal dichalcogenide monolayers and effect of dielectric environment. *Phys. Rev. B*, **92**, 205418.

[204] Langbein, W. and Hvam, J. M. (2000). Dephasing in the quasi-two-dimensional exciton-biexciton system. *Phys. Rev. B*, **61**, 1692–1695.

[205] Langbein, W., Meier, T., Koch, S. W., and Hvam, J. M. (2001). Spectral signatures of $\chi(5)$ processes in four-wave mixing of homogeneously broadened excitons. *J. Opt. Soc. Am. B*, **18**, 1318–1325.

[206] Langbein, Wolfgang and Patton, Brian (2005). Microscopic measurement of photon echo formation in groups of individual excitonic transitions. *Phys. Rev. Lett.*, **95**, 017403.

[207] Langbein, Wolfgang and Patton, Brian (2006). Heterodyne spectral interferometry for multidimensional nonlinear spectroscopy of individual quantum systems. *Opt. Lett.*, **31**, 1151–1153.

[208] Leegwater, Jan A. and Mukamel, Shaul (1994). Self-broadening and exciton line shifts in gases: Beyond the local-field approximation. *Phys. Rev. A*, **49**, 146–155.

[209] Leguy, Aurélien M. A., Goñi, Alejandro R., Frost, Jarvist M., Skelton, Jonathan, Brivio, Federico, Rodríguez-Martínez, Xabier, Weber, Oliver J., Pallipurath, Anuradha, Alonso, M. Isabel, Campoy-Quiles, Mariano, Weller, Mark T., Nelson, Jenny, Walsh, Aron, and Barnes, Piers R. F. (2016). Dynamic disorder, phonon lifetimes, and the assignment of modes to the vibrational spectra of methylammonium lead halide perovskites. *Phys. Chem. Chem. Phys.*, **18**, 27051–27066.

[210] Lenihan, A. S., Dutt, M. V. Gurudev, Steel, D. G., Ghosh, S., and Bhattacharya, P. (2004). Biexcitonic resonance in the nonlinear optical response of an InAs quantum dot ensemble. *Phys. Rev. B*, **69**, 045306.

[211] Leo, K., Wegener, M., Shah, J., Chemla, D. S., Göbel, E. O., Damen, T. C., Schmitt-Rink, S., and Schäfer, W. (1990). Effects of coherent polarization interactions on time-resolved degenerate four-wave mixing. *Phys. Rev. Lett.*, **65**, 1340–1343.

[212] Lepetit, L., Chériaux, G., and Joffre, M. (1995). Linear techniques of phase measurement by femtosecond spectral interferometry for applications in spectroscopy. *J. Opt. Soc. Am. B*, **12**, 2467.

[213] Lewis, E.L. (1980). Collisional relaxation of atomic excited states, line broadening and interatomic interactions. *Phys. Rep.*, **58**, 1–71.

[214] Li, Hebin, Bristow, Alan D, Siemens, Mark E, Moody, Galan, and Cundiff, Steven T (2013). Unraveling quantum pathways using optical three-dimensional Fourier-transform spectroscopy. *Nat. Commun.*, **4**, 1390.

[215] Li, Hebin, Moody, Galan, and Cundiff, Steven T (2013). Reflection optical two-dimensional Fourier-transform spectroscopy. *Opt. Express*, **21**, 1687–1692.

[216] Li, Hebin, Sautenkov, Vladimir A, Rostovtsev, Yuri V, and Scully, Marlan O (2009). Excitation dependence of resonance line self-broadening at different atomic densities. *J. Phys. B: At., Mol. Opt. Phys.*, **42**, 65203.

[217] Li, Hebin, Spencer, Austin P., Kortyna, Andrew, Moody, Galan, Jonas, David M., and Cundiff, Steven T. (2013). Pulse propagation effects in optical

2D Fourier-transform spectroscopy: Experiment. *J. Phys. Chem. A*, **117**, 6279–6287.

[218] Li, H, Varzhapetyan, T.S., Sautenkov, V.A., Rostovtsev, Y.V., Chen, H, Sarkisyan, D, and Scully, M.O. (2008). Improvement of spectral resolution by using the excitation dependence of dipole–dipole interaction in a dense atomic gas. *Appl. Phys. B*, **91**, 229–231.

[219] Li, Xiaoqin, Wu, Yanwen, Steel, Duncan, Gammon, D., Stievater, T. H., Katzer, D. S., Park, D., Piermarocchi, C., and Sham, L. J. (2003). An all-optical quantum gate in a semiconductor quantum dot. *Science*, **301**, 809–811.

[220] Li, Xiaoqin, Zhang, Tianhao, Borca, Camelia N., and Cundiff, Steven T. (2006). Many-body interactions in semiconductors probed by optical two-dimensional Fourier transform spectroscopy. *Phys. Rev. Lett.*, **96**, 057406.

[221] Li, Xiaoqin, Zhang, Tianhao, Mukamel, Shaul, Mirin, Richard P., and Cundiff, Steven T. (2009). Investigation of electronic coupling in semiconductor double quantum wells using coherent optical two-dimensional fourier transform spectroscopy. *Solid State Commun.*, **149**, 361–366.

[222] Liebel, Matz, Toninelli, Costanza, and van Hulst, Niek F. (2018). Room-temperature ultrafast nonlinear spectroscopy of a single molecule. *Nat. Photonics*, **12**, 45–49.

[223] Lindberg, M., Hu, Y. Z., Binder, R., and Koch, S. W. (1994). $\chi^{(3)}$ formalism in optically excited semiconductors and its applications in four-wave-mixing spectroscopy. *Phys. Rev. B*, **50**, 18060–18072.

[224] Liu, A., Almeida, D. B., Bae, W. K., Padilha, L. A., and Cundiff, S. T. (2019, Aug). Non-markovian exciton-phonon interactions in core-shell colloidal quantum dots at femtosecond timescales. *Phys. Rev. Lett.*, **123**, 057403.

[225] Liu, Albert, Almeida, Diogo B., Bae, Wan-Ki, Padilha, Lazaro A., and Cundiff, Steven T. (2019, sep). Simultaneous existence of confined and delocalized vibrational modes in colloidal quantum dots. *J. Phys. Chem. Lett.*, **10**, 6144–6150.

[226] Liu, Albert, Almeida, Diogo B., Bonato, Luiz G., Nagamine, Gabriel, Zagonel, Luiz F., Nogueira, Ana F., Padilha, Lazaro A., and Cundiff, S. T. (2021). Multidimensional coherent spectroscopy reveals triplet state coherences in cesium lead-halide perovskite nanocrystals. *Science Advances*, **7**, eabb3594.

[227] Liu, Albert and Cundiff, Steven T. (2020). Spectroscopic signatures of electron-phonon coupling in silicon-vacancy centers in diamond. *Phys. Rev. Materials*, **4**, 055202.

[228] Liu, Albert, Cundiff, Steven T., Almeida, Diogo B., and Ulbricht, Ronald (2021). Spectral broadening and ultrafast dynamics of a nitrogen-vacancy center ensemble in diamond. *Mater. Quantum Technol.*, **1**, 025002.

[229] Liu, Albert, Nagamine, Gabriel, Bonato, Luiz G., Almeida, Diogo B., Zagonel, Luiz F., Nogueira, Ana F., Padilha, Lazaro A., and Cundiff, Steven T. (2021). Toward engineering intrinsic line widths and line broadening in perovskite nanoplatelets. *ACS Nano*, **15**, 6499–6506.

[230] Lomsadze, Bachana (2018). Saturated absorption spectroscopy for diode laser locking. In *Encyclopedia of Modern Optics (Second Edition)* (Second Edition edn) (ed. B. D. Guenther and D. G. Steel), pp. 227 – 232. Elsevier, Oxford.

[231] Lomsadze, Bachana and Cundiff, Steven T. (2017). Frequency comb-based four-wave-mixing spectroscopy. *Opt. Lett.*, **42**, 2346.

[232] Lomsadze, Bachana and Cundiff, Steven T. (2017). Frequency combs enable rapid and high-resolution multidimensional coherent spectroscopy. *Science*, **357**, 1389–1391.

[233] Lomsadze, Bachana and Cundiff, Steven T. (2017). Multi-heterodyne two dimensional coherent spectroscopy using frequency combs. *Sci. Rep.*, **7**, 14018.

[234] Lomsadze, Bachana and Cundiff, Steven T. (2018). Frequency-comb based double-quantum two-dimensional spectrum identifies collective hyperfine resonances in atomic vapor induced by dipole-dipole interactions. *Phys. Rev. Lett.*, **120**, 233401.

[235] Lomsadze, Bachana and Cundiff, Steven T. (2019). Frequency comb-based multidimensional coherent spectroscopy. In *Coherent Multidimensional Spectroscopy* (ed. M. Cho), pp. 339–354. Springer Singapore.

[236] Lomsadze, B. and Cundiff, S. T. (2019). Tri-comb multidimensional coherent spectroscopy. *IEEE Photon. Technol. Lett.*, **31**, 1886–1889.

[237] Lomsadze, Bachana and Cundiff, Steven T. (2020). Line-shape analysis of double-quantum multidimensional coherent spectra. *Phys. Rev. A*, **102**, 043514.

[238] Lomsadze, B., Fehrenbach, C. W., and DePaola, B. D. (2012). Calculation of ionization in direct-frequency comb spectroscopy. *Phys. Rev. A*, **85**, 043403.

[239] Lomsadze, B., Fehrenbach, C. W., and DePaola, B. D. (2013). Measurement of ionization in direct frequency comb spectroscopy. *J. Appl. Phys.*, **113**, 103105.

[240] Lomsadze, Bachana, Fradet, Kelly M., and Arnold, Richard S. (2020). Elastic tape behavior of a bi-directional kerr-lens mode-locked dual-comb ring laser. *Opt. Lett.*, **45**, 1080–1083.

[241] Lomsadze, Bachana, Smith, Brad C., and Cundiff, Steven T. (2018). Tri-comb spectroscopy. *Nat. Photonics*, **12**, 676–680.

[242] Lorenz, V. and Cundiff, S. (2005). Non-markovian dynamics in a dense potassium vapor. *Phys. Rev. Lett.*, **95**, 163601.

[243] Lorenz, V O, Dai, X, Green, H, Asnicar, T R, and Cundiff, S T (2008). High-density, high-temperature alkali vapor cell. *Rev. Sci. Instrum.*, **79**, 123104.

[244] Lorenz, V O, Mukamel, S, Zhuang, W, and Cundiff, S T (2008). Ultrafast optical spectroscopy of spectral fluctuations in a dense atomic vapor. *Phys. Rev. Lett.*, **100**, 013603.

[245] Love, A. P. D., Krizhanovskii, D. N., Whittaker, D. M., Bouchekioua, R., Sanvitto, D., Rizeiqi, S. Al, Bradley, R., Skolnick, M. S., Eastham, P. R., André, R., and Dang, Le Si (2008). Intrinsic decoherence mechanisms in the microcavity polariton condensate. *Phys. Rev. Lett.*, **101**, 067404.

[246] Lucas, E., Lihachev, G., Bouchand, R., Pavlov, N. G., Raja, A. S., Karpov, M., Gorodetsky, M. L., and Kippenberg, T. J. (2018). Spatial multiplexing of soliton microcombs. *Nat. Photonics*, **12**, 699–705.

[247] Ludlow, A D, Zelevinsky, T, Campbell, G K, Blatt, S, Boyd, M M, de Miranda, M H G, Martin, M J, Thomsen, J W, Foreman, S M, Ye, Jun, Fortier, T M, Stalnaker, J E, Diddams, S A, Le Coq, Y, Barber, Z W, Poli, N, Lemke, N D, Beck, K M, and Oates, C W (2008). Sr lattice clock at $1 \times 10^{(-16)}$ fractional uncertainty by remote optical evaluation with a Ca clock. *Science*, **319**, 1805–1808.

[248] Madelung, O. (2012). *Semiconductors: Other than Group IV Elements and III–V Compounds*. Data in Science and Technology. Springer Berlin Heidelberg.

[249] Mai, Cong, Barrette, Andrew, Yu, Yifei, Semenov, Yuriy G., Kim, Ki Wook, Cao, Linyou, and Gundogdu, Kenan (2014). Many-body effects in valleytronics: Direct measurement of valley lifetimes in single-layer MoS_2. *Nano Lett.*, **14**, 202–206.

[250] Maialle, M. Z., Andrada e Silva, E. A., and Sham, L. J. (1993). Exciton spin dynamics in quantum wells. *Phys. Rev. B*, **47**, 15776.

[251] Mak, Kin Fai, Lee, Changgu, Hone, James, Shan, Jie, and Heinz, Tony F. (2010). Atomically thin mos2: A new direct-gap semiconductor. *Phys. Rev. Lett.*, **105**, 0136805.

[252] Maker, P. D., Terhune, R. W., Nisenoff, M., and Savage, C. M. (1962). Effects of dispersion and focusing on the production of optical harmonics. *Phys. Rev. Lett.*, **8**, 21–22.

[253] Maki, Jeffery J., Malcuit, Michelle S., Sipe, J. E., and Boyd, Robert W. (1991). Linear and nonlinear optical measurements of the Lorentz local field. *Phys. Rev. Lett.*, **67**, 972–975.

[254] Malý, Pavel and Mančal, Tomáš (2018). Signatures of exciton delocalization and exciton–exciton annihilation in fluorescence-detected two-dimensional coherent spectroscopy. *J. Phys. Chem. Lett.*, **9**, 5654–5659.

[255] Marian, Adela, Stowe, Matthew C., Lawall, John R., Felinto, Daniel, and Ye, Jun (2004). United time-frequency spectroscopy for dynamics and global structure. *Science*, **306**, 2063–2068.

[256] Martin, Eric W. (2018). *Coherent Spectroscopy at the Diffraction Limit*. phdthesis, The University of Michigan.

[257] Martin, Eric W. and Cundiff, Steven T. (2018). Inducing coherent quantum dot interactions. *Phys. Rev. B*, **97**, 081301(R).

[258] Martin, Eric W. and Cundiff, Steven T. (2018, Feb). Inducing coherent quantum dot interactions. *Phys. Rev. B*, **97**, 081301.

[259] Martin, Eric W., Horng, Jason, Ruth, Hanna G., Paik, Eunice, Wentzel, Michael-Henr, Deng, Hui, and Cundiff, Steven T. (2020). Encapsulation narrows and preserves the excitonic homogeneous linewidth of exfoliated monolayer MoSe2. *Phys. Rev. Appl.*, **14**, 021002.

[260] Masia, Francesco, Accanto, Nicolò, Langbein, Wolfgang, and Borri, Paola (2012). Spin-flip limited exciton dephasing in CdSe/ZnS colloidal quantum dots. *Phys. Rev. Lett.*, **108**, 087401.

[261] Mazurenko, Anton, Chiu, Christie S., Ji, Geoffrey, Parsons, Maxwell F., Kanász-Nagy, Márton, Schmidt, Richard, Grusdt, Fabian, Demler, Eugene, Greif, Daniel,

and Greiner, Markus (2017). A cold-atom Fermi-Hubbard antiferromagnet. *Nature*, **545**, 462–466.

[262] Meek, Samuel A., Hipke, Arthur, Guelachvili, Guy, Hänsch, Theodor W., and Picqué, Nathalie (2017). Doppler-free Fourier transform spectroscopy. *Opt. Lett.*, **43**, 162.

[263] Mehlenbacher, Randy D., McDonough, Thomas J., Grechko, Maksim, Wu, Meng-Yin, Arnold, Michael S., and Zanni, Martin T. (2015). Energy transfer pathways in semiconducting carbon nanotubes revealed using two-dimensional white-light spectroscopy. *Nat. Commun.*, **6**, 6732.

[264] Mehlenbacher, Randy D., McDonough, Thomas J., Kearns, Nicholas M., Shea, Matthew J., Joo, Yongho, Gopalan, Padma, Arnold, Michael S., and Zanni, Martin T. (2016). Polarization-controlled two-dimensional white-light spectroscopy of semiconducting carbon nanotube thin films. *J. Phys. Chem. C*, **120**, 17069–17080.

[265] Mehlenbacher, Randy D., Wang, Jialiang, Kearns, Nicholas M., Shea, Matthew J., Flach, Jessica T., McDonough, Thomas J., Wu, Meng-Yin, Arnold, Michael S., and Zanni, Martin T. (2016). Ultrafast exciton hopping observed in bare semiconducting carbon nanotube thin films with two-dimensional white-light spectroscopy. *J. Phys. Chem. Lett.*, **7**, 2024–2031.

[266] Mermillod, Q., Wigger, D., Delmonte, V., Reiter, D. E., Schneider, C., Kamp, M., S., Höfling, Langbein, W., Kuhn, T., Nogues, G., and Kasprzak, J. (2016). Dynamics of excitons in individual inas quantum dots revealed in four-wave mixing spectroscopy. *Optica*, **3**, 377–384.

[267] Miyata, K., Atallah, T. L., and Zhu, X.-Y. (2017). Lead halide perovskites: Crystal-liquid duality, phonon glass electron crystals, and large polaron formation. *Sci. Adv.*, **3**, e1701469.

[268] Miyata, K., Meggiolaro, D., Trinh, M. T., Joshi, P. P., Mosconi, E., Jones, S. C., De Angelis, F., and Zhu, X.-Y. (2017). Large polarons in lead halide perovskites. *Sci. Adv.*, **3**, e1701217.

[269] Mohler, Kathrin J., Bohn, Bernhard J., Yan, Ming, Mélen, Gwénaëlle, Hänsch, Theodor W., and Picqué, Nathalie (2017). Dual-comb coherent Raman spectroscopy with lasers of 1-GHz pulse repetition frequency. *Opt. Lett.*, **42**, 318.

[270] Monahan, D. M., Guo, L., Lin, J., Dou, L., Yang, P., and Fleming, G. R. (2017). Room-temperature coherent optical phonon in 2D electronic spectra of $CH_3NH_3PbI_3$ perovskite as a possible cooling bottleneck. *J. Phys. Chem. Lett.*, **8**, 3211–3215.

[271] Moody, G, Akimov, I A, Li, H, Singh, R, Yakovlev, D R, Karczewski, G, Wiater, M, Wojtowicz, T, Bayer, M, and Cundiff, S T (2014). Coherent coupling of excitons and trions in a photoexcited CdTe/CdMgTe quantum well. *Phys. Rev. Lett.*, **112**, 97401.

[272] Moody, Galan and Cundiff, Steven T. (2017). Advances in multi-dimensional coherent spectroscopy of semiconductor nanostructures. *Adv. Phys-X*, **2**, 641–674.

[273] Moody, Galan, Kavir Dass, Chandriker, Hao, Kai, Chen, Chang-Hsiao, Li, Lain-Jong, Singh, Akshay, Tran, Kha, Clark, Genevieve, Xu, Xiaodong, Berghauser,

Gunnar, Malic, Ermin, Knorr, Andreas, and Li, Xiaoqin (2015). Intrinsic homogeneous linewidth and broadening mechanisms of excitons in monolayer transition metal dichalcogenides. *Nat. Commun.*, **6**, 8315.

[274] Moody, Galan, Schaibley, John, and Xu, Xiaodong (2016). Exciton dynamics in monolayer transition metal dichalcogenides. *J. Opt. Soc. Am. B*, **33**, C39.

[275] Moody, G., Siemens, M. E., Bristow, A. D., Dai, X., Bracker, A. S., Gammon, D., and Cundiff, S. T. (2011). Exciton relaxation and coupling dynamics in a GaAs/AlGaAs quantum well and quantum dot ensemble. *Phys. Rev. B*, **83**, 245316.

[276] Moody, G., Siemens, M. E., Bristow, A. D., Dai, X., Karaiskaj„ D., Bracker, A. S., Gammon, D., and Cundiff, S. T. (2011). Exciton-exciton and exciton-phonon interactions in an interfacial GaAs quantum dot ensemble. *Phys. Rev. B*, **83**, 115324.

[277] Moody, G., Singh, R., H., Li., Akimov, I. A., Bayer, M., Reuter, D., Wieck, A. D., and Cundiff, S. T. (2013). Correlation and dephasing effects on the non-radiative coherence between bright excitons in an InAs QD ensemble measured with 2d spectroscopy. *Solid State Commun.*, **163**, 65.

[278] Moody, G., Singh, R., Li, H., Akimov, I. A., Bayer, M., Reuter, D., Wieck, A. D., Bracker, A. S., Gammon, D., and Cundiff, S. T. (2013). Influence of confinement on biexciton binding in semiconductor quantum dot ensembles measured with two-dimensional spectroscopy. *Phys. Rev. B*, **87**, 041304(R).

[279] Moody, G., Singh, R., Li, H., Akimov, I. A., Bayer, M., Reuter, D., Wieck, A. D., and Cundiff, S. T. (2013). Fifth-order nonlinear optical response of excitonic states in an inas quantum dot ensemble measured with two-dimensional spectroscopy. *Phys. Rev. B*, **87**, 045313.

[280] Mukamel, S. (1999). *Principles of Nonlinear Optical Spectroscopy*. Oxford University Press.

[281] Mukamel, S (2000). Multidimensional femtosecond correlation spectroscopies of electronic and vibrational excitations. *Annu. Rev. Phys. Chem.*, **51**, 691–729.

[282] Mukamel, Shaul and Rahav, Saar (2010). Ultrafast nonlinear optical signals viewed from the molecule's perspective. In *Adv. At. Mol. Opt. Phys.*, Volume 59, pp. 223–263. Elsevier.

[283] Mukherjee, Sudipta S., Skoff, David R., Middleton, Chris T., and Zanni, Martin T. (2013). Fully absorptive 3D IR spectroscopy using a dual mid-infrared pulse shaper. *J. Chem. Phys.*, **139**, 144205.

[284] Munoz, Maria F., Medina, Adam, Autry, Travis M., Moody, Galan, Siemens, Mark E., Bristow, Alan D., Cundiff, Steven T., and Li, Hebin (2020). Fast phase cycling in non-collinear optical two-dimensional coherent spectroscopy. *Opt. Lett.*, **45**, 5852–5855.

[285] Nardin, Gaël, Autry, Travis M, Silverman, Kevin L, and Cundiff, S T (2013). Multidimensional coherent photocurrent spectroscopy of a semiconductor nanostructure. *Opt. Express*, **21**, 28617.

[286] Nardin, Gaël, Moody, Galan, Singh, Rohan, Autry, Travis M., Li, Hebin, Morier-Genoud, François, and Cundiff, Steven T. (2014, Jan). Coherent excitonic cou-

pling in an asymmetric double ingaas quantum well arises from many-body effects. *Phys. Rev. Lett.*, **112**, 046402.

[287] Nataf, Pierre and Ciuti, Cristiano (2010). No-go theorem for superradiant quantum phase transitions in cavity QED and counter-example in circuit QED. *Nat. Commun.*, **1**, 72.

[288] Nemeth, Alexandra, Milota, Franz, Mančal, T., Pullerits, Tonu, Sperling, Jaroslaw, Hauer, Jürgen, Kauffmann, Harald F., and Christensson, Niklas (2010). Double-quantum two-dimensional electronic spectroscopy of a three-level system: Experiments and simulations. *J. Chem. Phys.*, **133**, 094505.

[289] Nesmeyanov, A. N. (1963). *Vapor pressure curve of chemical elements*. Elsevier, New York.

[290] Nish, Adrian, Hwang, Jeong-Yuan, Doig, James, and Nicholas, Robin J. (2007). Highly selective dispersion of single-walled carbon nanotubes using aromatic polymers. *Nat. Nanotechnol.*, **2**, 640–646.

[291] Nishida, Jun, Breen, John P., Lindquist, Kurt P., Umeyama, Daiki, Karunadasa, Hemamala I., and Fayer, Michael D. (2018). Dynamically disordered lattice in a layered Pb-I-SCN perovskite thin film probed by two-dimensional infrared spectroscopy. *J. Am. Chem. Soc.*, **140**, 9882–9890.

[292] Nowakowski, Paweł J., Khyasudeen, M. Faisal, and Tan, Howe-Siang (2018). The effect of laser pulse bandwidth on the measurement of the frequency fluctuation correlation functions in 2d electronic spectroscopy. *Chemical Physics*, **515**, 214–220.

[293] Oberli, D. Y., Shah, Jagdeep, Damen, T. C., Kuo, J. M., Henry, J. E., Lary, Jenifer, and Goodnick, Stephen M. (1990). Optical phonon-assisted tunneling in double quantum well structures. *Appl. Phys. Lett.*, **56**, 1239–1241.

[294] Ohnesorge, B., Albrecht, M., Oshinowo, J., Forchel, A., and Arakawa, Y. (1996). Rapid carrier relaxation in self-assembled InGaAs/GaAs quantum dots. *Phys. Rev. B*, **54**, 11532.

[295] Pakoulev, Andrei V., Rickard, Mark A., Kornau, Kathryn M., Mathew, Nathan A., Yurs, Lena A., Block, Stephen B., and Wright, John C. (2009). Mixed frequency-/time-domain coherent multidimensional spectroscopy: Research tool or potential analytical method? *Acc. Chem. Res.*, **42**, 1310–1321.

[296] Panitchayangkoon, Gitt, Voronine, Dmitri V., Abramavicius, Darius, Caram, Justin R., Lewis, Nicholas H. C., Mukamel, Shaul, and Engel, Gregory S. (2011). Direct evidence of quantum transport in photosynthetic light-harvesting complexes. *Proc. Nat. Acad. Sci. U.S.A.*, **108**, 20908–20912.

[297] Pantke, K.-H., Oberhauser, D., Lyssenko, V. G., Hvam, J. M., and Weimann, G. (1993). Coherent generation and interference of excitons and biexcitons in gaas/al$_x$ga$_{1-x}$as quantum wells. *Phys. Rev. B*, **47**, 2413–2416.

[298] Panzer, Fabian, Baderschneider, Sebastian, Gujar, Tanaji P., Unger, Thomas, Bagnich, Sergey, Jakoby, Marius, Bässler, Heinz, Hüttner, Sven, Köhler, Jürgen, Moos, Ralf, Thelakkat, Mukundan, Hildner, Richard, and Köhler, Anna (2016). Reversible laser-induced amplified spontaneous emission from coexisting tetrag-

onal and orthorhombic phases in hybrid lead halide perovskites. *Adv. Opt. Mater.*, **4**, 917–928.

[299] Park, Kisam and Cho, Minhaeng (2000). Theoretical description of the nonlinear response functions associated with eight distinctive three-dimensional vibrational spectroscopies. *J. Chem. Phys.*, **112**, 5021.

[300] Park, M., Neukirch, A. J., Reyes-Lillo, S. E., Lai, M., Ellis, S. R., Dietze, D., Neaton, J. B., Yang, P., Tretiak, S., and Mathies, R. A. (2018). Excited-state vibrational dynamics toward the polaron in methylammonium lead iodide perovskite. *Nat. Commun.*, **9**, 2525.

[301] Parzuchowski, Kristen M., Mikhaylov, Alexander, Mazurek, Michael D., Wilson, Ryan N., Lum, Daniel J., Gerrits, Thomas, Camp, Charles H., Stevens, Martin J., and Jimenez, Ralph (2021, apr). Setting bounds on entangled two-photon absorption cross sections in common fluorophores. *Phys. Rev. Applied*, **15**, 044012.

[302] Patton, B., Langbein, W., Woggon, U., Maingault, L., and Mariette, H. (2006). Time- and spectrally-resolved four-wave mixing in single CdTe/ZnTe quantum dots. *Phys. Rev. B*, **73**, 235354.

[303] Patton, B., Woggon, U., and Langbein, W. (2005). Coherent control and polarization readout of individual excitonic states. *Phys. Rev. Lett.*, **95**, 266401.

[304] Paul, J., Stevens, C. E., Liu, C., Dey, P., McIntyre, C., Turkowski, V., Reno, J. L., Hilton, D. J., and Karaiskaj, D. (2016). Strong quantum coherence between fermi liquid mahan excitons. *Phys. Rev. Lett.*, **116**, 157401.

[305] Perakis, Fivos, Borek, Joanna A., and Hamm, Peter (2013). Three-dimensional infrared spectroscopy of isotope-diluted ice ih. *J. Chem. Phys.*, **139**, 014501.

[306] Perdomo-Ortiz, Alejandro, Widom, Julia R., Lott, Geoffrey A., Aspuru-Guzik, Alán, and Marcus, Andrew H. (2012). Conformation and electronic population transfer in membrane-supported self-assembled porphyrin dimers by 2D fluorescence spectroscopy. *J. Phys. Chem. B*, **116**, 10757–10770.

[307] Perlík, Václav, Hauer, Jürgen, and Šanda, František (2017). Finite pulse effects in single and double quantum spectroscopies. *J. Opt. Soc. Am. B*, **34**, 430–439.

[308] Plechinger, Gerd, Nagler, Philipp, Kraus, Julia, Paradiso, Nicola, Strunk, Christoph, Schüller, Christian, and Korn, Tobias (2015). Identification of excitons, trions and biexcitons in single-layer WS_2. *Phys. Status Solidi RRL*, **9**, 457–461.

[309] Prevedel, R., Cronenberg, G., Tame, M. S., Paternostro, M., Walther, P., Kim, M. S., and Zeilinger, A. (2009). Experimental realization of Dicke states of up to six qubits for multiparty quantum networking. *Phys. Rev. Lett.*, **103**, 020503.

[310] Price, M. B., Butkus, J., Jellicoe, T. C., Sadhanala, A., Briane, A., Halpert, J. E., Broch, K., Hodgkiss, J. M., Friend, R. H., and Deschler, F. (2015). Hot-carrier cooling and photoinduced refractive index changes in organic-inorganic lead halide perovskites. *Nat. Commun.*, **6**, 8420.

[311] Prineas, J. P., Ell, C., Lee, E. S., Khitrova, G., Gibbs, H. M., and Koch, S. W. (2000). Exciton-polariton eigenmodes in light-coupled $in_{0.04}ga_{0.96}As/GaAs$ semiconductor multiple-quantum-well periodic structures. *Phys. Rev. B*, **61**, 13863–13872.

[312] Protesescu, Loredana, Yakunin, Sergii, Bodnarchuk, Maryna I., Krieg, Franziska, Caputo, Riccarda, Hendon, Christopher H., Yang, Ruo Xi, Walsh, Aron, and Kovalenko, Maksym V. (2015). Nanocrystals of cesium lead halide perovskites (CsPbX$_3$, X = Cl, Br, and I): Novel optoelectronic materials showing bright emission with wide color gamut. *Nano Lett.*, **15**, 3692–3696.

[313] Qiu, Diana Y., da Jornada, Felipe H., and Louie, Steven G. (2013). Optical spectrum of mos$_2$: Many-body effects and diversity of exciton states. *Phys. Rev. Lett.*, **111**, 216805.

[314] Raymer, Michael G., Landes, Tiemo, Allgaier, Markus, Merkouche, Sofiane, Smith, Brian J., and Marcus, Andrew H. (2021, may). How large is the quantum enhancement of two-photon absorption by time-frequency entanglement of photon pairs? *Optica*, **8**, 757.

[315] Raymer, M. G., Marcus, Andrew H., Widom, Julia R., and Vitullo, Dashiell L. P. (2013, oct). Entangled photon-pair two-dimensional fluorescence spectroscopy (EPP-2dfs). *J. Phys. Chem. B*, **117**, 15559–15575.

[316] Reutzel, Marcel, Li, Andi, and Petek, Hrvoje (2019, Mar). Coherent two-dimensional multiphoton photoelectron spectroscopy of metal surfaces. *Phys. Rev. X*, **9**, 011044.

[317] Reuven, A.B. (1975). Spectral line shapes in gases in the binary-collision approximation. *Adv. Chem. Phys.*, **33**, 235.

[318] Richter, J. M., Branchi, F., Valduga de Almeida Camargo, F., Zhao, B., Friend, R. H., Cerullo, G., and Deschler, F. (2017). Ultrafast carrier thermalization in lead iodide perovskite probed with two-dimensional electronic spectroscopy. *Nat. Commun.*, **8**, 376.

[319] Richter, Marten and Mukamel, Shaul (2010). Ultrafast double-quantum-coherence spectroscopy of excitons with entangled photons. *Phys. Rev. A*, **82**, 013820.

[320] Richter, Marten, Singh, Rohan, Siemens, Mark, and Cundiff, Steven T. (2018). Deconvolution of optical multidimensional coherent spectra. *Sci. Adv.*, **4**, eaar7697.

[321] Rodt, S., Schliwa, A., Pötschke, K., Guffarth, F., and Bimberg, D. (2005). Correlation of structural and few-particle properties of self-organized InAs/GaAs quantum dots. *Phys. Rev. B*, **71**, 155325.

[322] Röhlsberger, Ralf, Schlage, Kai, Sahoo, Balaram, Couet, Sebastien, and Rüffer, Rudolf (2010). Collective lamb shift in single-photon superradiance. *Science*, **328**, 1248–1251.

[323] Rowland, Clare E, Fedin, Igor, Zhang, Hui, Gray, Stephen K, Govorov, Alexander O, Talapin, Dmitri V, and Schaller, Richard D (2015). Picosecond energy transfer and multiexciton transfer outpaces Auger recombination in binary CdSe nanoplatelet solids. *Nat. Mater.*, **14**, 484–489.

[324] Rudin, S., Reinecke, T. L., and Segall, B. (1990). Temperature-dependent exciton linewidths in semiconductors. *Phys. Rev. B*, **42**, 11218–11231.

[325] Sackett, C. A., Kielpinski, D., King, B. E., Langer, C., Meyer, V., Myatt, C. J., Rowe, M., Turchette, Q. A., Itano, W. M., Wineland, D. J., and Monroe, C. (2000). Experimental entanglement of four particles. *Nature*, **404**, 256–259.

[326] Sanders, Jacob N., Saikin, Semion K., Mostame, Sarah, Andrade, Xavier, Widom, Julia R., Marcus, Andrew H., and Aspuru-Guzik, Alán (2012). Compressed sensing for multidimensional spectroscopy experiments. *J. Phys. Chem. Lett.*, **3**, 2697–2702.

[327] Sanguinetti, S., Watanabe, K., Tateno, T., Gurioli, M., Werner, P., Wakaki, M., and Koguchi, N. (2003). Modified droplet epitaxy GaAs/AlGaAs quantum dots grown on a variable thickness wetting layer. *J. Cryst. Growth*, **253**, 71.

[328] Sargent III, Murray, Scully, Marlan O., and Lamb, Jr., Willis E. (2018). *Laser Physics*. CRC Press.

[329] Sautenkov, V. A., van Kampen, H., Eliel, E. R., and Woerdman, J. P. (1996). Dipole-dipole broadened line shape in a partially excited dense atomic gas. *Phys. Rev. Lett.*, **77**, 3327–3330.

[330] Savona, V. and Langbein, W. (2006). Realistic heterointerface model for excitonic states in growth-interrupted GaAs quantum wells. *Phys. Rev. B*, **74**, 075311.

[331] Schaibley, John R., Yu, Hongyi, Clark, Genevieve, Rivera, Pasqual, Ross, Jason S., Seyler, Kyle L., Yao, Wang, and Xu, Xiaodong (2016). Valleytronics in 2D materials. *Nat. Rev. Mater.*, **1**, Valleytronics in2D material.

[332] Schlau-Cohen, G. S., Dawlaty, J. M., and Fleming, G. R. (2012). Ultrafast multidimensional spectroscopy: Principles and applications to photosynthetic systems. *IEEE J. Sel. Topics Quantum Electron.*, **18**, 283–295.

[333] Schliwa, Andrei, Winkelnkemper, Momme, and Bimberg, Dieter (2009). Few-particle energies versus geometry and composition of $in_x ga_{1-x} As$/GaAs self-organized quantum dots. *Phys. Rev. B*, **79**, 075443.

[334] Schlosser, Felix, Knorr, Andreas, Mukamel, Shaul, and Richter, Marten (2013). Using localized double-quantum-coherence spectroscopy to reconstruct the two-exciton wave function of coupled quantum emitters. *New J. Phys.*, **15**, 025004.

[335] Schnedermann, Christoph, Lim, Jong Min, Wende, Torsten, Duarte, Alex S., Ni, Limeng, Gu, Qifei, Sadhanala, Aditya, Rao, Akshay, and Kukura, Philipp (2016). Sub-10 fs time-resolved vibronic optical microscopy. *J. Phys. Chem. Lett.*, **7**, 4854–4859.

[336] Schröter, Marco, Pullerits, Tõnu, and Kühn, Oliver (2018). Using fluorescence detected two-dimensional spectroscopy to investigate initial exciton delocalization between coupled chromophores. *J. Chem. Phys.*, **149**, 114107.

[337] Schweigert, Igor V. and Mukamel, Shaul (2008). Simulating multidimensional optical wave-mixing signals with finite-pulse envelopes. *Phys. Rev. A*, **77**, 033802.

[338] Scully, Marlan O. (2009). Collective lamb shift in single photon dicke superradiance. *Phys. Rev. Lett.*, **102**, 143601.

[339] Scully, Marlan O., Fry, Edward S., Ooi, C. H. Raymond, and Wódkiewicz, Krzysztof (2006). Directed spontaneous emission from an extended ensemble of n atoms: Timing is everything. *Phys. Rev. Lett.*, **96**, 010501.

[340] Scully, Marlan O. and Svidzinsky, Anatoly A. (2009). The super of superradiance. *Science*, **325**, 1510–1511.

[341] Scully, Marlan O. and Zubairy, M. Suhail (1997). *Quantum Optics*.

[342] Selig, Ulrike, Langhojer, Florian, Dimler, Frank, Löhrig, Tatjana, Schwarz, Christoph, Gieseking, Björn, and Brixner, Tobias (2008). Inherently phasestable coherent two-dimensional spectroscopy using only conventional optics. *Opt. Lett.*, **33**, 2851.

[343] Shacklette, Justin M. and Cundiff, Steven T. (2002). Role of excitation-induced shift in the coherent optical response of semiconductors. *Phys. Rev. B*, **66**, 045309.

[344] Shacklette, J. M. and Cundiff, S. T. (2003). Nonperturbative transient fourwave-mixing line shapes due to excitation-induced shift and excitation-induced dephasing. *J. Opt. Soc. Am. B*, **20**, 764–769.

[345] Shamsi, Javad, Urban, Alexander S, Imran, Muhammad, De Trizio, Luca, and Manna, Liberato (2019). Metal halide perovskite nanocrystals: Synthesis, postsynthesis modifications, and their optical properties. *Chem. Rev.*, **119**, 3296–3348.

[346] Shang, Jingzhi, Shen, Xiaonan, Cong, Chunxiao, Peimyoo, Namphung, Cao, Bingchen, Eginligil, Mustafa, and Yu, Ting (2015). Observation of excitonic fine structure in a 2D transition-metal dichalcogenide semiconductor. *ACS Nano*, **9**, 647–655.

[347] Shen, Boqiang, Chang, Lin, Liu, Junqiu, Wang, Heming, Yang, Qi-Fan, Xiang, Chao, Wang, Rui Ning, He, Jijun, Liu, Tianyi, Xie, Weiqiang, Guo, Joel, Kinghorn, David, Wu, Lue, Ji, Qing-Xin, Kippenberg, Tobias J., Vahala, Kerry, and Bowers, John E. (2020). Integrated turnkey soliton microcombs. *Nature*, **582**, 365–369.

[348] Shen, F., Gao, J., Senin, A. A., Zhu, C. J., Allen, J. R., Lu, Z. H., Xiao, Y., and Eden, J. G. (2007). Many-body dipole-dipole interactions between excited Rb atoms probed by wave packets and parametric four-wave mixing. *Phys. Rev. Lett.*, **99**, 143201.

[349] Shen, Y. R. (2003). *The Principles of Nonlinear Optics*. Wiley-Interscience.

[350] Shim, S.-H., Strasfeld, David B., Ling, Yun L., and Zanni, Martin T. (2007). Automated 2D IR spectroscopy using a mid-IR pulse shaper and application of this technology to the human islet amyloid polypeptide. *Proc. Natl. Acad. Sci. U.S.A.*, **104**, 14197–14202.

[351] Shornikova, Elena V., Biadala, Louis, Yakovlev, Dmitri R., Sapega, Victor F., Kusrayev, Yuri G., Mitioglu, Anatolie A., Ballottin, Mariana V., Christianen, Peter C. M., Belykh, Vasilii V., Kochiev, Mikhail V., Sibeldin, Nikolai N., Golovatenko, Aleksandr A., Rodina, Anna V., Gippius, Nikolay A., Kuntzmann, Alexis, Jiang, Ye, Nasilowski, Michel, Dubertret, Benoit, and Bayer, Manfred (2018). Addressing the exciton fine structure in colloidal nanocrystals: the case of CdSe nanoplatelets. *Nanoscale*, **10**, 646–656.

[352] Sie, Edbert J., Frenzel, Alex J., Lee, Yi-Hsien, Kong, Jing, and Gedik, Nuh (2015). Intervalley biexcitons and many-body effects in monolayer mos₂. *Phys. Rev. B*, **92**, 125417.

[353] Siegman, Anthony E. (1986). *Lasers*. University Science Books, Sausalito, CA.

[354] Siemens, M. E., Moody, G., Li, H., Bristow, A. D., and Cundiff, S. T. (2010). Resonance lineshapes in two-dimensional Fourier transform spectroscopy. *Opt. Express*, **18**, 17699.

[355] Singh, Akshay, Moody, Galan, Tran, Kha, Scott, Marie E., Overbeck, Vincent, Berghäuser, Gunnar, Schaibley, John, Seifert, Edward J., Pleskot, Dennis, Gabor, Nathaniel M., Yan, Jiaqiang, Mandrus, David G., Richter, Marten, Malic, Ermin, Xu, Xiaodong, and Li, Xiaoqin (2016). Trion formation dynamics in monolayer transition metal dichalcogenides. *Phys. Rev. B*, **93**, 041401.

[356] Singh, Akshay, Moody, Galan, Wu, Sanfeng, Wu, Yanwen, Ghimire, Nirmal J., Yan, Jiaqiang, Mandrus, David G., Xu, Xiaodong, and Li, Xiaoqin (2014). Coherent electronic coupling in atomically thin mose$_2$. *Phys. Rev. Lett.*, **112**, 216804.

[357] Singh, Rohan, Autry, Travis M, Nardin, Gaël, Moody, Galan, Li, Hebin, Pierz, Klaus, Bieler, Mark, and Cundiff, Steven T (2013). Anisotropic homogeneous linewidth of the heavy-hole exciton in (110)-oriented GaAs quantum wells. *Phys. Rev. B*, **88**, 45304.

[358] Singh, Rohan, Moody, Galan, Siemens, Mark E., Li, Hebin, and Cundiff, Steven T. (2016). Quantifying spectral diffusion by the direct measurement of the correlation function for excitons in semiconductor quantum wells. *J. Opt. Soc. Am. B*, **33**, C137–C143.

[359] Singh, Rohan, Richter, Marten, Moody, Galan, Siemens, Mark E., Li, Hebin, and Cundiff, Steven T. (2017). Localization dynamics of excitons in disordered semiconductor quantum wells. *Phys. Rev. B*, **95**, 235307.

[360] Singh, Rohan, Suzuki, Takeshi, Autry, Travis M., Moody, Galan, Siemens, Mark E., and Cundiff, Steven T. (2016). Polarization-dependent exciton linewidth in semiconductor quantum wells: A consequence of bosonic nature of excitons. *Phys. Rev. B*, **94**, 081304(R).

[361] Skribanowitz, N., Herman, I. P., MacGillivray, J. C., and Feld, M. S. (1973). Observation of Dicke superradiance in optically pumped HF gas. *Phys. Rev. Lett.*, **30**, 309–312.

[362] Smallwood, Christopher L., Autry, Travis M., and Cundiff, Steven T. (2017). Analytical solutions to the finite-pulse bloch model for multidimensional coherent spectroscopy. *J. Opt. Soc. Am. B*, **34**, 419–429.

[363] Smallwood, Christopher L. and Cundiff, Steven T. (2018). Multidimensional coherent spectroscopy of semiconductors. *Laser Photonics Rev.*, **12**, 1800171.

[364] Smallwood, Christopher L., Ulbricht, Ronald, Day, Matthew W., Schröder, Tim, Bates, Kelsey M., Autry, Travis M., Diederich, Geoffrey, Bielejec, Edward, Siemens, Mark E., and Cundiff, Steven T. (2021). Hidden silicon-vacancy centers in diamond. *Phys. Rev. Lett.*, **126**, 213601.

[365] Smith, Brad C., Lomsadze, Bachana, and Cundiff, Steven T. (2018). Optimum repetition rates for dual-comb spectroscopy. *Opt. Express*, **26**, 12049.

[366] Smith, Earl W, Cooper, J., and Roszman, Larry J. (1973). An analysis of the unified and scalar additivity theories of spectral line broadening. *J. Quant. Spectrosc. Radiat. Transfer*, **13**, 1523–1538.

[367] Somma, Carmine, Folpini, Giulia, Reimann, Klaus, Woerner, Michael, and Elsaesser, Thomas (2016). Phase-resolved two-dimensional terahertz spectroscopy including off-resonant interactions beyond the $\chi^{(3)}$ limit. *J. Chem. Phys.*, **144**, 184202.

[368] Somma, Carmine, Folpini, Giulia, Reimann, Klaus, Woerner, Michael, and Elsaesser, Thomas (2016). Two-phonon quantum coherences in indium antimonide studied by nonlinear two-dimensional terahertz spectroscopy. *Phys. Rev. Lett.*, **116**, 177401.

[369] Song, Yin, Konar, Arkaprabha, Sechrist, Riley, Roy, Ved Prakash, Duan, Rong, Dziurgot, Jared, Policht, Veronica, Matutes, Yassel Acosta, Kubarych, Kevin J., and Ogilvie, Jennifer P. (2019). Multispectral multidimensional spectrometer spanning the ultraviolet to the mid-infrared. *Rev. Sci. Instrum.*, **90**, 013108.

[370] Spencer, Austin P., Li, Hebin, Cundiff, Steven T., and Jonas, David M. (2015). Pulse Propagation Effects in Optical 2D Fourier-Transform Spectroscopy: Theory. *J. Phys. Chem. A*, **119**, 3936–3960.

[371] Splendiani, Andrea, Sun, Liang, Zhang, Yuanbo, Li, Tianshu, Kim, Jonghwan, Chim, Chi-Yung, Galli, Giulia, and Wang, Feng (2010). Emerging photoluminescence in monolayer MoS_2. *Nano Lett.*, **10**, 1271–1275.

[372] Stievater, T. H., Li, X., Cubel, T., Steel, D. G., Gammon, D., Katzer, D. S., and Park, D. (2002). Measurement of relaxation between polarization eigenstates in single quantum dots. *Appl. Phys. Lett.*, **81**, 4251.

[373] Stievater, T. H., Li, X., Steel, D. G., Gammon, D., Katzer, D. S., Park, D., Piermarocchi, C., and Sham, L. J. (2001). Rabi oscillations of excitons in single dots. *Phys. Rev. Lett.*, **87**, 133603.

[374] Stone, Katherine W., Gundogdu, Kenan, Turner, Daniel B., Li, Xiaoqin, Cundiff, Steven T., and Nelson, Keith A. (2009). Two-quantum 2d ft electronic spectroscopy of biexcitons in GaAs quantum wells. *Science*, **324**, 1169–1173.

[375] Stone, Katherine W., Turner, Daniel B., Gundogdu, Kenan, Cundiff, Steven T., and Nelson, Keith A. (2009). Exciton–exciton correlations revealed by two-quantum, two-dimensional fourier transform optical spectroscopy. *Acc. Chem. Res.*, **42**, 1452–1461.

[376] Stranks, S. D., Eperon, G. E., Grancini, G., Menelaou, C., Alcocer, M. J. P., Leijtens, T., Herz, L. M., Petrozza, A., and Snaith, H. J. (2013). Electron-hole diffusion lengths exceeding 1 micrometer in an organometal trihalide perovskite absorber. *Science*, **342**, 341–344.

[377] Suh, Myoung-Gyun and Vahala, Kerry J. (2018). Soliton microcomb range measurement. *Science*, **359**, 884–887.

[378] Suh, Myoung-Gyun, Yang, Qi-Fan, Yang, Ki Youl, Yi, Xu, and Vahala, Kerry J. (2016). Microresonator soliton dual-comb spectroscopy. *Science*, **354**, 600–603.

[379] Suter, Dieter, Klepel, Harald, and Mlynek, Jürgen (1991). Time-resolved two-dimensional spectroscopy of optically driven atomic sublevel coherences. *Phys. Rev. Lett.*, **67**, 2001–2004.

[380] Suzuki, Takeshi, Singh, Rohan, Bayer, Manfred, Ludwig, Arne, Wieck, Andreas D., and Cundiff, Steven T. (2016). Coherent control of the exciton-biexciton

system in an inas self-assembled quantum dot ensemble. *Phys. Rev. Lett.*, **117**, 157402.

[381] Suzuki, Takeshi, Singh, Rohan, Bayer, Manfred, Ludwig, Arne, Wieck, Andreas D., and Cundiff, Steven T. (2018). Detuning dependence of Rabi oscillations in an InAs self-assembled quantum dot ensemble. *Phys. Rev. B*, **97**, 161301.

[382] Suzuki, Takeshi, Singh, Rohan, Moody, Galan, Aßmann, Marc, Bayer, Manfred, Ludwig, Arne, Wieck, Andreas D., and Cundiff, Steven T. (2018). Dephasing of InAs quantum dot *p*-shell excitons studied using two-dimensional coherent spectroscopy. *Phys. Rev. B*, **98**, 195304.

[383] Swagel, E., Paul, J., Bristow, A. D., and Wahlstrand, J. K. (2021). Analysis of complex multidimensional optical spectra by linear prediction. *Opt. Express*, **29**, 37525–37533.

[384] Swallows, Matthew D., Bishof, Michael, Lin, Yige, Blatt, Sebastian, Martin, Michael J, Rey, Ana Maria, and Ye, Jun (2011). Suppression of collisional shifts in a strongly interacting lattice clock. *Science*, **331**, 1043–1046.

[385] Szudy, J. and Baylis, W.E. (1975). Unified Franck-Condon treatment of pressure broadening of spectral lines. *J. Quant. Spectrosc. Radiat. Transfer*, **15**, 641–668.

[386] Takagahara, T. (1999). Theory of exciton dephasing in semiconductor quantum dots. *Phys. Rev. B*, **60**, 2638.

[387] Takemura, N., Trebaol, S., Anderson, M. D., Kohnle, V., Léger, Y., Oberli, D. Y., Portella-Oberli, M. T., and Deveaud, B. (2015). Two-dimensional Fourier transform spectroscopy of exciton-polaritons and their interactions. *Phys. Rev. B*, **92**, 125415.

[388] Tan, Zhi-Kuang, Moghaddam, Reza Saberi, Lai, May Ling, Docampo, Pablo, Higler, Ruben, Deschler, Felix, Price, Michael, Sadhanala, Aditya, Pazos, Luis M., Credgington, Dan, Hanusch, Fabian, Bein, Thomas, Snaith, Henry J., and Friend, Richard H. (2014). Bright light-emitting diodes based on organometal halide perovskite. *Nat. Nanotechnol.*, **9**, 687–692.

[389] Tanimura, Yoshitaka and Mukamel, Shaul (1993). Two-dimensional femtosecond vibrational spectroscopy of liquids. *J. Chem. Phys.*, **99**, 9496.

[390] Tavis, Michael and Cummings, Frederick W. (1968). Exact solution for an *n*-molecule—radiation-field Hamiltonian. *Phys. Rev.*, **170**, 379–384.

[391] Tekavec, P. F., Lott, G. A., and Marcus, A. H. (2007). Fluorescence-detected two-dimensional electronic coherence spectroscopy by acousto-optic phase modulation. *J. Chem. Phys.*, **127**, 214307.

[392] Thämer, Martin, De Marco, Luigi, Ramasesha, Krupa, Mandal, Aritra, and Tokmakoff, Andrei (2015). Ultrafast 2D IR spectroscopy of the excess proton in liquid water. *Science*, **350**, 78–82.

[393] Thouin, Félix, Neutzner, Stefanie, Cortecchia, Daniele, Dragomir, Vlad Alexandru, Soci, Cesare, Salim, Teddy, Lam, Yeng Ming, Leonelli, Richard, Petrozza, Annamaria, Kandada, Ajay Ram Srimath, and Silva, Carlos (2018). Stable biexcitons in two-dimensional metal-halide perovskites with strong dynamic lattice disorder. *Phys. Rev. Mater.*, **2**, 034001.

[394] Tian, Peifang F, Keusters, Dorine, Suzaki, Yoshifumi, and Warren, Warren S. (2003). Femtosecond phase-coherent two-dimensional spectroscopy. *Science*, **300**, 1553–1555.

[395] Titze, Michael, Fei, Chengbin, Munoz, Maria, Wang, Xuewen, Wang, He, and Li, Hebin (2019). Ultrafast carrier dynamics of dual emissions from the orthorhombic phase in methylammonium lead iodide perovskites revealed by two-dimensional coherent spectroscopy. *J. Phys. Chem. Lett.*, **10**, 4625–4631.

[396] Titze, Michael, Li, Bo, Zhang, Xiang, Ajayan, Pulickel M., and Li, Hebin (2018). Intrinsic coherence time of trions in monolayer $MoSe_2$ measured via two-dimensional coherent spectroscopy. *Phys. Rev. Mater.*, **2**, 054001.

[397] Titze, Michael and Li, Hebin (2017). Interpretation of optical three-dimensional coherent spectroscopy. *Phys. Rev. A*, **96**, 032508.

[398] Tiwari, V., Peters, W. K., and Jonas, D. M. (2013). Electronic resonance with anticorrelated pigment vibrations drives photosynthetic energy transfer outside the adiabatic framework. *Proc. Natl. Acad. Sci. U.S.A.*, **110**, 1203.

[399] Tollerud, Jonathan and Davis, Jeffrey A. (2016, Jul). Two-dimensional double-quantum spectroscopy: peak shapes as a sensitive probe of carrier interactions in quantum wells. *J. Opt. Soc. Am. B*, **33**, C108–C114.

[400] Tollerud, Jonathan and Davis, Jeffrey A. (2016). Two-dimensional double-quantum spectroscopy: peak shapes as a sensitive probe of carrier interactions in quantum wells. *J. Opt. Soc. Am. B*, **33**, C108.

[401] Tollerud, Jonathan O., Cundiff, Steven T., and Davis, Jeffrey A. (2016). Revealing and characterizing dark excitons through coherent multidimensional spectroscopy. *Phys. Rev. Lett.*, **117**, 097401.

[402] Tollerud, Jonathan O. and Davis, Jeffrey A. (2017). Separating pathways in double-quantum optical spectroscopy reveals excitonic interactions. *Laser Photonics Rev.*, **11**, 1600249.

[403] Tollerud, Jonathan O, Hall, Christopher R, and Davis, Jeffrey A (2014). Isolating quantum coherence using coherent multi-dimensional spectroscopy with spectrally shaped pulses. *Opt. Express*, **22**, 6719.

[404] Tommila, J., Schramm, A., Hakkarainen, T. V., Dumitrescu, M., and Guina, M. (2013). Size-dependence properties of single InAs quantum dots grown in nanoimprint lithography patterned GaAs pits. *Nanotechnology*, **24**, 235204.

[405] Tong, Yu, Bladt, Eva, Aygüler, Meltem F, Manzi, Aurora, Milowska, Karolina Z, Hintermayr, Verena A, Docampo, Pablo, Bals, Sara, Urban, Alexander S, Polavarapu, Lakshminarayana, and Feldmann, Jochen (2016). Highly luminescent cesium lead halide perovskite nanocrystals with tunable composition and thickness by ultrasonication. *Angew. Chem. Int. Ed.*, **55**, 13887–13892.

[406] Trebino, R., DeLong, K. W., Fittinghoff, D. N., Sweetser, J. N., Krumbügel, M. A., and Kane, D. J. (1997). Measuring ultrashort laser pulses in the time-frequency domain using frequency-resolved optical gating. *Rev. Sci. Instrum.*, **68**, 3277.

[407] Trinh, M. T., Wu, X., Niesner, D., and Zhu, X.-Y. (2015). Many-body interactions in photo-excited lead iodide perovskite. *J. Mater. Chem. A*, **3**, 9285–9290.

[408] Trocha, P., Karpov, M., Ganin, D., Pfeiffer, M. H. P., Kordts, A., Wolf, S., Krockenberger, J., Marin-Palomo, P., Weimann, C., Randel, S., Freude, W., Kippenberg, T. J., and Koos, C. (2018). Ultrafast optical ranging using microresonator soliton frequency combs. *Science*, **359**, 887–891.

[409] Trotzky, S., Chen, Y. A., Flesch, A., McCulloch, I. P., Schollwöck, U., Eisert, J., and Bloch, I. (2012). Probing the relaxation towards equilibrium in an isolated strongly correlated one-dimensional Bose gas. *Nat. Phys.*, **8**, 325–330.

[410] Turner, Daniel B., Hassan, Yasser, and Scholes, Gregory D. (2012). Exciton superposition states in CdSe nanocrystals measured using broadband two-dimensional electronic spectroscopy. *Nano Lett.*, **12**, 880–886.

[411] Turner, Daniel B. and Nelson, Keith A. (2010). Coherent measurements of high-order electronic correlations in quantum wells. *Nature*, **466**, 1089–1092.

[412] Turner, D B, Stone, K W, Gundogdu, K, and Nelson, K A (2009). Three-dimensional electronic spectroscopy of excitons in GaAs quantum wells. *J. Chem. Phys.*, **131**, 144510.

[413] Turner, D. B., Stone, K. W., Gundogdu, K. G., and Nelson, K. A. (2011). The coherent optical laser beam recombination technique (COLBERT) spectrometer: coherent multidimensional spectroscopy made easier. *Rev. Sci. Instrum.*, **82**, 081301.

[414] Turner, Daniel B., Wen, Patrick, Arias, Dylan H., Nelson, Keith A., Li, Hebin, Moody, Galan, Siemens, Mark E., and Cundiff, Steven T. (2012). Persistent exciton-type many-body interactions in GaAs quantum wells measured using two-dimensional optical spectroscopy. *Phys. Rev. B*, **85**, 201303.

[415] Utsunomiya, S., Tian, L., Roumpos, G., Lai, C. W., Kumada, N., Fujisawa, T., Kuwata-Gonokami, M., Löffler, A., Höfling, S., Forchel, A., and Yamamoto, Y. (2008). Observation of Bogoliubov excitations in exciton-polariton condensates. *Nat. Phys.*, **4**, 700–705.

[416] van Kampen, H., Sautenkov, V. A., Shalagin, A. M., Eliel, E. R., and Woerdman, J. P. (1997). Dipole-dipole collision-induced transport of resonance excitation in a high-density atomic vapor. *Phys. Rev. A*, **56**, 3569–3575.

[417] Vaughan, Joshua C., Hornung, T., Stone, K. W., and Nelson, Keith A. (2007). Coherently controlled ultrafast four-wave mixing spectroscopy. *J. Phys. Chem. A*, **111**, 4873–4883.

[418] Victor, K., Axt, V. M., Bartels, G., Stahl, A., Bott, K., and Thomas, P. (1996). Microscopic foundation of the phenomenological few-level approach to coherent semiconductor optics. *Z. Phys. B*, **99**, 197–205.

[419] Virk, Kuljit S. and Sipe, J. E. (2009). Multidimensional Fourier spectroscopy of semiconductors. i. Nonequilibrium Green function approach. *Phys. Rev. B*, **80**, 165318.

[420] Virk, Kuljit S. and Sipe, J. E. (2009). Multidimensional Fourier spectroscopy of semiconductors. ii. Decoherence effects. *Phys. Rev. B*, **80**, 165319.

[421] Vöhringer, P., Arnett, D. C., Westervelt, R. A., Feldstein, M. J., and Scherer, N. F. (1995). Optical dephasing on femtosecond time scales: Direct measurement and calculation from solvent spectral densities. *J. Chem. Phys.*, **102**, 4027–4036.

[422] Volder, Michael F. L. De, Tawfick, Sameh H., Baughman, Ray H., and Hart, A. John (2013). Carbon nanotubes: Present and future commercial applications. *Science*, **339**, 535–539.

[423] Volkov, V., Schanz, R., and Hamm, P. (2005). Active phase stabilization in fourier-transform two-dimensional infrared spectroscopy. *Opt. Lett.*, **30**, 2010.

[424] Wagner, Wolfgang, Li, Chunqiang, Semmlow, John, and Warren, Warren S. (2005). Rapid phase-cycled two-dimensional optical spectroscopy in fluorescence and transmission mode. *Opt. Express*, **13**, 3697.

[425] Wang, F. (2005). The optical resonances in carbon nanotubes arise from excitons. *Science*, **308**, 838–841.

[426] Wang, Hailin, Ferrio, Kyle, Steel, Duncan G., Hu, Y. Z., Binder, R., and Koch, S. W. (1993). Transient nonlinear optical response from excitation induced dephasing in gaas. *Phys. Rev. Lett.*, **71**, 1261–1264.

[427] Wang, H., Ferrio, K. B., Steel, D. G., Berman, P. R., Hu, Y. Z., Binder, R., and Koch, S. W. (1994). Transient 4-wave-mixing line-shapes - effects of excitation-induced dephasing. *Phys. Rev. A*, **49**, R1551–R1554.

[428] Wang, Hailin, Shah, Jagdeep, Damen, T.C., and Pfeiffer, L.N. (1994). Polarization-dependent coherent nonlinear optical response in GaAs quantum wells: Dominant effects of two-photon coherence between ground and biexciton states. *Solid State Commun.*, **91**, 869–874.

[429] Wang, He, Valkunas, Leonas, Cao, Thu, Whittaker-Brooks, Luisa, and Fleming, Graham R. (2016). Coulomb screening and coherent phonon in methylammonium lead iodide perovskites. *J. Phys. Chem. Lett.*, **7**, 3284–3289.

[430] Wang, Lili, Griffin, Graham B., Zhang, Alice, Zhai, Feng, Williams, Nicholas E., Jordan, Richard F., and Engel, Gregory S. (2017). Controlling quantum-beating signals in 2d electronic spectra by packing synthetic heterodimers on single-walled carbon nanotubes. *Nat. Chem.*, **9**, 219–225.

[431] Wang, Qing Hua, Kalantar-Zadeh, Kourosh, Kis, Andras, Coleman, Jonathan N., and Strano, Michael S. (2012). Electronics and optoelectronics of two-dimensional transition metal dichalcogenides. *Nat. Nanotechnol.*, **7**, 699–712.

[432] Wang, Yixian, Shan, Xiaonan, and Tao, Nongjian (2016). Emerging tools for studying single entity electrochemistry. *Faraday Discuss.*, **193**, 9–39.

[433] Wegener, M., Chemla, D. S., Schmitt-Rink, S., and Schäfer, W. (1990). Line shape of time-resolved four-wave mixing. *Phys. Rev. A*, **42**, 5675.

[434] Wehrenfennig, Christian, Liu, Mingzhen, Snaith, Henry J., Johnston, Michael B., and Herz, Laura M. (2014). Charge carrier recombination channels in the low-temperature phase of organic-inorganic lead halide perovskite thin films. *APL Materials*, **2**, 081513.

[435] Weidman, Mark C., Seitz, Michael, Stranks, Samuel D., and Tisdale, William A. (2016). Highly tunable colloidal perovskite nanoplatelets through variable cation, metal, and halide composition. *ACS Nano*, **10**, 7830–7839.

[436] Weiner, A. M., De Silvestri, S., and Ippen, E. P. (1985). Three-pulse scattering for femtosecond dephasing studies: theory and experiment. *J. Opt. Soc. Am. B*, **2**, 654.

[437] Weiner, John, Bagnato, Vanderlei S., Zilio, Sergio, and Julienne, Paul S. (1999). Experiments and theory in cold and ultracold collisions. *Rev. Mod. Phys.*, **71**, 1–85.

[438] Weisbuch, C., Nishioka, M., Ishikawa, A., and Arakawa, Y. (1992). Observation of the coupled exciton-photon mode splitting in a semiconductor quantum microcavity. *Phys. Rev. Lett.*, **69**, 3314–3317.

[439] Wen, P, Christmann, G, Baumberg, J J, and Nelson, Keith A (2013). Influence of multi-exciton correlations on nonlinear polariton dynamics in semiconductor microcavities. *New J. Phys.*, **15**, 025005.

[440] Wieczorek, Witlef, Krischek, Roland, Kiesel, Nikolai, Michelberger, Patrick, Tóth, Géza, and Weinfurter, Harald (2009). Experimental entanglement of a six-photon symmetric Dicke state. *Phys. Rev. Lett.*, **103**, 020504.

[441] Wilmer, Brian L., Passmann, Felix, Gehl, Michael, Khitrova, Galina, and Bristow, Alan D. (2015). Multidimensional coherent spectroscopy of a semiconductor microcavity. *Phys. Rev. B*, **91**, 201304(R).

[442] Wong, Cathy Y. and Scholes, Gregory D. (2011, Apr). Biexcitonic fine structure of cdse nanocrystals probed by polarization-dependent two-dimensional photon echo spectroscopy. *J. Phys. Chem. A*, **115**, 3797–3806.

[443] Wright, John C (2011). Multiresonant coherent multidimensional spectroscopy. *Annu. Rev. Phys. Chem.*, **62**, 209–230.

[444] Wu, Kewei, Bera, Ashok, Ma, Chun, Du, Yuanmin, Yang, Yang, Li, Liang, and Wu, Tom (2014). Temperature-dependent excitonic photoluminescence of hybrid organometal halide perovskite films. *Phys. Chem. Chem. Phys.*, **16**, 22476–22481.

[445] Xiao, Di, Liu, Gui-Bin, Feng, Wanxiang, Xu, Xiaodong, and Yao, Wang (2012). Coupled spin and valley physics in monolayers of MoS_2 and other group-VI dichalcogenides. *Phys. Rev. Lett.*, **108**, 196802.

[446] Xing, Guichuan, Mathews, Nripan, Lim, Swee Sien, Yantara, Natalia, Liu, Xinfeng, Sabba, Dharani, Grätzel, Michael, Mhaisalkar, Subodh, and Sum, Tze Chien (2014). Low-temperature solution-processed wavelength-tunable perovskites for lasing. *Nat. Mater.*, **13**, 476–480.

[447] Xing, G., Mathews, N., Sun, S. Y., Lim, S. S., Lam, Y. M., Gratzel, M., Mhaisalkar, S., and Sum, T. C. (2013). Long-range balanced electron- and hole-transport lengths in organic-inorganic $CH_3NH_3PbI_3$. *Science*, **342**, 344–347.

[448] Xu, Xiaodong, Yao, Wang, Xiao, Di, and Heinz, Tony F. (2014). Spin and pseudospins in layered transition metal dichalcogenides. *Nat. Phys.*, **10**, 343–350.

[449] Yajima, Tatsuo and Taira, Yoichi (1979). Spatial optical parametric coupling of picosecond light pulses and transverse relaxation effect in resonant media. *J. Phys. Soc. Jpn.*, **47**, 1620–1626.

[450] Yampolsky, Steven, Fishman, Dmitry A., Dey, Shirshendu, Hulkko, Eero, Banik, Mayukh, Potma, Eric O., and Apkarian, Vartkess A. (2014). Seeing a single molecule vibrate through time-resolved coherent anti-Stokes Raman scattering. *Nat. Photonics*, **8**, 650–656.

[451] Yang, J., Wen, X., Xia, H., Sheng, R., Ma, Q., Kim, J., Tapping, P., Harada, T., Kee, T. W., and Huang, F. Z., et al. (2017). Acoustic-optical phonon up-

conversion and hot-phonon bottleneck in lead-halide perovskites. *Nat. Commun.*, **8**, 14120.

[452] Yang, Lijun and Mukamel, Shaul (2008). Revealing exciton-exciton couplings in semiconductors using multidimensional four-wave mixing signals. *Phys. Rev. B*, **77**, 075335.

[453] Yang, Lijun and Mukamel, Shaul (2008). Two-dimensional correlation spectroscopy of two-exciton resonances in semiconductor quantum wells. *Phys. Rev. Lett.*, **100**, 057402.

[454] Yang, Lijun, Zhang, Tianhao, Bristow, Alan D., Cundiff, Steven T., and Mukamel, S. (2008). Isolating excitonic Raman coherence in semiconductors using two-dimensional correlation spectroscopy. *J. Chem. Phys.*, **129**, 234711.

[455] Yang, Y., Ostrowski, D. P., France, R. M., Zhu, K., van de Lagemaat, J., Luther, J. M., and Beard, M. C. (2016). Observation of a hot-phonon bottleneck in lead-iodide perovskites. *Nat. Photonics*, **10**, 53–59.

[456] Ye, Jun and Cundiff, Steven T. (ed.) (2005). *Femtosecond optical frequency comb : principle, operation, and application*. Springer.

[457] Yetzbacher, M K, Belabas, N, Kitney, K A, and Jonas, D M (2007). Propagation, beam geometry, and detection distortions of peak shapes in two-dimensional Fourier transform spectra. *J. Chem. Phys.*, **126**, 44511.

[458] You, Yumeng, Zhang, Xiao-Xiao, Berkelbach, Timothy C., Hybertsen, Mark S., Reichman, David R., and Heinz, Tony F. (2015). Observation of biexcitons in monolayer WSe_2. *Nat. Phys.*, **11**, 477–481.

[459] Yu, Shaogang, Titze, Michael, Zhu, Yifu, Liu, Xiaojun, and Li, Hebin (2019). Long range dipole-dipole interaction in low-density atomic vapors probed by double-quantum two-dimensional coherent spectroscopy. *Opt. Express*, **27**, 28891.

[460] Yu, Shaogang, Titze, Michael, Zhu, Yifu, Liu, Xiaojun, and Li, Hebin (2019). Observation of scalable and deterministic multi-atom Dicke states in an atomic vapor. *Opt. Lett.*, **44**, 2795.

[461] Zhang, Tianhao, Borca, Camelia N., Li, Xiaoqin, and Cundiff, S. T. (2005). Optical two-dimensional Fourier transform spectroscopy with active interferometric stabilization. *Opt. Express*, **13**, 7432.

[462] Zhang, Tianhao, Kuznetsova, Irina, Meier, Torsten, Li, Xiaoqin, Mirin, Richard P., Thomas, Peter, and Cundiff, Steven T. (2007). Polarization-dependent optical 2D Fourier transform spectroscopy of semiconductors. *Proc. Natl. Acad. Sci. U.S.A.*, **104**, 14227–14232.

[463] Zhang, T., Kuznetsova, I., Meier, T., Li, X., Mirin, R. P., Thomas, P., and Cundiff, S. T. (2007). Polarization-dependent optical two-dimensional fourier transform spectroscopy of semiconductors. *Proc. Natl. Acad. Sci. U.S.A.*, **104**, 14227.

[464] Zhang, Zhengyang, Wells, Kym Lewis, and Tan, Howe-Siang (2012, DEC 15). Purely absorptive fifth-order three-dimensional electronic spectroscopy. *Opt. Lett.*, **37**, 5058–5060.

[465] Zhao, Wei and Wright, John C. (2000, Feb). Doubly vibrationally enhanced four wave mixing: The optical analog to 2d nmr. *Phys. Rev. Lett.*, **84**, 1411–1414.

[466] Zheng, Junrong, Kwak, Kyungwon, Asbury, John, Chen, Xin, Piletic, Ivan R., and Fayer, M. D. (2005). Ultrafast dynamics of solute-solvent complexation observed at thermal equilibrium in real time. *Science*, **309**, 1338–1343.

[467] Zhu, X.-Y. and Podzorov, V. (2015). Charge carriers in hybrid organic-inorganic lead halide perovskites might be protected as large polarons. *J. Phys. Chem. Lett*, **6**, 4758–4761.

Index